Imagining New Human-Animal Futures in Australia

AUSTRALIAN STUDIES: INTERDISCIPLINARY PERSPECTIVES

Series Editor
Anne Brewster
Associate Professor,
University of New South Wales

Volume 5

PETER LANG
Oxford • Bern • Berlin • Bruxelles • New York • Wien

Imagining New Human-Animal Futures in Australia

Jane Mummery and Debbie Rodan

PETER LANG

Oxford • Bern • Berlin • Bruxelles • New York • Wien

Bibliographic information published by Die Deutsche Nationalbibliothek.
Die Deutsche Nationalbibliothek lists this publication in the Deutsche
National-bibliografie; detailed bibliographic data is available on the
Internet at http://dnb.d-nb.de.

A catalogue record for this book is available from the British Library.

Library of Congress Cataloging-in-Publication Data

Names: Mummery, Jane, 1970- author. | Rodan, Debbie, author.
Title: Imagining new human-animal futures in Australia / Jane Mummery,
   Debbie Rodan.
Description: Oxford ; New York : Peter Lang, [2022] | Series: Australian
   studies. Interdisciplinary perspectives, 2297-8194 ; vol no. 5 |
   Includes bibliographical references and index.
Identifiers: LCCN 2022019705 (print) | LCCN 2022019706 (ebook) | ISBN
   9781789973143 (hardback) | ISBN 9781789973150 (ebook) | ISBN
   9781789973167 (epub)
Subjects: LCSH: Human-animal relationships--Australia. | Animal
   welfare--Australia.
Classification: LCC QL85 .M86 2022 (print) | LCC QL85 (ebook) | DDC
   591.94--dc23/eng/20220713
LC record available at https://lccn.loc.gov/2022019705
LC ebook record available at https://lccn.loc.gov/2022019706

Cover image: Photo courtesy of David Smyth.

Cover design by Peter Lang Ltd.

ISSN 2297-8194
ISBN 978-1-78997-314-3 (print)
ISBN 978-1-78997-315-0 (ePDF)
ISBN 978-1-78997-316-7 (ePub)

© Peter Lang AG 2022

Published by Peter Lang Ltd, International Academic Publishers,
Oxford, United Kingdom
oxford@peterlang.com, www.peterlang.com

Jane Mummery and Debbie Rodan have asserted their right under the Copyright, Designs
and Patents Act, 1988, to be identified as Authors of this Work.

All rights reserved.
All parts of this publication are protected by copyright.
Any utilisation outside the strict limits of the copyright law, without the
permission of the publisher, is forbidden and liable to prosecution.
This applies in particular to reproductions, translations, microfilming,
and storage and processing in electronic retrieval systems.

This publication has been peer reviewed.

Contents

List of Figures — vii

Acknowledgements — xi

List of Abbreviations — xiii

Introduction — 1

CHAPTER 1
Imagining Human-Animal Relations in the Australian Anthropocene — 19

CHAPTER 2
Riding the Sheep's Back: From Live Stock to Sentience — 45

CHAPTER 3
Transitioning Australia's Meat-Eating Culture — 75

CHAPTER 4
Making Friends and Kin with Companion Animals: Skippy to Red Dog — 111

CHAPTER 5
Neither Food nor Friend: Cohabiting with the Animals Out There — 153

CHAPTER 6
Stewardship, Sustainability, and Protecting the Great Barrier Reef — 197

CHAPTER 7
Different/Together: Building More-Than-Human Communities   227

POSTSCRIPT
Making Change   263

Bibliography   273

Index   339

Figures

Figure 1. An illustration of the impact of bauxite mining in mature Jarrah forest near Dwellingup, Western Australia. This image, taken in May 2021, shows the three phases of mining in one frame: (a) clearing and stacking of mature Jarrah before burning (lower area); (b) strip mining (upper left and centre); (c) replanting (top right). Key issues are that restoration cannot reinstate the full ecological functionality and resilience of old growth forest and it is uncertain in a drying climate whether restoration areas will achieve old growth status at all. *Source:* Photo courtesy of Dave Osborne, WalkGPS.com.au. 28

Figure 2. 'Aussie Spirit' mural created by Melbourne's Murals at Black Rock, Victoria, to pay tribute to the many volunteers who worked to support Australia's wildlife after the Black Summer fires of 2019–2020. *Source:* Photo courtesy of Kevin Rennie. 32

Figure 3. Intensive poultry farming for meat at a South Australian farm in 2016. *Source:* Photo courtesy of the Farm Transparency Project. 47

Figure 4. RSPCA, Animals Australia, leading politicians, and thousands of concerned Australians gathered in Australian capital cities to rally for a kinder future free from the cruelty of live export. Image from Ban Live Export National Rally in 2011 in Sydney, New South Wales. *Source:* Photo courtesy of James Morgan and Nikki To. 63

Figure 5. Vegan activist groups such as Vegan Australia strive to remind Australians that there is no substantive

	difference between the animals we choose as companions and those we treat as resources for our consumption – both are sentient. This image is from a 'Choose Vegan' campaign carried out by Vegan Australia in Melbourne, Victoria, in 2017. *Source:* Photo courtesy of Vegan Australia.	81
Figure 6.	*Make It Possible* campaign poster from Animals Australia. *Source:* Photo by author.	85
Figure 7.	Many Australian birds such as these wild sulphur crested cockatoos in Sherbrooke, Victoria, are intensely curious, intelligent, social and prepared to interact with humans. Such transient interactions could be considered as a building block for the extension of kinship outside of companion species. *Source:* Photo by Laya Clode on Unsplash.	136
Figure 8.	Kangaroo Island, South Australia, has been described as an Australian equivalent of the Galapagos. As this satellite image released by NASA shows, over half a million acres – almost half – of the island was burnt during the 2019–2020 Black Summer bushfires. This included 96 per cent of Flinders Chase National Park, one of Australia's oldest national parks, and a home to a vibrant diversity of species, including many endangered animals and birds. *Source:* Photo from NASA Earth Observatory images by Lauren Dauphin.	163
Figure 9.	A volunteer in Canberra topping up water containers to support wildlife on Black Mountain, Canberra, the Australian Capital Territory (ACT), while guarding against smoke inhalation from the 2019–2020 Black Summer bush fires. *Source:* Photo courtesy of Josephine Mummery.	165

Figure 10.	Brumbies in the Snowy Mountains, Kosciuszko National Park, Australia. *Source*: Photo by Christine Mendoza on Unsplash.	181
Figure 11.	Land clearing in north Queensland. Queensland remains a deforestation front in Australia with most clearing in the state linked to the cattle industry. *Source:* Photo courtesy of Bill Laurance.	205
Figure 12.	Coral bleaching severity survey on Orpheus Island 2017, Great Barrier Reef, Queensland. *Source:* Photo courtesy of ARC Centre of Excellence for Coral Reef Studies Flickr account (CC BY-ND 2.0).	216
Figure 13.	Dog agility courses have been designed to allow dogs and humans to play in ways that provide dogs with mental and physical stimulation. Agility courses can be either run competitively or just for fun. Dogs would typically run and play in such spaces off-leash. *Source:* Photo courtesy of Sylvia Hamilton.	237
Figure 14.	Nesting boxes installed at a height of ten metres in trees at the Fremantle Arts Centre, Western Australia, during an artist's residency in the spring of 2009. The residency culminated in an exhibition entitled 'to hear the language of birds'. Nesting boxes were in use by galahs when the artist visited the site a year later. *Source:* Photo courtesy of Paul Uhlmann.	247

Acknowledgements

As is the case for every author with every book, we are indebted to and thankful for the contributions and support offered to us by colleagues, friends, and family, as well as by our series editor and Peter Lang's excellent staff. We are particularly thankful to Anne Brewster, editor of the series *Australian Studies: Interdisciplinary Perspectives*, at Peter Lang Oxford, and the anonymous reviewers for their valuable feedback. We are also appreciative of the support that Edith Cowan University and Federation University have given to this project. We would like to thank Claire Dolling, Terry Eyssens, Cheryl Lange, Josephine Mummery, Marnie Nolton, Laurel Plapp, Jessica Reeves, and David Smyth for all their varied assistance in helping us bring this book into reality as well as helping us avoid some real bloopers. We would also like to thank Sylvia Hamilton, Bill Laurance, Greg McFarlane, Josephine Mummery (again), David Osbourne, Kevin Rennie, David Smyth, and Paul Uhlmann for generously sharing some of their photographs with us. Without these various contributions this book would be significantly less rich. We would finally like to note our appreciation for the various nonhuman members of our respective households and broader communities who share our space, demand our time and care, and draw us away from the computer.

Abbreviations

AAWS	Australian Animal Welfare Strategy
ABC	Australian Broadcasting Corporation
ACF	Australian Conservation Foundation
ACT	Australian Capital Territory
AMCS	Australian Marine Conservation Society
ANZAC	Australian and New Zealand Army Corps
API	Animal Protection Index
CGBR	Citizens of the Great Barrier Reef
CSIRO	Commonwealth Scientific and Industrial Research Organisation
Cth	Commonwealth
EPBC Act	Environmental Protection and Biodiversity Conservation Act
FP2P	From Paddock to Plate
GBR	Great Barrier Reef
GBRF	Great Barrier Reef Foundation
GBRMPA	Great Barrier Reef Marine Park Authority
IPBES	Intergovernmental Science-Policy Platform for Biodiversity and Ecosystem Services
IPCC	Intergovernmental Panel on Climate Change
MLA	Meat & Livestock Australia
NACCHO	National Aboriginal and Community Controlled Health Organisation

NAILSMA	North Australian Indigenous Land and Sea Management Alliance
NFF	National Farmers' Federation
NGO	Non-governmental organization
NSW	New South Wales
OECD	The Organisation for Economic Co-operation and Development
RSPCA	Royal Society for the Prevention of Cruelty to Animals
UNESCO	United Nations Educational, Scientific and Cultural Organization
WIRES	Wildlife Information and Rescue Service
WWF	World Wide Fund for Nature

Introduction

Australians are lucky. We live in cities and towns with green spaces – parks, reserves, botanical gardens, coastal foreshores. Many of us also have green spaces – gardens – attached to our homes. Our urban and suburban spaces are inhabited by a multitude of nonhuman species: animals as well as plants. These spaces teem not just with ubiquitous insects, spiders, and other non-vertebrate life – some of which we admittedly consider undesirable living companions (cockroaches, termites, silverfish, various wasp species, poisonous spiders, to name a few) – but with the raucous birds native to Australia, multiple species of lizards and snakes, companion as well as other introduced animals. Some of Australia's native animals might also reside in our communities and in our backyards. Living by the coast or near other water sources – our rivers and lakes – makes other species visible to us too: fish, sharks, dolphins and whales, tortoises and turtles, rays, and other water-based species. We see and hear what is in our proximity – we marvel at the visitations of whales and dolphins, fear the visitations of sharks, groan at the noisiness of birdsong early in the morning, grumble at the ruckus and mess from cockatoos and galahs descending on our flowering shrubs and trees, and plan walking and running routes to avoid being swooped by magpies during spring nesting times. Some of us might go fishing or crabbing in nearby waterways. Many of us share our lives with companion animals, but we might also participate in backyard bird counts, take up birdwatching, swim with seals, dolphins, whale sharks or even sharks, go snorkelling, surfing or diving, or visit wildlife parks and zoos. This ubiquity of nonhuman life close to us is, however, also a problem. It can mean we do not notice the losses in biodiversity that are increasing throughout Australia.

The Intergovernmental Science-Policy Platform for Biodiversity and Ecosystem Services (IPBES) has assessed that nearly a million species face extinction across the world (IPBES, 2019; Leahy, 2019), and the World Wide Fund for Nature's *Living Planet Report 2020* has reported that the past fifty years (1970 to 2020) has seen the loss of almost 70 per cent of the animals on Earth (WWF, 2020b). Through global surveying, Australia has been identified as one of the highest contributors to species extinction and biodiversity loss. Climate change is further exacerbating the vulnerability of Australia's biodiversity, particularly through Australia's increased rates of catastrophic bushfire. These issues are expected to give further impetus to what is being labelled an extinction crisis. Unfortunately, this has been a largely invisible crisis for many Australians – perhaps because we continue to see a variety of nonhuman life in our backyards and suburbs and thus assume the health of other ecosystems. At the same time, Australians loudly proclaim their care for animals, with substantial public engagement in a wide variety of animal welfare debates. Some of these examine the fates of farmed animals within agriculture and Australia's live animal export industry; of Australia's native animals whose habitat is destroyed through land development and agricultural use, as well as the disaster of catastrophic fire; of introduced animals perceived to threaten the remnant habitat of native species, agricultural interests or native species directly; and of those animals used and managed in entertainment industries and hunting. Such debates also extend to how conflicts between human and animal interests should be managed, whether this is to do with the tensions between preserving animal habitat and our development interests, or how to manage the threat to human life perceived posed by sharks. Australia is also a country known for its iconic and *charismatic* species (usually taken to mean animals with popular appeal and some symbolic value, see, e.g., Ducarme et al., 2013),[1] and for the natural environment experiences – including animal experiences – it offers visitors. It is a country, in other words, where the divergent and competing interests of humans and of other species are in ongoing and sometimes very visible tension, and where there is little agreement about how to manage these tensions.

This, then, is the context of this book which is aiming to identify and map some of the main assumptions and narratives of human-animal relations in common use within Australia. More specifically, this book is driven by two overlapping objectives. The first is to contextualize and outline some of the key features of human-animal relations in Australia as told through common and shared narratives of social life. The second is to explore some of the ways these relations might be – and are starting to be – imagined differently, with these imaginings slowly gaining some public visibility via the circulation of counter-narratives. Considering these imaginings together, we suggest, might lead to different, more respectful ways of living with and among animals, and new – better – futures in Australia with regards to human-animal relations and communities. As this book will endeavour to show, common Australian narratives regarding human-animal relations have generated and maintained some very particular perspectives with regards to (a) what is taken for granted in human-animal relations, and (b) what is considered normal treatment and management of animals across various contexts. These narratives not only highlight and share several assumptions and value judgements regarding nonhuman animals and human-animal relations, but further normalize individual as well as institutional understandings and practices. At the same time, such narratives need to be recognized as able to be reshaped and/or countered, and indeed, as actively being challenged through the circulation and promotion of a variety of counter-narratives. Mainstream narratives may describe normative views and practices, but they do not thereby entail the continuation of these views and practices, unchanged, into the future.

Social Imaginaries and the Functions of Narrative

In order to examine both the current dynamics and possible developments of human-animal relations in a broad Australian context, we engage the concept of *social imaginary*. Basically, as defined by the philosopher Charles Taylor (2004), a social imaginary refers to a very broad collective understanding about how the world should be and how members of a society should live together. It is the prevailing common sense of a society, the 'invisible cement' that binds a given society together (Castoriadis, 1975/1987, p. 143). More specifically, social imaginary refers to the shared, even paradigmatic, understandings that make possible the common practices, models for relationships, and widely shared senses of legitimacy that inform a society. It stands for the way the people of a given society envision their relations to others, their shared norms and ideas of existence, and how that collective sense provides legitimacy to their self-understandings, surroundings, and their everyday practices. These shared understandings of self and normative social practices are intertwined with what Taylor refers to as the 'moral order' of a society, by which he means 'the rights and obligations we have as individuals in regard to each other' in the broad societal context (2004, p. 4). More broadly, a society's moral order refers to the models and parameters of relationships that are seen as normal by members of that society and typically reflected in that society's legal system. These would include relationships between people, between people and their surroundings, between people and animals, and even between people and their symbolic systems (such as God). To use Taylor's (p. 23) words, a social imaginary refers to:

> something much broader and deeper than the intellectual schemes people may entertain when they think about reality in a disengaged mode. I am thinking, rather, of the ways people imagine their social existence, how they fit together with each other, how things go on between them and their fellows, the expectations that are normally met, and the deeper normative notions and images that underlie these expectations.

A social imaginary is, in other words, what allows members of a society to generally make sense of and justify both individual and collective actions, whether these involve participating in democratic processes or activist actions, or accepting the use of nonhuman animals as resource. Because a society makes sense of itself and, indeed, institutes itself, via its social imaginary, members of that society are all broadly able to understand such actions as valid or legitimate collective decisions and carry them out without difficulty. It is also via the normativity it celebrates – and legislates – that a social imaginary provides the resources for critiquing any deviations in the face of these norms. This is the point that a society's collective social imaginary 'endures through time and so becomes increasingly embedded in all our institutions, our judicial systems, our national narratives, our founding fictions, our cultural traditions' (Gatens & Lloyd, 1999, p. 143).

The social imaginary thus provides the background legitimacy for any given society's daily routines and social repertoires. That is, a social imaginary is structured by the social dynamics that produces it while at the same time also structuring those forces. Social imaginaries can consequently be understood as products of history that 'generate individual and collective practices in accordance with the schemes generated by history' (Bourdieu, 1980/1990, pp. 54–5). As such, the social imaginary of our society does not just describe how we see the world in our heads; it *makes* that world. The key point here is that social imaginaries are both symbolically and materially produced and maintained. They are produced and maintained by our beliefs, assumptions, and practices – and, as noted above, produce and maintain those beliefs, assumptions, and practices. At the same time, because the concept of the social imaginary is inextricably connected to the imagination (see Bohmann & Montero, 2014), it has the capacity to explain not only the dynamics of the reproduction of the social status quo, but the dynamics of resistance and transformation (Adams et al., 2015). Social imaginaries thus draw attention to both the paradigmatic shaping and reproduction of social norms (often through normative social narratives), and the potential of their reimagining and reshaping. In this way, social imaginaries are not just paradigmatic but can also draw attention to a 'paradigm-in-the-making' (Adams et al., 2015, p. 15).

Social imaginaries achieve these reproductive and productive capacities through their connection to forms of imagining – such as narrative and myth – which are 'visibly rooted in everyday life' (Anderson, 2006, pp. 35–6) and both reflect and create reality (Davies, 2002). Drawn from 'social, cultural and, perhaps, unconscious imperatives, which [they] at the same time reveal' (Andrews et al., 2003, p. 8), narratives legitimize and challenge both individual and collective understandings of the world. They illustrate 'the inclusion or exclusion of social groups, the enactment of institutional routines, the perpetration of social roles, etc.' (De Fina & Georgakopoulou, 2008, p. 382). Insofar, then, as narratives can variously circulate, legitimate, and challenge a society's institutions and norms, they can provide important insights into the operations of that society's social imaginary (Bohmann & Montero, 2014, p. 7). There are, however, two key points that need to be unpacked here. The first is that while there are many narratives associated with any given social imaginary, and, further, many different imaginaries – capitalist, constitutional, cosmopolitan, democratic, ecological, economic, feminist, global, historical, hypermodern, humanitarian, nationalist, political, politico-juridical, populist, and religious, for instance (see, e.g., Adams & Smith, 2019) – we take all of this diversity to be nested within and contributing to the broader social imaginary that binds a particular society together. The second key point is that while the various dimensions of a social imaginary may not be universally ascribed to or identified with by all of those considered members of that society – particularly the case in a diverse society such as Australia[2] – this lack of universal acceptance does not by itself undermine that social imaginary. Such aspects of non-identification may generate senses of not fully belonging, of being minoritized, with reference to that social imaginary, as well as enabling counter-narratives that may articulate dynamics of resistance and transformation.

It is through these dynamics of resistance and transformation – and the diversity of associated narratives – that a given social imaginary can be understood as both paradigmatic and malleable, normative but able to be reimagined and reshaped. This is the reminder that narratives are outstanding at describing how the world appears and what it feels like from different perspectives (Hall, 1996; Herman, 2009; Mar & Oatley, 2008; Stadler, 2003). Narratives do not simply endorse existing values and forms of relation but they also open up novel ways of looking at things and new possibilities for action. To clarify, insofar as narratives both draw on and evoke imagination, they also invite us to think *from what is* to *what if* or *what might happen next* (Sools, 2012). In helping to open 'the black box of what we think is possible' (Wittmayer et al., 2015, p. 7), narratives thus have the capability to extend or reshape a given social imaginary, along with its norms and restrictions. That is, whilst they can enforce the tenets of a social imaginary they can also – via the circulation of counter-narratives – reach beyond the assumptions and limits of that social imaginary and suggest or promote its reshaping. Given the significance of these points for our consideration of the broad Australian social imaginary and its openness towards social change regarding human-animal relations, this book engages and is informed by an approach that is mostly drawn from considerations and analyses of the diverse forms of social narratives and storytelling that are circulated throughout the Australian public sphere. To focus on how storytelling can spark people's imagination with regards to social change, both individually and collectively, we will thus be considering how the circulation of diverse narratives can help align people with proposed changes to the collective, conventional practices and norms that comprise a society's social imaginary.

Narratives and the Generation of Social Change

In their exploration of the 'hypothetical, the possible, and the actual' (Brockmeier, 2009, p. 228), it is clear that narratives can be important contributors to the co-creation of new social narratives and the reshaping of the social imaginary. Enabling the critique of established norms and the spread of new ideas, narratives and storytelling thus comprise a fundamental strategy through which processes of change can be introduced and dealt with in community (Herman, 2009). These possibilities play out through the different ways narratives can construct meaning through organizing events, concepts, characters and situations into 'sequences of causation and consequence' (Stadler, 2003, p. 87). These can offer new ideas of collective and individual identity and agency (Squire et al., 2008). Because both the production and consumption of narrative matter in this context – as well as a narrative's *traction* (its success in attaining broad-based support) – it is worth identifying some of the main factors that may make a particular narrative effective in its call for change. Here a first point is that for narratives to be successful in gaining traction and thus promoting social change, they need to connect into personal narratives and recognizable everyday interactions (Epstein et al., 2014). They further need to be effective in their use of emotional modes of appeal that can fuel identification, empathy, and the acceptance of a need for change.

Scholars interested in the role of narratives in social change argue that engaging an audience's emotion is central to them learning from that narrative and being persuaded by its proposals (Chattoo Borum & Feldman, 2017). This is the view that new understandings of events, life or our world can be both introduced and gain traction through our emotional and creative responses to stories. Indeed, stories that resonate with us can 'serve as springboards for ethical action' (Witherell & Noddings, 1991, p. 8; see also Winskell & Enger, 2014). In particular, it is our capacities for identification and empathy with various characters, plots, and storylines, and to be *transported into a narrative world* (Green & Brock, 2000, 2002) that matter. It is these capacities that allow us to connect with, experience, and learn from stories, and as such envision, compare and connect with alternative constructions of reality (Green et al., 2004; Tal-Or & Cohen, 2016). It is this work of connection – of building traction – that thus also allows audiences to find narrative 'visions of possible futures' legitimate and/or desirable (Sefcovic & Condit, 2001, p. 289). While the connective work of narrative is enhanced through the use of images and other visual media, particularly those that add realist credibility (Aiello & Parry, 2020; Rose, 2016), it is also worth noting that the impact of stories is not dependent on whether readers perceive them as either fictional or nonfictional (Małecki et al., 2019; see also Green & Brock, 2000). Both can inspire and generate social change. It is this work of connection and collective revisioning through the development and circulation of narratives that will thus be explored throughout this book, insofar as it is this work that will be instrumental in the possible reshaping of Australia's social imaginary with regards to human-animal relations.

Telling This Story

This brings us finally to a few pragmatic points. First, as we have noted, this is a story of developments and possibilities regarding the presentation and legitimation of human-animal relations in Australia throughout its social imaginary. This is not to say that such possibilities are only taking shape in Australia but exploring cross-country and global connections would take more space than we have available. Regarding the Australian context and social imaginary, we also need to emphasize that our interest in this book lies with the diversity of narratives and counter-narratives about human-animal relations that are either already widely circulated and shared throughout the Australian public sphere or are emerging and gaining traction. More specifically, this means that our focus is on narratives that are either already in common circulation and endorsed by existing institutions, or ones that are being actively circulated as alternatives in an effort to increase their traction. These, then, are the narratives developed and endorsed by animal industries and political and economic institutions, by the biological and environmental conservation sciences, by cultural intermediaries and opinion leaders, as well as by animal activist groups and social movements. These might be narratives about animal welfare,[3] meat-eating and alternative diets, biodiversity loss, hunting, the management of introduced species, the preservation of habitat, and so forth. Such narratives are thus variously circulated throughout policy and associated scientific literatures, throughout multimedia industry and activist campaigns, as well as through the mainstream press and social media. They are also visible in cultural and creative storytelling circulated through screen media (film and television) and other forms of publishing. Each of these domains provides rich resources for our consideration, and our analyses of the narratives being made visible and debated within these domains are used to underpin and illustrate our discussions in each chapter. This broad consideration is essential because human-animal relations are played out and debated across multiple spheres of thought and practice.

A second point concerns language and terminology. As all writers understand, the selection of terminology can be fraught. Although the relationship between reality and its linguistic representation is extremely complex, it is clear that language encodes what is important to people and provides the means for them to articulate their understanding of what is significant. Language, then, is normative, and thus an arena of political struggle. This means that the linguistic categories people use carry baggage in the form of assumptions and normative viewpoints, and this is particularly the case when discussing nonhuman animals and our relations to them. A discussion of *livestock*, for instance, carries with it an array of human-oriented economic viewpoints wherein animals have their diverse lived experiences and their needs and interests reduced to their function as units of a resource. They are live *stock* or *stock* animals and the *stock* part of this category is given the most value. Similarly, the attribution of *pest* as a descriptor of some wild living animals carries with it connotations of being of no value, of being a threat to established (human) interests. Indeed, the standard definition of a *pest* animal or *vertebrate pest* is thus 'an animal that has a significant net deleterious impact on a valued resource' (Braysher et al., 1996, p. 18). Such a definition and the use of such a label, however, is of course subjective, reflecting a particular perception, bias or expectation regarding the weighting of what is of value (Marks, 1999). As with the *livestock* label, the *pest* label is typically used on the basis of human-oriented economic viewpoints which are seen as normative. What we are noting here is that so-called category terms for animals such as livestock or pest (or feral, native, wild-living, pet, companion, for example) reflect value judgements. Even using the term *animal* in contrast to *human* is a marker of a value judgement, in this case marking humans as being distinctly different from and more valuable than other animals.

In the context of this book, such linguistic categories are both constructed and contested across and through the narratives we examine. This is because many of the commonplace Australian narratives regarding non-human animals and human-animal relations are still very much informed by human-oriented economic viewpoints. This is not surprising given the historical role of agricultural practices and industries in Australia, both with and prior to European occupation and settlement.[4] Although the relative economic importance of agriculture has declined in Australia since it was seen as the foundation for national prosperity, agriculture continues to be 'a pillar of the Australian economy' and central to narratives about Australian identity and prosperity (Agribusiness Australia, 2020, p. 2; Lockie, 2015).[5] Among other things, this means that terms such as *livestock* and *pest* animals remain in common usage in many common narratives concerning animals. These terms may (or may not) be challenged in counter-narratives, and may be replaced by those considered (for now at least) to carry less reductive and human-oriented judgements and values. For instance, the term *pet* with its connotations of ownership and human indulgence might be substituted by *companion animal* with its emphasis on the values of companionship and friendship. With regards to our own practice throughout this book, because these struggles of terminology are *live* in the narratives we are examining, we have chosen to use the terms that are in commonplace use even though they are often problematic – these include *livestock, pet, pest, feral, native*. We have also mainly talked in this book about *humans* and *animals*, although we acknowledge that humans themselves are also animals with overlapping attributes and capacities. In this instance, whilst we acknowledge that there are increasing attempts in some fields to distinguish *human animals* from *other animals, nonhuman animals,* or *animals other than humans*, these terms are not commonplace in the social narratives we are examining. However, insofar as we also examine a variety of counter-narratives throughout the forthcoming chapters we do also acknowledge and explore these terms as sites of struggle. Considering a diversity of narratives, both those with existing public traction and those striving to build public traction, is important as they also help us remember to account for at least some of the normative viewpoints built into our own positionalities, including the ongoing traction of anthropocentric assumptions (a point we discuss in detail in the next chapter).

A third key point concerns the orientations and interconnection of the chapters of this book. As has been noted, each chapter is oriented by some of the current points of concern and debate with regards to human-animal relations in Australia. These include: live export, industrialized animal agriculture, food norms, the status of companion animals, the Black Summer fires, the loss of native animal habitat, the treatment of native animals and so-called pest animals, protecting the Great Barrier Reef, and aims of rewilding Australian communities for example. These debates are informed by intersecting ideas and narratives about Australians' relations with animals which orient around a range of key concepts: anthropocentrism, sentience, care, friendship, responsibility, stewardship, and community. These concepts are also explored with reference to some of the values that underpin Australia's First Nations Peoples' understandings of human-animal relations. Because neither of us are of First Nations descent, we strive to engage with these ideas and values cautiously and respectfully, recognizing that we can only understand the topmost and public layers of these ways of being.[6] At the same time, we honour that these ways of understanding and being with animals and land have not only enabled complex multispecies communities in Australia for millennia – and continue to do so – but that they can also offer important narratives for non-First Nations Australians for coming to imagine and live with nonhuman animals in less anthropocentric ways into our shared future.

All of these concepts and associated narratives will be examined and traced both individually and via their points of intersection throughout the forthcoming chapters. A further point to be noted here is that while many of these debates and narratives intersect and share key ideals, many of them remain disconnected from each other. This is also the case with regards to many of the campaigns calling for improvements in the treatment of animals in Australia. That is, the majority of the advocacy and campaigning work in Australia with regards to animals and human-animal relations is issue-specific and/or species-specific. While it has been argued that the utilization of issue-specific advocacy diverts attention from contending with the root cause of injustice towards animals (see, e.g., Francione, 2010b, 2010c, 2010d), meaning that considerations of animal treatment are rarely systemic, systemic approaches towards animal injustice are simply not the norm within Australia at this point in time.[7] What this means is that the resulting stories of developments and possibilities we draw attention to throughout these chapters remain fragmentary, reflecting the nature of current advocacy work in Australia. Fragments of thought and practice that resist normalized shapes and practices can, however, trigger 'a sense of awareness and reflection' (Caso, 2010, p. 109) and can provide important resources for creating alternative narratives and reshaping the social imaginary – perhaps to the point that systemic address to the root causes of injustice towards animals will become the norm.

To situate these possibilities and imaginings, Chapter 1 outlines the key contexts that inform legacy and current human-animal relations in Australia. First and foremost, these include the framework and narrative of the *Anthropocene* and associated longstanding tendencies towards *anthropocentrism*. Together these make a highly influential view of the world and of human-animal relations that has taken root in both scientific and popular discourse. Also considered are the commonplace anthropocentric values of *human exceptionalism* and *speciesism* given their influence on both conventional economic ideas and everyday practices of capitalism and consumerism, and Australians' typical forms of interaction with animals. In

Chapter 2, we explore recent increases in public concern within Australia for livestock welfare, particularly the way such concern is being developed and shaped through reference to ideas of sentience. Examining a major point of current contention in Australia – live animal export – we consider the ways animal activist and advocacy groups are entreating the Australian community to make animals' individual capacities for sentience the focal point for welfare and to thus avoid framing animals as simply a resource or units of *live stock*. Continuing these considerations, Chapter 3 explores the paradox that while Australia is one of the world's big meat-eating cultures, increasing awareness of and care for animal sentience has meant that veganism and other dietary habits that minimize or avoid meat are also rapidly gaining in social prevalence and status. Having traced two responses to the animal welfare narrative of sentience, those relating first to the framing and farming of livestock animals and secondly to the development of new food norms, Chapter 4 maps a different kind of revaluation which rather takes the recognition of sentience as entailing the possibility of friendship between humans and animals. This chapter also shifts focus from examining our relations with those animals conventionally known – and valued – as live stock, to examining our relations concerning the animals we consider our companions, and the value we give them as being independent entities with their own interests.

Chapter 5 examines our relations with and responsibilities for those animals which escape conventional categorization – and subsequent consideration – as food or friend. In this chapter we thus examine Australians' relations with brumbies, kangaroos, crocodiles, and a range of other species that live outside of our standard senses of responsibility. In particular we consider the basis for our notions of responsibility with regards to these other animals. Where do our responsibilities lie? With the lives and living of all animals? With members of *desirable* species? With ecosystems? Such a focus is continued in Chapter 6. Continuing a consideration of the scope of our responsibility for animal lives and living, this chapter considers the capacities of a narrative of stewardship and sustainability in the context of protecting Australia's Great Barrier Reef. More specifically, this chapter marks a change in focus from mapping the course of the narrative of sentience in potentially reshaping human-animal relations, to exploring ideas and mechanisms that could require more equitable forms for the consideration and protection of both human and animal interests. Chapter 7 takes a broader look again at what is circulating in the mainstream media and Australian public sphere with regards to initiatives towards developing and supporting multispecies communities. This chapter is followed by a brief postscript in which we return to the key themes, arguments, and achievements of the book – in particular, assessing the capacity of the issues, debates, possibilities, and trends discussed to deliver new ways of living with animals that might enable the reshaping of Australia's social imaginary.

While different issues and possibilities are foregrounded in each chapter, readers can also expect to see key insights, ideas, and focuses emerge and re-emerge across different contexts and chapters, each time being sketched out to show a different facet or application. Such cross-hatching and points of intersection are reflective of trying to consider something the scale of a social imaginary. After all, the point about social imaginaries is that they stand for a framework of understanding that applies throughout all of a society's relational dimensions. There is, for example, no single illustration of the relationality – and narrativity – demonstrative of a social imaginary, nor any single source for its framework. Rather such relationality and narrativity inform multiple norms in multiple contexts. Certainly, in many of the legacy narratives informing the Australian social imaginary since European occupation, only human life and human activity have been conventionally given value in the moral terms necessary for their protection. Human-animal relations, as well as human-ecosystem relations, have thus been based on and tied to the use and usefulness of the nonhuman. Social imaginaries are, however, not set in stone, and while they are totalizing, they can change with the development and uptake of new social narratives. As such, this book is actually a story of change and hope. It is a story that suggests change is happening – as will be outlined through the forthcoming chapters – and which suggests that such change can reshape the social imaginary and initiate and make normative new models for Australian social life that are able to support more than human interests. This would not be for a society without humans, but a society of humans and animals that responds to and promotes the interests of multiple species and that does so up front and explicitly.

Notes

1. The iconic and symbolic status of many of Australia's native species – along with the concurrent status of many of these species as threatened or endangered – is also a driver behind their memorialization and display through the vehicle of public or community art, particularly in the forms of sculpture and street art. Artists and other advocates, for instance, talk of such works as deliberately featuring 'native flora and fauna species found locally, some of which are threatened, endangered or vulnerable' (Claire Foxton as cited in City of Parramatta, 2021), as serving 'as a reminder of the responsibility we share as guardians of our natural environment' (Ann-Maree Greaney as cited in City of Townsville, 2022), and as a form of 're-populating the cityscape with animals, as a way to have them re-enter the contemporary landscape that was once theirs' (Engelen, 2011).
2. Australia is one of the most ethnically diverse societies in the world today, with the population of Australia consisting of more than 270 ethnic groups from nearly 200 different countries, along with Australia's First Nations Peoples who comprise over 500 different clan groups or *nations* around the continent. It is estimated that the identity of the Australian nation in the twenty-first century consists of 58 per cent of Australians with an Anglo-Celtic background, 21 per cent with a non-European background, 18 per cent with a European background and 3 per cent with an Australian First Nations Peoples background, specifically Australian Aboriginal and/or Torres Strait Islander (Arvanitakis et al., 2020). Until the mid-twentieth century, however, Australian society was, with some accuracy, regarded in the wider world as essentially British – or at any rate Anglo-Celtic – due to its original settlement by the British (an act that disregarded Australia's existing First Nations inhabitants) and its subsequent strong promotion of pro-British and European immigration, particularly between 1901 and 1973 – the period of Australia's *Immigration Restriction Act 1901*. Known as the *White Australia policy*, this policy aimed to not only restrict numbers of non-White migrants to Australia, but also to deport *undesirable* migrants who were already in the country. What this history means is that Australia's social imaginary has been very strongly informed by Anglo-Celtic settler views – to the point, as noted above, that post-settlement Australia was considered Anglo-Celtic for well over a century. Australia's current cultural diversity does mean, of course, that this social imaginary is under constant tension from a verity of social narratives. Our focus on human-animal interactions and relations, however, also makes clear the essential anthropocentrism of not just Australia's White settler culture but global industrialized modernity. Challenging and countering anthropocentrism is fundamental work for all human societies – work that is given further impetus by the encroaching climate crisis and collapse of biodiversity.

3   Since its genesis, the animal advocacy movement has been broadly divided into two camps, one advocating for the *humane* treatment of animals based on consideration for their welfare and the other for the complete abolition of human (ab)use of animals. These have been known respectively as animal welfarism and the animal rights (sometimes animal liberation) based approach. These two positions have usually been understood as irreconcilable due to contrasting positions regarding human use of animals (De Villiers, 2017). Animal welfarists contend that that as long as animals are treated humanely, human use of animals is morally acceptable. Animal rights advocates contend, in contrast, that animals should never be regarded as property and any *use* of animals by humans is unacceptable (see, e.g., Regan, 1986). Animal welfarists thus prioritize a reform agenda which seeks the development and implementation of legislation to improve the lives of the animals, while animal rights advocates call for a complete change in the way humans relate to animals, including the extension of rights to animals (De Villiers, 2017). These differences have seen some animal rights proponents argue that animal welfarism facilitates continued and increased exploitation of animals, although others see the increasing concern for animal welfare as marking the taking of incremental steps towards animal rights (see, e.g., De Villiers, 2017). As will be illustrated throughout the following chapters, Australian legislation and the majority of activism is strongly oriented towards an animal welfare approach.

4   While it is extremely difficult to generalize about Australia's First Nations cultures, Aboriginal and Torres Strait Islander Peoples were established hunters and gatherers in a landscape of which they had a very detailed and local knowledge. While First Nations Peoples would certainly have had a significantly less instrumentalist view of animals than did Europeans at the end of the eighteenth century, the social and cultural use of animals was still a norm in First Nations cultures. Animal use practices included extensive fishing in mainland coastal areas and wetlands as well as hunting, and animal products were essential to the meeting of human physical needs (Rose, 1996). Animal use practices were, however, situated in a broader understanding of human and animal interconnectedness and care.

5   Although Australia's agricultural and rural sectors face a number of inter-related social and economic stressors, agricultural interests continue to hold political weight in Australia with the wellbeing of farmers still seen as crucial to the wellbeing of all (see, e.g., Lockie, 2015). An important voice for these interests has been the National Party of Australia, which first entered Australian politics in the early 1920s as the Country Party of Australia. Set up as a voice for rural Australians, the party grew out of farm interest groups that had been established from the mid-nineteenth century (Botterill, 2009). The party has maintained an unbroken presence in the Commonwealth Parliament to this day and, after almost 100 years, remains the most influential of all political parties in representing the needs and interests of Australians living and working beyond the capital cities. In particular,

the party presents itself as a specialist party, concentrating on improving the lifestyle and livelihood of people across regional Australia and increasing the competitiveness of regional business, industry and tourism, and the sustainable development of agriculture and mining (The Nationals, n.d.). Political influence is also held in Australia by farm interest groups, represented since 1979 by the National Farmers' Federation (NFF). A federation of state farm organizations and commodity councils, the NFF has played a central role in guiding Australian agricultural policy, and economic policy more generally, since its formation (Halpin, 2004; Trebeck, 1990). The NFF is still the recognized voice of Australian primary producers, and represents Australian agriculture on national and foreign policy issues including workplace relations, trade and natural resource management (NFF, 2019).

6   Here we acknowledge and respect the limitations set by the Gay'Wu Group of Women (a deep collaboration between five Yolŋu women from North East Arnhem Land and three non-Aboriginal women) in their sharing of Yolŋu women's wisdom. 'We share songspirals with you and we ask that you treat them with respect. [...] The words in this book are our knowledge, our property. You can talk about it, but don't think you can become the authority on it. [...] You need to honour the context of our songspirals, acknowledge the layers of our knowledge. You can talk about the very top layer but you need to be respectful and aware of the limits of what we are sharing and what you in turn can share' (2019, pp. xxv-xxvi). As the Group continues, such sharing is nonetheless integral: 'We want people from different backgrounds, different cultures to walk with us, to learn our culture [...] We want you to touch, and hear, our world. [...] We invite you to sit on the ground with us. We can balance both cultures, we can share' (p. xxvi). In our own way and in our own context, we hope to respectfully show some of the pathways that might arise from such sharing.

7   Although we stand by our point that single-issue campaigning can be effective in the reshaping of Australia's social imaginary, we do take the point that such a campaigning model can lead to conflict between campaigns and thus confusion as to desired ends. Such problems were clearly demonstrated in Australia with the launch and collapse of the Animal Effect app. Launched in May 2013, Animal Effect was a multi-platform social networking app that aimed to deliver animal advocacy news, events, and content directly to subscribers. Its vision was to amplify the collective voices of animal lovers and advocates who, united by Animal Effect, are creating a world in which all animals are treated with care and respect (Doherty, 2013; Mann, 2013). Explicitly designed as a capacity-builder and mobiliser, Animal Effect aimed to facilitate the articulation both of individuals into communities of people with shared interests, and of multiple disparate activist organizations into new and powerful configurations. Unfortunately, what the release of the app actually made visible was the range of tensions between different activists and advocacy

groups, with different groups unwilling to act collaboratively (C. Mann, personal communication, July 16, 2016).

CHAPTER 1

Imagining Human-Animal Relations in the Australian Anthropocene

Type *Animal News* into a search engine, and results pop up from news sites and animal advocacy organizations sharing stories of animals and of human interactions with them from around the world. Search results include news stories about a wide range of animals, viral animal videos, new insights into animal behaviour published in scientific journals, and updates on activities posted by activist and advocacy groups. Digging deeper shows a focus on the curation of content by associated institutions to draw and engage reader attention. One Australian news site, for example, contextualizes its curation of content under 'Wildlife News Headlines' as including the 'Latest wildlife news stories and videos including Australian native animals, endangered species and conservation, zoos, sea life, African animals and other nature news stories' (Wildlife News Headlines, 2021). Opening this site (during an internet search dated September 1, 2021) provided a plethora of animal news stories from this date back until August 27, 2020. Similar kinds of content are collected and curated by other news organizations. Typical content for such sites includes a mix of happy animal stories (rescues, births, reunions, and so on) and stories about problem human-animal interactions. This latter content includes acts of human cruelty to animals, animal attacks on humans, and animal-based problems for society such as insect or rodent plagues. This mix of content is significant insofar as it foregrounds the issues that human-animal interactions and relations are not all benign, and that there are few broadly accepted guidelines for managing such interactions.

This diversity of stories is also a cogent reminder that we share our spaces with nonhuman others and that this sharing of space opens a range of questions regarding our understanding of these others – of their behaviours, their needs, their preferences – and what responsibilities we might have towards them. Such questions drive this book as a whole, but have also taken on enhanced importance as we find ourselves living in a period shaped by and driving increased understanding of nonhuman agency and by escalating and mutually reinforcing processes of biosocial destruction (including mass species extinctions and climate change). Many names have been proposed for this condition, but the one with the most traction in both scientific and popular discourse is the *Anthropocene*. It is this, then, that marks the key context and narrative that informs not just Australian but global understandings of human-animal interaction. To thus situate the narratives that strive to reshape Australia's social imaginary with regards to human-animal relations, it is first necessary to outline the framework and narrative of the Anthropocene and its associated longstanding tendencies towards *anthropocentrism*. Also considered in this chapter are the commonplace anthropocentric values of *human exceptionalism* and *speciesism* given their influence on both conventional economic ideas and everyday practices of capitalism and consumerism, and our relations with animals. As this chapter will endeavour to show, these various values and ideas have generated and maintained some very particular perspectives with regards to (a) what is taken for granted in human-animal relations, and (b) what is considered normal treatment and management of animals across various contexts. Given that the ideas of anthropocentrism, human exceptionalism and speciesism are by no means specific to Australia, some consideration will also be given in this chapter as to how they have taken shape in and shaped the Australian social imaginary. These are all issues that will of course be returned to and explored in more detail in later chapters.

The Anthropocene and the Narrative of Anthropocentrism

The Anthropocene, a term coined by Paul J. Crutzen and Eugene F. Stoermer in 2000 (see also Crutzen, 2002), is the suggestion that because the last 300 years or so of human activities have had planet-wide effects, the current epoch should be described as a *human age* (Monastersky,

2015). That is, the Anthropocene – highlighting *anthropos* (*human* or *man* in Greek) – describes an era in which human presence and activity can be recognized as working alongside *natural* global ecological drivers (e.g. volcanoes, earthquakes, and tsunamis) in shaping and reshaping the planet's environments (Clark et al., 2005). The impact of human activities can be identified in three main phenomena: (a) the exponential increase of human population growth (from 1 billion in 1800 to 7.7 billion in 2020); (b) human transformation of the planet's ecosystems; and (c) global industrialization, large-scale fossil fuel consumption and greenhouse gas emission. When considered in the context of nonhuman life and interests, these activities have brought about a series of detrimental outcomes. These have included: (a) the destruction and pollution of land, freshwater and marine habitats and ecosystems; (b) the confinement of non-domesticated animals to ever more restricted areas; (c) the introduction of multiple animal species (including companion and livestock animals) into environments previously uninhabited by them; and (d) the general breeding and use of animals to suit human demands with little concern for other, nonhuman interests or needs (Burns & Paterson, 2014). The impacts of such activities are particularly evident in Australia, and mark the site of multiple debates as to the future of Australia's biodiversity. Indeed, it has been estimated that Australia's introduced species have come to not only greatly outnumber but exert more ecological influence than do our native species (Woinarski, 2014).

It is worth considering some of the broader impacts of these activities. For instance, research on the Earth's *biomass* – namely the combined weight of all living things, including humans – has shown that while humans currently account for about 36 per cent of the biomass of all mammals on the planet, domesticated livestock accounts for 60 per cent, and the entirety of wild mammals together only 4 per cent (Bar-On et al., 2018). The same research calculates that our activity across the planet has reduced the biomass of wild marine and terrestrial mammals by six times. The same holds true for birds – the biomass of poultry is now about three times higher than that of wild birds. The point here is that, instead of wild animals, a small number of farmed animal species (mainly cows and pigs) now dominate global biomass (Benton et al., 2021). Even more telling is the point again that while human impact is considered to have caused the loss of 83 per cent of all wild mammals, 70 per cent of the Earth's animals are thought to have been lost in the last fifty years (Carrington, 2018; WWF, 2020b). Comparing the respective biomass of humans and wild mammals between 1900 and 2000 illustrates the rapidity of these changes. For instance, in 1900, humans were estimated as comprising 22 per cent of the global biomass of land mammals, with wild mammals estimated at 17 per cent. By 2000, the numbers were humans 31 per cent and wild mammals 3 per cent (Smil, 2011). As such figures show, whereas the biomass of humans in 1900 was relatively close to that of all wild mammals, by 2000 it is ten times greater. It is also worth noting that the biomass of plant matter has also been reduced by half through human activity. That is, while modern agriculture is certainly using an increasing land area for growing crops – an area that is furthermore still growing – the total mass of domesticated crops is itself vastly outweighed by the loss of plant mass resulting from our practices of deforestation, forest management and other land-uses (Elhacham et al., 2020).

To sum these points up, then, the stress on human activities and interests that is illustrated by the Anthropocene has seen the natural world and nonhumans treated as either little more than as resources for humans, or as not possessing enough value to be protected. At the same time, this naming of the Anthropocene is drawing new attention to what has been normalized to the point of invisibility. For example, it should first be evident that the idea of the Anthropocene draws attention to the breadth and impact of our *anthropocentrism*. This refers to both a kind of conceptual separation of humanity from the natural world, and to our tendency to not look beyond our own needs and interests. More specifically, anthropocentrism refers to human centredness and, typically, an assumption of human superiority, according to which only human interests count in any morally significant way:

> From an anthropocentric viewpoint [...] animals are constructed as radically other, and humans are empathically separated from them. Nature is seen as being lower order and as lacking continuity with humans. (Burns, 2014, p. 11)

More broadly, anthropocentrism carries with it an assumption of human superiority across a range of criteria, such as intelligence, creativity, freedom, morality, and reasoning. All of these attribute agency or the capacity for reasoned, intentional action 'only to humans' (Boyd et al., 2015, p. x) – a view which in turn further strengthens and justifies anthropocentric perspectives and practices. It is these perspectives and practices that have resulted not just in an exponential growth in human population and environmental degradation, but mass species extinction, and treatments of nonhumans that are not monitored by the ethical standards required of human-human interactions (Warkentin, 2010). Indeed, anthropocentrism has even normalized the idea that the 'destruction of natural systems and domination and extinction of other animals' is an inevitable, even if regrettable, result of human development (Boyd et al., 2015, p. ix).

Nevertheless, consideration of the Anthropocene can lead to questions concerning our easy prioritizing of human interests over those of other species and ecosystems, and even over the health and resilience of global environmental processes and systems. Indeed, such questions are gaining in scope as it becomes clear that our unbounded anthropocentrism is endangering the ecological processes that support all life on the planet (Rockström et al., 2009). This endangerment is often summed up with reference to the phenomenon of anthropogenic or human-caused climate change. With the basic idea of climate change referring to 'a significant and lasting change in the statistical distribution of weather patterns over periods ranging from decades to millions of years' (Okoye, 2012, p. 136), anthropogenic climate change refers to the scientific contention that the levels of climate change visible in the twenty-first century are the result of human activity, specifically the long-term and widespread use of fossil fuels, changes in agriculture and other land-uses, and the associated and ongoing release of high levels of greenhouse gases into the Earth's atmosphere through these activities. The significant point here is that these greenhouse gas emissions are driving up global temperatures. This will see larger proportions of the Earth being affected by drought and extreme weather events, and increases in desertification. Also to be expected are rises in sea level due to the loss of sea ice, coastal erosion with losses of coastal wetlands, mangroves and coral reef systems, and persistent changes to oceanic currents, acidification and levels of salinity. These effects in turn entail others, such as famine, reductions in fresh water supplies, changes in the growth and distribution of plants, animals and insects, and losses of biodiversity due to species being unable to adapt to changing conditions, habitat collapse and ecological mistiming (IPCC, 2018). These effects, it must be stressed, are already observable throughout the world, and have been seen as consequences of the processes associated with the Anthropocene and the patterns of thinking typical of anthropocentrism.

While descriptions of the Anthropocene highlight on the one hand how our anthropocentrism has come to threaten the ecological processes that support all life, they also draw attention to humanity's 'inescapable dependency' on these same ecological processes (Boyd et al., 2015, p. ix; see also Baskin, 2015; Chiew, 2015). The concept of the Anthropocene thus both highlights the longstanding hierarchical dualisms of anthropocentrism – our prioritizing of (our) culture over nature, humans over animals – whilst also questioning them. Hence, although it has been suggested that the concept of the Anthropocene might further entrench some of the norms of anthropocentrism (Boyd et al., 2015; Creed, 2017), these ideas still offer a useful device for drawing attention to the results of an unbounded anthropocentrism. We thus agree with other analysts that the importance of the Anthropocene is that it invites – even requires us (Blue, 2015) – to rethink our conventional assumptions regarding human-animal relations, as well as our broader relationship with our planetary environment (Chakrabarty, 2012). The invitation of the Anthropocene is thus to come to think and live differently, with a key aspect of this being to imagine a new future of living much more respectfully alongside other species, of consciously building and maintaining multispecies communities even in urban and other built-up settings. It is this invitation that lies at the heart of our objectives for this book.

What Drives the Anthropocene?

If the designation of the Anthropocene provides a productive rubric for reexamining past, present and future understandings of the relations between humanity and nature (see, e.g., Haraway, 2016), questions concerning the nature of human-animal relations are important here. It is worth noting that in Crutzen and Stoermer's (2000) original formulation of the Anthropocene, three sites of human-animal interaction were highlighted by the authors as significant markers of human impacts within the Anthropocene. Illustrative of the pervasiveness and impact of anthropocentrism, these were: the growth in global cattle populations, species extinction, and the expansion of industrialized fishing (2000, p. 17). More specifically, it has been evaluated that the intensifying of global agricultural practices over the past fifty years – particularly with regards to the conversion of natural ecosystems for crop production or pasture for livestock – has been a principal cause in changing global species distribution and reducing biodiversity (see Benton et al., 2021; FAO, 2006). Despite recognition of the increasingly urgent need to redress such anthropocentric practices and reduce biodiversity loss, recent attempts to arrest the decline have been unsuccessful (Convention on Biological Diversity, 2020).

Responding to these issues of biodiversity loss in the Anthropocene is a task that asks us, among other things, to consider how we could better live with and alongside animals (Boyd et al., 2015). To properly consider these issues does, however, mean needing to understand past and present assumptions as to our ordering of human-animal relations. This requires identifying and examining several influential concepts (and associated practices) that interconnect with the anthropocentric viewpoint: namely, *speciesism*, *human exceptionalism*, and *capitalism*. Of these, human exceptionalism and speciesism go hand in hand with each other and anthropocentrism, with each concept contending that only humans and human interests matter. Human exceptionalism, then, is the belief that human beings have special status based on our capacities, typically taken to mean that only humans possess moral status and value in their own right.[1] Certainly, other species and objects may be attributed some moral status and value but this is due to human interests in them – for instance, we might value members of other species but only because of their cultural, economic or personal importance to us. Concerning the capacities typically used to identify humans as the kinds of beings with this special status, several candidates have been proposed. These have included certain kinds of cognitive and behavioural capacities such as using language, thinking abstractly, developing family ties, solving social problems, expressing emotions, starting wars, or having sex for pleasure, as well as some religious assumptions with regards to the possession of a soul, or being chosen by God. What is important with reference at least to each of the secular capacities, is that none of these activities appear unique to human beings. That is, both scholarly and popular work on animal behaviour suggest that many of the activities that have been thought to be distinct to humans also occur in at least some other animals (see, e.g., de Waal & Tyack, 2003; King, 2013). It is worth noting here that because human behaviour and cognition share deep roots with the behaviour and cognition of other animals, approaches that try to find sharp behavioural or cognitive boundaries between humans and other animals will remain controversial and contested.

Speciesism, coined by Richard Ryder in the 1970s but popularized by Peter Singer (1974, 1975/1990), refers to the practice of favouring one's own species and interests while exploiting or harming members of other species for 'morally arbitrary reasons' (Hayward, 1997, p. 52). Defined in the *The New Shorter Oxford English Dictionary* (1993, p. 2,972) as meaning '[d]iscrimination against or exploitation of certain animal species by humans, based on an assumption of human superiority', speciesism basically involves assigning different values or rights to individuals on the basis of species membership. Humans, Singer (1974, 1975/1990) argues, are speciesist when we give less weight to the interests of animals than we would give to the similar interests of humans, but we are also speciesist when we use our own interests as a basis for giving preference to members of one animal species over another. An Australian example would be the practice of fencing kangaroos off from water sources because we want to keep such resources for the exclusive use of our livestock, or trying to fence dingoes out of agricultural areas – or of feeling justified in killing members of both species as pests. More generally, the term *speciesism* has been used to describe practices of human domination over animals and the exclusion of all nonhuman animals from the rights and freedoms that are conventionally granted to humans. Speciesism, in other words, assumes and is dependent on human exceptionalism in order to construct divisions between species, thereby disregarding within-species variation that can often be more marked than between-species differences. It is, in effect, a form of human chauvinism, which ultimately values humans simply '*because they are humans*' (Hayward, 1997, pp. 56–7, original emphasis).

Although, as has been shown, neither human exceptionalism nor speciesism stand up to critical examination, they have both – alongside anthropocentrism – been fundamental influences in the understandings of human-animal relations that have been part and parcel of the Anthropocene. Significantly, such perspectives have permitted and endorsed using animals – and the natural world more broadly – as resources, as 'unmanufactured commodities' (Smith, 1994, p. 214). Indeed, anthropocentric thinking, together with human exceptionalism, speciesism, and human chauvinism, have enabled the 'subordination of nonhuman organisms' on a planet-wide scale, in tandem with denying that either the natural world or nonhuman animals could have moral status or value in their own right (Kopnina et al., 2018, p. 115). This, of course, is the very framing of the natural world that the Anthropocene illustrates. Such views have further been instrumental to the capitalist market-based economic system, which itself rests on the assumptions that humans primarily act in their own self-interest under the force of competition, and that the land – to use Adam Smith's phrasing from 1776 – is a major source of human revenue and wealth, and thus an essential resource for the economic pursuit of self-interest and the circulation of capital (Smith, 1994). This idea of nature as a source from which to extract and generate capital also affirms Karl Marx's analysis in *The Communist Manifesto* (first published 1848), which posits that, given the rise of the market economy and industry, the natural world will be subjected to the use of humanity (Marx & Engels, 1955). A further related assumption has been that there is such an abundance of natural resources that there is no need to consider their protection or conservation (Sookram, 2013). Such views have meant that economic transactions have typically focused on extraction and product development costs, ignoring any possible payment to nature (Kilbourne & Polonsky, 2005). Nature, in other words, is considered to be a free input that is able to be exploited in the pursuit of economic growth (Mundt, 1993). A detailed examination of the role of anthropocentrism in the development and operation of capitalism, and of how capitalist development has further shaped anthropocentric understandings of human-nonhuman relations is beyond the scope of this chapter. What, however, is important here is that the anthropocentric idea of nature – in which the nonhuman world is largely taken as a form of resource awaiting extraction and use (Smith, 2009) – was central to the development of our contemporary economic and political systems, undergirding projects of modernism, capitalism, and colonialism.

Under this anthropocentric capitalist lens, then, the environment and animals become 'economic commodities' with 'market value' (Barua, 2019; Francione & Garner, 2010, p. 27). Considered as resources to be extracted or managed for profit, trees become lumber, wilderness is converted to farms (or tourist destinations), animals and insects are categorized as *livestock* or *pests*, and the land itself becomes something to manipulate, use, and manage (e.g. Figure 1).

Figure 1. An illustration of the impact of bauxite mining in mature Jarrah forest near Dwellingup, Western Australia. This image, taken in May 2021, shows the three phases of mining in one frame: (a) clearing and stacking of mature Jarrah before burning

(lower area); (b) strip mining (upper left and centre); (c) replanting (top right). Key issues are that restoration cannot reinstate the full ecological functionality and resilience of old growth forest and it is uncertain in a drying climate whether restoration areas will achieve old growth status at all. *Source:* Photo courtesy of Dave Osborne, WalkGPS.com.au.

Economic actors under capitalism as such position themselves as 'conquerors' of the natural world (Leopold, 1987, p. 204), a view which when informed by the capitalist desire for profit has supported not just the 'exploitation' of the environment and animals, but has led to models of increasingly 'intensive processing; and/or reduced costs' (McCausland, 2014, p. 211). Under such a framework, animal welfare may be considered but only because poor animal health may impede profits or otherwise impact logistical systems. However, by effectively placing both ecosystem *health* and animal welfare in a state of tension with productivity and profit, such models tend to lead to environmental degradation and the acceptance of practices impacting animals that would be considered barbaric if applied to humans (Gunderson, 2013). Economic growth has thus been identified as one of the most significant causes of environmental degradation and losses in biodiversity (Crist, 2015; Sookram, 2013). The toxic entanglement of anthropocentrism with economics has further led to human relationships with animals being diagnosed as essentially violent (Wadiwel, 2015). Under this framework, after all, the intensive *production* and slaughter of agricultural animals, the hunting of wild animals, the culling of *managed* wildlife, lethal *control* of pest species, the use and *sacrifice* of laboratory animals, the *euthanasia* of unwanted companion animals, and so on, are all not just permissible but pervasive. They are practices that have become seen as fundamentally normal. All such practices, however, display a disregard for 'what[ever] sort of life the animal may hold, what[ever] potential this life may possess, what[ever] sort of value the animal's own life may have for itself' (Wadiwel, 2015, p. 165). These practices illustrate how human exceptionalism and speciesism have come to contextualize and inform all of our human relationships and interactions with animals. That is, whether we kill and consume or welcome animals into our homes as companions, we are still operating under the auspices of human exceptionalism and a 'logic of [human] sovereignty' (Calarco, 2015, p. ix). This logic and our associated assumptions of 'property relations' indeed 'frame and structure nearly all of our relations with animals, from the most hostile to the most pacific' (p. ix).

The frameworks and practices that shape animal lives (and ecosystem health) that are underpinned by these core concepts of human exceptionalism and speciesism are thus always going to be ones of their use and management to fulfil human interests. This is the case even when animal welfare is supposedly being taken into consideration – an issue that will be examined in later chapters. Indeed, it is worth briefly noting here that countries such as Australia that supposedly provide regulation in support of animal welfare, nonetheless continue to legitimize and protect 'the industries that carry out the largest-scale and most intrusive exploitation of animals ever undertaken' (Armstrong, 2017, p. 240). Not only is there no legislation protecting animals generally against human exploitation – impossible given their accepted status as resource – but the legislation protecting animals against cruelty itself remains differentially applicable (Ellis, 2010). As we outline in the following chapters, Australian ideas regarding the proper treatment of animals are also diverse and rife with inconsistencies (Zuolo, 2020). That is to say, the species and type of animal – more specifically, their type of usefulness – determines the kinds and levels of protection potentially made available to them, with animals counted as *livestock* or *pests* open to management in ways that are considered too inhumane for those animals we value as *companions* (Bagaric & Akers, 2012; Cao, 2010; Chen, 2016; White, 2008). These issues will be discussed in more detail in later chapters. It is, however, worth more clearly identifying here the kinds of assumptions and narratives regarding human-animal relations that have become normalized in Australia (and elsewhere) through the entrenchment of the conceptual framework of anthropocentrism.

Human-Animal Relations in the Australian Anthropocene

As noted earlier, Australia provides an interesting basis from which to consider the question of human-animal relations in the Anthropocene. As subsequent chapters will show, Australia's multiple and entangled cultures and histories of First Nations and European settler (and, later, migrant) occupations have generated diverse and often conflicting views about and interests in the continent's (native and introduced) animal populations. The following provides an initial glimpse into some of this complexity of attitude as it plays out in the twenty-first century. To begin with, although most countries use animals variously as companions, for production, and for sport and entertainment, the scale and variety of the presence and influence of animals in Australia is unparalleled. Indeed, 'Australian society is characterized by a pervasive influence of animals in all aspects of human life' (Coleman, 2018, p. 14). This presence and influence can be noted across a variety of domains. To begin with, given that many of the plants, mammals, reptiles and frogs that inhabit the continent are endemic to Australia,[2] this has seen Australian wildlife increasingly featuring in Australian tourism marketing (e.g. koalas and quokkas), with some native animals (kangaroos and emus) also considered symbolic of Australian nationhood (Franklin, 1996, 2007, 2008; Gressier, 2016). In addition, the catastrophic impact of the wildfires in 2019–20 on Australian native animals saw strong upsurges of public concern for these animal populations (e.g. Figure 2; Mathews, 2020; Simo, 2020).

Figure 2. 'Aussie Spirit' mural created by Melbourne's Murals at Black Rock, Victoria, to pay tribute to the many volunteers who worked to support Australia's wildlife after the Black Summer fires of 2019–2020. *Source:* Photo courtesy of Kevin Rennie.

At the same time, the Australian economy has historically been supported strongly by its meat and livestock industries (Franklin, 2007), with Australia described as having ridden to prosperity in the 1950s *on the sheep's back*.[3] Over half of the Australian agricultural industry is still devoted to livestock production, with animal agriculture continuing to make a significant contribution to the national GDP (Australian Bureau of Statistics, 2017, 2021; Johnson, 2017). Australia has also been and remains one of the world's largest exporters of beef, lamb, mutton and goat meat (Meat & Livestock Australia, 2020).

In addition, Australians show very strong commitments to the keeping of companion animals (Franklin, 2007). A Newgate Research quantitative study of Australian households and the state of pet ownership has shown that, as of 2019, over 60 per cent of Australian households have a pet, and that 90 per cent of Australians have had a pet at some time in their lives.

Of Australian households without a current pet, over 60 per cent express interest in the idea of getting a pet. This same study estimates that there are almost 29 million companion animals in Australia today – more than the estimated human population of 25 million (Animal Medicines Australia, 2019). Australia furthermore has a history of concern for animal welfare, and has even been identified as a 'bastion' of animal rights groups (Bauer & English, 2011, p. 228). Not only do the majority of Australians generally show a low tolerance for what they perceive as animal mistreatment (Chen, 2016), but levels of community opposition to a range of animal management techniques have been shown to increase with their lethalness (Burgin et al., 2015). This concern extends beyond the realm of companion animals, with a report commissioned by Australia's federal government finding, for instance, that 95 per cent of Australians considered the welfare of livestock animals to be insufficiently protected within standard industry practice (Futureye, 2018). However, although most Australians might express concern for livestock welfare (Coleman, 2007, 2018; Mazur, 2006; Southwell et al., 2006; Taylor & Signal, 2009), this concern does not seem to translate consistently into practices of purchase and consumption of animal products, showing a lack of attention to livestock welfare (Chen, 2016). In other words, while Australians might reveal a high level of interest in, and emotional engagement with, the topic of animal welfare, this is inconsistently translated into practices that support high animal welfare. There are also discrepancies apparent with regards to Australians' expressions of concern and preparedness to take action across diverse species. That is, research has shown that many Australians still show rapid drops in levels of concern for animal welfare as 'we move from animals that live in close proximity to most Australians to those that do not' (Chen, 2016, p. 51). Such a 'hierarchy of concern' (p. 52) for different animal species is, as noted above, also evidenced in the discrepancies in legislated protection.

What is key is that these different kinds of human-animal relationships are all informed by a broader narrative of anthropocentrism which normalizes any discrepancies regarding the consideration of animal welfare. Resting on the anthropocentric view that only humans possess moral status and value in their own right, this is a narrative that takes for granted four major assumptions and their associated practices. These are as follows.

The first is that the human exploitation of animals for products (food and fibre, for instance), for labour and for human entertainment is justifiable. The second is that human development interests and the responsible management of ecosystems justify animal harm and animal deaths. The third is that all (nonhuman) animals should be regulated for their impact on the community and relevant ecosystems; and, finally, the fourth is that little to no attention needs to be given to animal interests and lives unless there is some legal obligation to do so. What these assumptions make clear is that the management of animals through reference to human interests is completely normal. It must also be noted, however, that these points can and do support a concern for animal welfare – but, as noted above, this is always a consideration dictated by human interests. That is, 'animals are generally protected only to the extent that their welfare coincides with human interests' (Nurse, 2016, p. 175). It is no surprise, then, that an anthropocentric narrative of animal life dismisses arguments for animal rights and freedoms as unreasonable, even extremist; an attack on human wellbeing and economic security (Munro, 2004; Sorenson, 2009, 2016).

Australia's Legislative Protection of Animals

Importantly, it is this kind of broadly anthropocentrist position with regards to animals that continues to inform their legislative protection in the Australian context. Not only are there still no rights attributed to any animals in Australia – meaning that animals are not considered in the broader society to possess moral status in the way that humans do – but there is even no national Australian law applying to animal welfare and setting out basic principles and protections for animals, such as are contained in New Zealand's Animal Welfare Act 1999. As has been summed up in the Animal Protection Index (API) 2020 report on Australia (in which Australia slipped from a previously held C to a D ranking, placing Australia behind New Zealand, Mexico and Malaysia's C ratings and the United Kingdom, Sweden, and Switzerland's ratings of B):

> The Australian Commonwealth does not have a Minister of Animal Welfare or Associate Minister of Animal Welfare and none of the eight states or territories has a Minister or Associate Minister dedicated to animal welfare. Australia does not have a Commissioner for Animals at either federal or state level. The website of the Department of Agriculture and Water Resources states that its Ministers are the Minister for Agriculture and Water Resources and the Assistant Minister for Agriculture and Water Resources. There is no reference to animal welfare. (as cited in World Animal Protection, 2020, p. 48)

As a result of these deficits, neither the Commonwealth Prime Minister nor other Commonwealth Ministers accept any responsibility for promoting animal welfare as a priority – a point that suggests that animal welfare is not considered to be a crucial national issue. Despite calls for the establishment of an Australian Commission for Animal Welfare or other such body (see, e.g., RSPCA, 2017; World Animal Protection, 2020), there are also no current national government bodies with the remit to provide advice, governance and leadership on animal welfare issues.[4] Rather, Australia's individual state and territory governments are responsible for setting and enforcing animal welfare standards through their administration of state legislation for animal welfare or the prevention of animal cruelty. This has resulted in, at minimum, eight separate pieces of legislation aimed at preventing cruelty to animals which differ in many of their specifics (Englefield et al., 2019; Geysen & White, 2009; Morton et al., 2021).[5] Within each Australian jurisdiction, enforcement of the relevant animal welfare law is shared by government agencies and the state body of the Royal Society for the Protection of Animals (RSPCA) (Morton et al., 2020). (Note, however, that as the RSPCA is a charitable organization, its ability to enforce anti-cruelty statutes is limited by strict budgetary constraints.) It is also worth noting that while the wording of the various pieces of legislation provides recognition of the ability for animals to be harmed or experience suffering as a result of either acts or omissions, there is an evidential burden in most Australian states and territories of proving in court that distress or injury has occurred (Cao, 2015). In other words, animals must endure some degree of suffering before animal cruelty provisions can be applied and enforced.

While state and territory legislation do all implicitly recognize that animals feel pain by providing for the prevention of pain in some circumstances, varying degrees of animal pain and psychological trauma are still legally justified in Australia in the contexts of farming, research and teaching, and entertainment (World Animal Protection, 2020). In general terms, while there are variations between state and territory legislation regarding animals and the protection of animal welfare,[6] there are also several common features with regards to animal protection regulation. Common features include: (a) the normative understanding that domesticated animals are units of personal property; (b) the adoption in legislation of a generic (and highly ambiguous) standard of *no unnecessary suffering* for the assessment of allegations of cruelty to animals; (c) the use of exemptions in specified circumstances (e.g. in research and animal agriculture) from legislated prohibitions against cruelty; and (d) the establishment of and reliance on one of the key institutional actors in the animal protection field, the RSPCA (White, 2016). As is evident, these understandings and standards are oriented at best to the protection of animal welfare.

As noted above, the Australian Commonwealth and, by extension, the state and territory governments continue to justify several exemptions with regards to the protection of animals from cruelty. For example, the welfare conditions of farm animals in Australia are not protected by anti-cruelty legislation but have rather been elaborated in the form of Model Codes of Practice for industry. Commissioned by the Primary Industries Standing Committee and endorsed by the Primary Industries Ministerial Council, these Model Codes of Practice provide guidance on the farming of a range of animal species – covering farming practices from rearing to transport and slaughtering. Although these codes are not legally binding and are best described as sets of guidelines detailing minimum (rather than *best*) welfare standards, they have been incorporated by most states and territories into their legislation. Within these jurisdictions, administration of animal welfare legislation is then delegated to the departments of agriculture (or equivalent). These departments vary greatly in their progress to implement the Codes into law and compliance with the Codes is not always mandatory nor well regulated. This has resulted in variations in laws, discrepancies between state laws and Codes, and current laws that still permit levels of animal suffering unacceptable to the broader community (World Animal Protection, 2016). It is also worth noting that the absence of national leadership in the field of farmed animal welfare is clearly at odds with the geography of the industries, which are ultimately national in their operation. Indeed, a national forum in 2015 hosted by the Australian Veterinary Association, National Farmers' Federation, and the RSPCA resulted in agreement from many participants that national leadership, coordination and consistency is 'required to promote strategic thinking, partnerships and shared investment rather than a patchwork of differing standards' (as cited in World Animal Protection, 2016, p. 6). Although the Commonwealth is working to replace these Model Codes of Practice with nationally agreed Australian Animal Welfare Standards and Guidelines (Australian Government, 2021), these standards and guidelines are not only developed, cooperatively, by government and livestock industries but jointly funded by both groups, and their development is proceeding extremely slowly. With regards to these standards and guidelines, the standards are outlined as the legal obligations which should be met by any person in charge of animals. The guidelines, conversely, outline the recommended practices to achieve the desired animal welfare outcomes. However, while they may exceed the legal welfare requirements specified in the standards, they are not definitions of best practice, and non-compliance does not constitute an offence (Manning et al., 2021). Indeed, RSPCA analysis of the standards and guidelines produced to date makes the point that they simply reflect current industry practice and fail to raise the bar on animal welfare standards (RSPCA, 2021b). (Native wildlife, feral and pest animals, as well as animals used in research, are faced with similar kinds of discrepancies with regards to their legislated protection; see Chapters 5 and 6 for discussion of the welfare considerations of native wildlife, and of feral and pest animals in Australia.)

Consideration of the legislation and standards regarding livestock animal welfare makes several inter-related failings visible (RSPCA, 2017; World Animal Protection, 2016). These include, first, a lack of independence and transparency in the standards development process with, for instance, a disproportionate level of influence continuing to be awarded to livestock industries in the development of national farm animal welfare standards

(World Animal Protection, 2016). Second is the failure to consider community values and expectations with regards to animal welfare in any systemic way (RSPCA, 2020). Third is inconsistency in the scientific basis for different standards, with different concepts in animal welfare science being adopted across various standards. Such diversity in concepts leads to the prioritizing of different measures of welfare – for instance, basic biological health and functioning as opposed to affective states (how animals feel) as opposed to so-called *natural* living – and thus the adoption of different methodologies and interpretations of scientific data (RSPCA, 2017). This is the point that while often agreeing in practice, these differing concepts represent views of welfare that may not necessarily coincide:

> A pig farmer using criteria based on biological functioning might conclude that the welfare of a group of confined sows is high because the animals are well fed, reproducing efficiently, and free from disease and injury. Critics using other criteria might conclude that the welfare of the same animals is poor because they are unable to lead natural lives, or because they show signs of frustration and discomfort. (Fraser, 2010, p. 48)

Finally, institutional conflicts of interest have long been noted. Such conflicts are very clear with regards to the case of livestock welfare given the influence of the livestock industry on the setting and enforcement of welfare standards. This is the point that 'animal welfare is likely to be of secondary importance when the primary objective of the agency responsible for livestock welfare is to promote a productive and profitable agricultural sector' (Productivity Commission, 2016). (This conflict of interest is very visible in the adherence of national, and state and territory governments to live animal export despite the cruelty recognized as inherent in the industry; see Chapter 2.)

What these various points make visible is that Australia's legislative framework for animals is, at best, oriented to welfarism as the humane treatment of animals. This welfarist orientation is further emphasized by the awarding of much of the work of animal protection in Australia to the RSPCA. This is the point that the RSPCA too is oriented to welfarism, a point made very clear throughout its websites:

> RSPCA Australia, as the federation's national body, is a leading source of animal welfare science, and works with governments and industries to progress animal welfare across a range of issues. In addition, RSPCA Australia runs the RSPCA Approved Farming Scheme, which works with farmers and brands to improve the lives of millions of farm animals. (RSPCA, 2021a)

The alignment of the RSPCA to welfarist principles is further emphasized by its objectives, which combine a stress on the protection of animals through legislative means with a stress on the need to improve animal welfare and educate stakeholders and the broader Australian community regarding the humane treatment of animals (RSPCA, 2019e). An RSPCA campaign from earlier this century makes this welfarist orientation very explicit, stating that the RSPCA 'does not oppose the farming of animals. We just think they should be given a fair go' (as cited in Glasgow, 2008, p. 186). There is an important point to note here. This concerns a key aspect of welfarist approaches to the prevention of cruelty to animals which is that meaningful improvements in the treatment of animals can be achieved through the existing legal framework and without any change to their foundational anthropocentric status as resource and property (Shyam, 2018). Such assumptions are very clear in the RSPCA Australia vision, objectives, and standard practices.

Such assumptions are also adhered to more generally throughout Australia's socio-political domains. Part of this is due to some animal activists and activist organizations gaining a reputation for extremism for engaging in more or less serious acts of unlawfulness to promote their cause. Such acts include various forms of surveillance, trespass and direct action (the latter including damage to public and private property, the infliction of mental or physical injury on participants in animal cruelty (or on their families), and economic sabotage) as well as information provision (Bagaric & Akers, 2012). The key point here is that not all forms of action are perceived as justifiable within the broader Australian society (Pedersen & White, 2021). Certainly, some such actions may be condoned if they bring to light obvious – and clearly gratuitous – acts of animal cruelty (O'Sullivan et al., 2017). For instance, the performance of covert surveillance by activist and advocacy organizations to expose practices of live animal baiting in the greyhound racing industry was seen as justifiable whistleblowing by most Australians once the story broke (Casey, 2015; De Brito, 2015). However, actions that are threatening or violent (to humans), or that condemn industries as a whole and those who work in them for their *use* of animals – where that *use* is not recognizably or legislatively defined as cruelty – are much less likely to be broadly accepted by the Australian public (Pedersen & White, 2021). For example, the Aussie Farms Map Project (now titled the Farm Transparency Project) is an Australian animal rights activist project that utilizes Google Maps to pinpoint the location of farms within Australia and, in some instances, includes contact details of farmers and captured images and/or footage of the conditions in which animals are kept (Barnes & White, 2020). The map created outrage in the farming community who saw it as invading their privacy and a threat to their safety (Moret, 2019). Such outrage – and the concurrent suggestion that farmers are being targeted, threatened, and bullied by activists – was received with some sympathy by the broader Australian community (Barnes & White, 2020; Pedersen & White, 2021). Such actions have also led to political responses in the form of new legislation being drafted to better 'to protect the privacy of Australian farmers and primary producers […] from the unlawful actions of animal activists' (as cited in Moraro, 2019).[7] The key point here again is that while there is strong sympathy in Australia for the protection of animal welfare, actions towards improvements in welfare are expected to take legal and/or socially acceptable forms.

Certainly, as the following chapters will show, these normative understandings and associated standards for welfare – and, indeed, for activism – are in contention and increasingly so across Australia, but the broadly anthropocentric narrative regarding animals continues to be influential. Our question is whether such a narrative, with its associated assumptions, values and practices, can be revised or replaced by something else that enables different kinds of human-animal relations. These would be relations that are not so determined by anthropocentric frameworks and that would have the potential to reshape Australia's social imaginary with regards to human-animal relations. The next chapters explore some of the narratives and counter-narratives that are being developed and used to enable this reshaping.

Notes

1. Moral status – also known as moral standing – is an important ethical concept that underpins all views and debates regarding the normal or most appropriate parameters for human-animal relations. It is, for instance, our beliefs about moral status that make legitimate (or illegitimate) our various actions toward and regarding animals. Basically, the criteria for having moral status are the criteria for being an entity towards which people accept they have moral obligations to act on behalf of that entity or for its interest (Warren, 1997). Put another way, moral status encompasses the degree to which we believe various living things deserve our consideration. Many ethical theories have, for instance, only awarded full moral status – meaning the category for receiving full moral consideration – to healthy, cognitively able human adults (a category that has also historically often been further restricted in accordance with then social and political norms). Given that the majority of human societies have relied – and still rely – on nonhuman animals not being recognized as possessing the same moral status as humans, arguments towards shifting the accepted moral status of animals are often contentious. We explore a range of these arguments and their reception within Australia throughout this book.

2. Due to the continent's long geographic isolation, Australia is one of the most important nations on Earth for biodiversity. As one of only seventeen *megadiverse* nations and home to more species than any other developed country, most of Australia's wildlife is found nowhere else in the world. More specifically, 87 per cent of Australia's mammal species, 93 per cent of reptiles, 94 per cent of frogs and 45 per cent of bird species are found only in Australia (Australian Wildlife Conservancy, n.d.).

3. Australia riding to prosperity *on the sheep's back* is a colloquial phrase referring to the prosperity that Australia derived from wool and sheep from the 1870s to the 1960s (Cashin & McDermott, 2002; Schedvin, 1979). It has been argued that the wool export industry gave Australia one of the highest living standards in the world during this period. In full the phrase is often expressed as *riding on the sheep's back*, and sometimes as *living off the sheep's back*.

4. In 2004 Australia adopted the Australian Animal Welfare Strategy (AAWS), which explicitly covered 'all sentient animals – that is, those with a capacity to experience suffering and pleasure' (Australian Government, 2011). The recognition of animal sentience underpinned the whole Strategy, as the Australian government recognized that 'sentience is the reason that welfare matters' (Australian Government, 2011). The Strategy was accompanied by a National Implementation Plan for the period 2010–14. However, the AAWS was defunded in 2013 and the Implementation Plan was not renewed post-2014. Following the dismantling of the AAWS, the Australian Animal Welfare Advisory Committee and the Animal

Welfare Committee were also disbanded. The Animal Welfare Task Group, comprising representatives from state and territorial governments, was to continue to progress some AAWS projects, but funding for both the AAWS and the Task Group ceased in June 2015 (World Animal Protection, 2016). This has left the Animal Welfare Task Group as a body with no Commonwealth funding, its membership limited to government officials, obliged to meet only twice a year and with no published minutes or actions (World Animal Protection, 2020).

5 It is worth noting that a recent review of animal protection law in Australia (see Morton et al., 2021) identified over forty current primary pieces of legislation – including both national (Commonwealth) statutes, and state and territory-based statutes – that incorporate provisions for both the protection of animal welfare or prevention of animal cruelty and their enforcement by penalty. Penalty was defined broadly to include monetary fines, custodial sentences, animal welfare directions or notices, court-mandated prohibitions or animal seizures. Note that this study focused only on statutes, meaning that delegated legislation such as regulations and codes of practice were not included. The authors also explained that they had excluded a range of other statutes. These included those put in place for the management of animals largely for public health reasons (e.g. dog and cat management acts), and any statute controlling humane killing methods (e.g. biosecurity acts). Provisions that included stealing or killing animals with the intention to steal were also excluded (sections of crimes acts), because they related to damaging personal property and were not focused on animal welfare. Statutes in force to regulate professions, such as the veterinary industry or research involving animal use, were excluded as they were largely administrative in nature (e.g. controlling licensing) and lack any enforceable animal welfare provisions. Emergency management statutes that could make reference to managing animals in natural disasters were excluded for the same reasons. Finally, any statutes that discussed the human effects of animal abuse were excluded (e.g. domestic violence acts), as these acts are in place to protect human suffering resulting from emotional abuse caused by animal cruelty. The fact that there are hundreds, if not thousands, of pieces of legislation in Australia dealing with animal welfare and protection – even if this is not explicitly the focus of all of them – further underscores the need for a national body tasked with ensuring a united front toward animal protection.

6 Key variations arise, for example, with regard to legislative definitions of *animals*. For instance, while the legislation of all Australian states and territories include mammals, reptiles, amphibians, and birds in their definitions of an animal, inconsistencies arise for aquatic species, such as fish, crustaceans and cephalopods. Many states and territories do not count them as animals at all, while others include provisions only for certain fish or crustaceans based on their use by humans. Variations also occur between Australian states and territories in definitions of *cruelty* and in lists of prohibited activities, and *animal welfare* itself is left undefined, even given

the almost universal naming of state animal protection acts as Animal Welfare Acts (see Morton et al., 2021).

7   Such legislation has become known as *Ag-gag* laws, meaning legislation designed to curb animal activist (and environmental) monitoring and investigative activities (Englezos, 2018). These laws have been used in Australia to pursue animal rights activists on criminal charges and, more recently, have prevented news outlets such as *The Guardian* from publishing footage of animal welfare violations gained covertly by animal activists (Knaus, 2021). These laws will be discussed in more detail in Chapter 2.

CHAPTER 2

Riding the Sheep's Back: From Live Stock to Sentience

As we have noted in the previous chapter, post-settlement Australia has famously envisaged itself as having ridden the sheep's back to prosperity. Indeed, livestock agriculture and related industries are still significant in the Australian economy and continue to hold political influence. This has meant narratives foregrounding the normalcy of the anthropocentric exploitation of livestock animals in animal agriculture have been told and accepted across Australia's social imaginary since settlement. Such narratives have been shaped by the four assumptions noted earlier: that human exploitation of animals is justifiable; that human development interests justify animal harm and animal deaths; that all (nonhuman) animals should be regulated for their impact on the (human) community and relevant ecosystems; and, that little to no attention needs to be given to animal interests unless there is some legal obligation to do so. Given our focus on examining how these assumptions and their associated narratives regarding animals have both informed human-animal relations in post-settlement Australia and are slowly being challenged by counter-narratives, and given the continuing influence of agricultural industries within Australia, this chapter begins our consideration of these issues by first outlining some of the ways these assumptions have played out with regards to considerations of livestock animal welfare in Australia. A second focus is on the ways these assumptions and narratives are starting to be reimagined and retold by some of Australia's peak animal welfare bodies so as to develop a stronger and more holistic orientation to not just animal welfare but animal wellbeing. As this chapter will show, activist and advocacy organizations including Animals Australia and the Royal Society for the Prevention of Cruelty to Animals (RSPCA) have been instrumental in not only (a) developing stronger and more holistic narratives prioritizing animal wellbeing; and (b) establishing such narratives in the Australian public sphere; but also (c) using the empathy generated by these narratives to mobilize the broader Australian community to recognize the moral status of animals used in agriculture instead of simply considering their economic value. Of particular focus here will be the way these organizations have used narrative approaches to entreat the Australian public to recognize the individual capacities of sentience regardless of species and thus to move away from seeing animals as simply a resource or units of *live stock* with no interests of their own deemed worthy of respect. Such a move, as this book will trace, is integral to the work of reimagining the shape of human-animal relations.

From *Live Stock* to Sentience

The first site for this work concerns the construction and use by a number of animal welfare groups of a counter-narrative of sentience, used particularly with regards to challenging anthropocentric economic narratives of *live stock*. As we have noted, under the anthropocentric narrative of justified animal use, livestock animals are defined as all animals farmed for use and profit, including poultry and aquatic animals. The narrative of animal use has thus seen livestock reduced to resources named in food, fabrics, and other materials including fertilizers, insulation, many cosmetics, beauty and household products, and so forth. This narrative always prioritizes human interests – in particular, our economic interests – over the interests and lives of animals. Indeed, the interests of individual animals are not *seen*, only an animal's fitness for use and potential economic value matters. This is a view that Bidda Jones, head of science and strategy at RSPCA Australia, acknowledges:

> It is a sad fact that farm animal welfare has always been compromised in Australia. When compared with many other developed countries, ensuring livestock are well treated is of lower priority. The phasing out or elimination of discredited agricultural practices like caged egg production, sow stalls, and other practices that prevent animals from moving freely or expressing their natural behaviour, is retarded here

compared with most nations in Western Europe and across the Tasman in New Zealand. (as cited in Jones & Davies, 2016, p. 62)

Following the trajectory of this anthropocentric assumption regarding justified animal use, animal agriculture in Australia continues to normalize the *managing* of farmed animals through drugs, surgical and other interventions, as well as the *culling* of unwanted animals. Such management is so that they *fit* easily within the space and timeframe given over to their growth into the resources that are valued more than them (e.g. Figure 3). Indeed, with successful animal farming being measured in terms of efficiency and profits per unit, a key normative focus has been the aim, for instance, of obtaining 'more edible muscle meat for less money' (Stephany, 2010, p. 361).[1]

Figure 3. Intensive poultry farming for meat at a South Australian farm in 2016.
Source: Photo courtesy of the Farm Transparency Project.

According to this anthropocentric narrative of the justifiability of human exploitation of animals, animal welfare might matter but only so long as animals' usefulness to us is still their primary value – and only so long as welfare measures will deliver a better *product* outcome. Some contemporary versions of these views may place a stronger stress on maintaining animal welfare, but never to the point of destabilizing the use framework for animals. And, indeed, some of these versions suggest that strengthening attention to animal welfare will in fact make those animals *more fit* for their purpose. Hence the arguments by many chefs that animals afforded free range and high welfare conditions during their lives (and their deaths), for example, will provide better, tastier and healthier meat and other products, thus enhancing our enjoyment in their use (see Chapter 3 for further discussion of these issues).

In the Australian context, this priority towards considering livestock animals via their use value is evident in three main ways. These are (a) the legislation used to regulate livestock animals (introduced in the previous chapter); (b) the negativity of many political and industry responses to calls to change such legislation and the treatments they permit; and (c) broader conceptual and psychological barriers to re-evaluating the prevailing narrative of anthropocentrism. To begin with, according to the legislation used to regulate livestock animals in Australia, these animals are legally property or stock with no rights to bodily integrity or liberty (Arbon & Duncalfe, 2014; Bagaric & Akers, 2012). Livestock animals' lives and deaths in Australia are regulated through a suite of Model Codes of Practice commissioned by the Primary Industries Standing Committee and endorsed by the Primary Industries Ministerial Council. (Note that the Commonwealth is slowly replacing the Model Codes of Practice with Australian Animal Welfare Standards and Guidelines, but that these are still being developed jointly by government and Australia's livestock industries.) In either case – Model Codes or the Standards and Guidelines – livestock are not covered by the provisions of anti-cruelty statutes that protect other cohorts of animals (such as our companion animals) from cruelty and welfare abuses. These Model Codes of Practice are best described as sets of guidelines detailing minimum (rather than *best*) welfare standards, and which are furthermore neither compulsory nor practically enforceable. Under this framework, anti-cruelty legislation can only be applied to a specific instance of animal cruelty if no Code of Practice covers that animal, and conversely the Code of Practice can actually be used as a legal defence by a farmer if he or she is brought up on charges of cruelty. And here it is worth noting that some actions toward animals that are forbidden under anti-cruelty legislation are permissible under the Codes of Practice – including the removal of very young animals from their mothers, the performance of surgeries without pain relief, and the keeping of animals in confinement with no or little *enrichment* of their living conditions.[2] As the volunteer-based organization Lawyers for Animals (2021) has stated, the Code of Practice *loophole* has resulted in some 500 million Australian livestock animals being excluded from the protection of existing animal welfare legislation (see also Bruce & Faunce, 2017). What these points make clear is that Australia's laws still 'fail to provide meaningful protections' to livestock animals (Bagaric & Akers, 2012; Sharman, 2009, p. 36). In addition, there is no coherent and adequately resourced strategy to enforce what little protection these animals might have.

This model of relying on industry-controlled Codes of Practice for animal welfare protection has come under increasing pressure from animal advocates and activists who have argued that there is an inherent conflict of interests between animal welfare and the production of animal-based resources. Such groups stress that it is inappropriate to have the requirements of livestock welfare decided and regulated by the same industries that produce animal-based resources for economic gain. Their point is that in such a context, any conflict of interests – between protecting animal welfare and increasing economic profits, for instance – will only rarely (if ever) be decided in favour of protecting or strengthening animal welfare. The broader issue pointed out by advocates and activists is that livestock animals are just as *sentient* as the animals protected under anti-cruelty legislation – but that livestock industries have been permitted to systematically deny the implications of livestock sentience for (human) economic benefit. Defined as the capacity to subjectively experience one's life and the world in both positive (life-affirming) and negative (life-degrading) ways (Bekoff, 2013), *sentience* is presented as the condition in ethical thinking for a being to have its own interests, and to prefer – and deserve – to have those interests respected and protected (Singer, 1975/1990). A sentient being, then, is 'a being who has interests; that is, a being who prefers, desires, or wants' (Francione, 2012). Given this capacity to have preferences, a key interest of all sentient beings is thus to live a life of wellbeing rather than of suffering. This was one of Peter Singer's main points in his 1975 book *Animal Liberation*: 'If a being suffers there can be no moral justification for refusing to take that suffering into consideration' (1975/1990, p. 8). He also stresses that all sentient beings – no matter their species – deserve to have their interests considered in any actions that would impact upon them. In his words: 'the interests of every being affected by an action' should always 'be taken into account and given the same weight as the like interests of any other being' (p. 5). As he also says, this recognition of sentience, of the possession of interests, and of the weighting of interests against each other, should not be automatically biased along species lines. This would be to practise what he calls *speciesism*. As with sexism or racism, this is defined to mean that we are ignoring or differentially weighting the similar interests of members of different but strongly related groups, such as preferencing men over women, members of one racialized group over others, or humans over nonhuman animals (p. 6). Here the point is that as sentient beings able to experience pain and suffering as well as their own wellbeing, livestock animals should also have their welfare requirements properly recognized and protected.

The argument here is that sentience should mark a basis for at least some moral consideration. That is, the 'capacity for feelings of pleasure and pain and for the forms of life of which animals are capable clearly imposes duties of compassion and humanity in their case' (Rawls, 1971, p. 512). The challenge, of course, is whether such duties can overrule the *use* of animals for human benefit. In Australia, calls to significantly change the treatment and regulation of livestock animals on the basis of their sentience and our corresponding moral obligation have met with significant resistance from industry and governmental bodies. Indeed, as we noted in the previous chapter, protests against the treatment of livestock animals have a history of being framed in Australia as attempts to damage the national interest and as ill-informed attacks on the practices and importance of Australian agriculture (see, e.g., Murphy, 2018; Wagstaff, 2015). For example, in opposition to activist challenges to Australia's live animal export industry regarding levels of animal suffering, the industry response has consistently been that any ban on live exports – or on broader industry-accepted and legal practices involving livestock animals – would 'have serious negative consequences for Australia' (Wagstaff, 2015; see also Jooste, 2016; Keogh, 2013; The Livestock Collective, 2021).[3] Such a view has given rise to three interconnected tendencies. First, as Bidda Jones from RSPCA Australia has noted, should activist organizations gain footage or other evidence of obvious welfare abuse, the ongoing tendency has been for it to be dismissed by both industry and political bodies 'as the exception rather than the norm' (Jones & Davies, 2016, p. 57). Framing issues in terms of *isolated instances of cruelty* means that livestock workers or companies specified in footage can be targeted rather than the industry as a whole, meaning that industry-accepted practices and the companies that carry out such practices remain unchallenged and unchanged (Englezos, 2018). A second set of industry and political techniques in the face of activist challenge has been to vilify activist aims. Thus the National Farmers Federation (NFF) has dismissed campaigning by Animals Australia and other activist and advocacy organizations to improve the welfare conditions of livestock animals as an attempt to make the country *go vegan*, demand the complete abolition of human use of these animals, and harm the national economy (Bettles, 2013b, 2017; Rodan & Mummery, 2018). Public protest activities have finally tended to be dismissed by livestock industries and indeed the government as 'reflections of a lack of community knowledge or understanding' (Coleman, 2018, p. 18).

The economic argument has been used to excuse ongoing livestock animal cruelty for many decades. For instance, the cruelty and animal welfare abuses endemic to live animal export have long been recognized, with a Senate Select Committee on Animal Welfare stating in a 1985 review that if the decision about the live export trade was based 'purely on animal welfare grounds, there is enough evidence to stop the trade' (as cited in Petrie, 2016). In Bidda Jones' reflection on this review, as head of science and strategy at RSPCA Australia, she notes that the report outlined the division between the values-based arguments of animal welfare advocates in seeking an end to the trade, and the monetary-based arguments of the industry and government agricultural economists. The report then stated outright that the committee 'has found it difficult to reconcile economic value with animal welfare' (as cited in Jones & Davies, 2016, p. 62). More recently, the 2018 Moss Review, which was commissioned in response to the multiple waves of exposé footage gathered on the welfare abuses of Australia's current live export industry, stated that although 'live animal exports present a high risk to animal health and welfare', and despite increasing levels of public opposition, the industry can still be continued (Moss, 2018, p. viii). This general argument – that animal welfare should remain 'divorced from economic and other considerations' (as cited in Petrie, 2016) – is common to successive Australian governments, industry bodies, and producers. It is clearly present in calls for Australia to 'fill the gap that New Zealand will leave' as New Zealand implements its ban on live animal export in 2021 on the basis of animal welfare concerns (Sullivan & Verley, 2021).

The Australian industry bodies administering and representing the live export trade have tended to justify the trade as a world food resource and production issue – as helping meet the global 'demand for essential red meat protein' (see, e.g., Centre for International Economics, 2010). According to this perspective, banning live export would do 'irreparable harm to communities around the world' (The Livestock Collective, 2021). Cases are further made for the significance of Australian live animal export for countries such as Indonesia and Vietnam (Burton et al., 2018; The Livestock Collective, 2021). Industry responses to activist campaigning with regards to animal welfare have in turn consistently downplayed welfare concerns, but concurrently contended that significantly strengthening welfare requirements for livestock, for instance, would decrease the profitability of agribusiness to the detriment of Australia's economy (Eyers, 2016).[4] All such attitudes exemplify the denial of sentience in livestock animals. Such a denial is further exemplified in the paradoxical situation that has arisen from beginning moves to recognize animal sentience in Australian legislation. For instance, despite amendments to the *Animal Welfare Act 1992* (ACT) in 2019 that have meant that animal sentience and the 'intrinsic value' of animals are now legally protected in the Australian Capital Territory (ACT), the new animal welfare offences created by the Act still do not apply where the relevant conduct is in accordance with an existing Code of Practice (as cited in Kotzmann, 2019). In other words, animal industry practices that might otherwise be considered instances of animal cruelty are once again exempted from the requirements of the legislation.

Political and industry resistance to calls for change in the livestock industry by animal activists has, as in other parts of the world, taken the form of attempting to legislate against activist action itself. Thus activists who engage in public protest activities or covertly access private property to record acts of animal cruelty are being framed in industry and some political contexts as economic saboteurs, 'un-Australian' and 'green collared criminals' (SBS News, 2019), even as 'terrorists' (ABC News, 2013; Greer, 2013; Sorenson, 2016). Protest activities have further led to proposals that Australia's federal government would be prepared to support farmers in any legal claims against protesters (SBS News, 2019), as well as attempts to institute within Australia what have come to be called *Ag-gag* laws. As was noted in Chapter 1, an *Ag-gag* law is the name given to legislation designed to curb animal activist (and environmental) monitoring and investigative activities. They are laws which effectively *gag* or reduce the possibility of public awareness and discussion of controversial issues in agriculture (Englezos, 2018). Proposals to introduce such legislation first arose in the United States in the late 1990s and, where these laws have been enacted, they typically contain one or more of the following provisions:

- a prohibition of taking photographs or video footage on or in an agricultural facility or property without the permission of the proprietor;
- a prohibition on seeking employment with an agricultural business under false pretences or without disclosing ties to animal rights organizations; and
- a requirement that any documentary evidence of animal mistreatment is reported to relevant authorities within a short time frame, often a 24 to 48 hour period. (RSPCA, 2019h)

Such laws seek to duplicate existing trespass laws, but with a twist. Seriously increasing penalties, making it illegal to distribute or broadcast images that have not been surrendered to the police within a specified period, and making it a crime to seek employment with the aim of exposing animal suffering would mean that animal activists would struggle to present instances of animal cruelty as systemic problems, as well as be unable to alert the broader community to otherwise invisible animal suffering (Englezos, 2018; Gelber & O'Sullivan, 2021; O'Sullivan, 2015; Shea, 2015).

Another factor can be identified alongside these forms of resistance to calls to see past the potential use value of livestock animals to their sentience and hence their need for protection against cruelty and welfare abuses. This is our deep saturation with anthropocentric narratives of speciesism and human superiority. According to such narratives, as discussed in the previous chapter, it is simply wrong-headed to equate human and animal interests, and it is *natural* that animal welfare concerns take a second place to human economic interests (Munro, 2004). As one response to the screening of an exposé of the 'shocking abuse of [Australian] cattle in an Egyptian abattoir' puts this view, the mistreatment of animals caught on this footage is 'horrific and appalling', but we must remember that 'animals are food' (Flint, 2013). Under this perspective, animal mistreatment may be shocking, but at the end of the day, animals are nothing more than their use value. Research into the prevalence and traction of such views – and into the attributes of those rejecting such views – has suggested that coming to take animal activists' views seriously actually entails a turning away from dominant cultural ideologies. That is, coming to care about animals in more than superficial ways requires in many situations a de-socialization in relation to dominant norm and value systems and a resocialization into an extended form of care wherein animal sentience matters (Pallotta, 2005; Rodan & Mummery, 2016). More specifically, it has been argued that becoming an animal activist requires a 're-engineering' of an individual's understanding of the world and of human-animal relations (Hansson & Jacobsson, 2014, p. 263). This is to the point that the individual not only cares about but actively strives to protect the interests of animals – becoming, as we have examined elsewhere, an 'everyday activist' (Rodan & Mummery, 2016). The argument here is that we need to learn how to care, and through this to realize that it is unethical to continue to weight human economic interests more heavily than animal suffering (Singer, 2018). It is this work of public re-education that is being undertaken by animal activist and advocacy organizations.

Storytelling with Sentience: Animals Australia and the RSPCA

> Sheep are not iron ore or wheat. They are sentient beings who cannot protest against what we are doing to them, except by dying.
>
> – Singer, 2018

It is consequently this aim of making visible and problematizing the levels of suffering perceived as acceptable for the animals who continue to be seen primarily as units of economic value, as *livestock*, that drives much of the campaigning work of animal advocacy and protection organizations. Thus, organizations such as Animals Australia and RSPCA Australia have developed a range of narratives to remind Australians that all animals, including livestock animals, are sentient beings and that sentience matters. Although the presence of such narratives can be identified in the advocacy work of the majority of these organizations, they are given particular prominence by Animals Australia. Very briefly, Animals Australia is a not-for-profit organization representing over 2 million individual supporters. It is recognized – along with RSPCA Australia and Voiceless[5] – as one of Australia's foremost national animal protection and advocacy organizations (Chen, 2016). Animals Australia prioritizes several missions. These are to (a) investigate, expose and raise community awareness of animal cruelty; (b) provide animals with the strongest representation possible to government and other decision-makers; (c) educate, inspire, empower and enlist the support of the community to prevent and prohibit animal cruelty; and (d) generally strengthen the animal protection movement (Animals Australia, n.d.-a). In delivery of these, the organization engages not only the platforms of public rallies and protests, government and industry submissions, corporate outreach, print and broadcast media (television, radio, newspapers, and billboards), as well as the web to present its campaigns, but also multiple forms of social media, including Facebook, YouTube and Twitter (Rodan & Mummery, 2014, 2018).

Most importantly, Animals Australia very explicitly uses narrative forms in its campaigning work in order to foster emotional resonance and empathy, and transportation and identification among audience members. As noted in the previous chapter, such capacities are considered integral for promoting changes in audience beliefs and behaviour (Witherell & Noddings, 1991). Although all of Animals Australia's campaigns use this format, two in particular engage narrative forms and devices very strongly. These are the *Make It Possible* campaign (and its sequel campaign, *Somewhere*, released in 2020) and the organization's broader campaign and exposé work against live export (Animals Australia, 2012, 2020c). Launched in 2012, and still active, *Make It Possible* tackled the welfare abuse of pigs and chickens within the Australian intensive farming industry. It was designed as a visually compelling multimedia campaign, using both still images and videos of not only animals within factory farm contexts but of celebrity and ordinary Australians – who are filmed expressing their horror, disgust, and determination to make a difference within this situation – to connect with, shock, and engage viewers in both personal and collective activisms both online and offline. Both the campaign narrative – the expression by livestock animals of their dreams for a world and life where freedom and kindness are the norm, and where they can live their lives in pursuit of their own interests (see the transcript, Animals Australia, 2012) – and images are used to drive a single message home across all of the campaign's platforms and genres. This is that factory farming is a major cause of animal cruelty; all factory-farmed animals experience a life of intolerable and unnecessary suffering; and each of us can and should work to end the factory farming of animals (see Lyn White as cited in ABC Landline, 2013). (This message is reiterated in Animals Australia's 2020 campaign, *Somewhere*.)

Combining *Babe* (1995) style animation effects with real footage from Australian factory farms – visual effects which won the makers a Mobius Award in 2012 – the *Make It Possible* YouTube video presents animals as being fundamentally *like* the viewers, possessing a similar interest in living a life of wellbeing. Engaging viewers' tacit knowledge of how it feels to be restricted, along with their empathy, animals in factory farms are presented in the campaign video (see the transcript, Animals Australia, 2012) as suffering subjects who are easily identified with. They are presented as yearning for a better life, for 'a new way of living'. They are described as 'living lives of abject misery', as 'waking up each day, just to suffer'. Such descriptions are reinforced with footage showing confinement, overcrowding, lightlessness, and industrial sterility. The campaign reminds viewers that the animals kept in these barren and constrictive conditions are 'no different to our pets at home'; that they are 'someone, not something'. Further driving this message home, animals in this campaign are explicitly anthropomorphized, given human voice and facial expressions. They are also individuated, with recurring close-ups of animal faces and eyes, directly challenging any tendency to see these animals as nothing more than a resource to be used, as *livestock*. With its final scene of a winged pig escaping confinement, the video also engages a powerful social narrative of exile, alienation, and hope. As Anthony Ritchie, one of the campaigners with Animals Australia, stressed with regards to the development of the *Make It Possible* video, this narrative focus on likeness and the facilitation of identification and empathy was integral:

> Pigs and chickens aren't animals that people instantly connect with or have empathy for so our first task was simply getting people to like them – to think about them in a different way and to understand that these animals share the same capacity to suffer and to feel love as our dogs and cats at home. The success of movies like *Charlotte's Web* and *Babe* gave us a great formula to work with and that's what we had in mind when we created our 'hero pig'. The rest of the TVC uses real footage from factory farms in Australia – it was critical that what we were showing reflected the current situation for most animals raised for food in Australia today. Finding the song 'Somewhere' and obtaining the rights to use it brought the vision together. We always knew that if animals could plead their own case for a kinder world then factory farming would have ended long ago and the words to 'Somewhere' so beautifully encapsulate our core message – that at the very least animals raised for food should be provided with a life worth living. (as cited in van Gurp, 2012)

Similar strategies are used throughout associated campaigns with regards to the treatment of livestock animals in Australia's live animal export industry. These campaigns have primarily been based on the forty plus exposés of the industry carried out by Animals Australia, sometimes in association with Animals International, RSPCA Australia and other activist organizations. In each of these campaigns, a harrowing series of video exposés and still images have captured the extreme cruelty and suffering experienced by these animals, whether during their transportation and/or in their last moments.[6] Many of these images have been broadcast through national television and shared over multiple forms of media. In these instances – see, for example, 'A Bloody Business' featured on *Four Corners* (Doyle, 2011) on May 30, 2011, and 'Sheep, Ships and Videotape' aired on *60 Minutes Australia* (Sacre, 2018) on April 8, 2018 – although only documented surveillance footage is used, animals are again individualized and personalized. It is impossible not to see the suffering of individuals in their final fights for life. This is particularly exemplified in the following analysis of footage screened as part of 'A Bloody Business' on *Four Corners*:

> For many people, one of the haunting and lingering images of the footage was of a black steer standing by as his fellows were brutally killed. As the steer watched what was happening, and just before he too met the same treatment, he trembled in fear. The eyes and body, the rippling black skin and demeanour, were vividly expressive of sheer physical terror, and perhaps even of an animal, bodily fear of death [...] (Coghlan, 2012)

Animal experience is thus made visible by and described in these exposés in terms of individual sensation. In 'A Bloody Business', for instance, animals are shown experiencing 'extreme pain'. They are 'kicked, slapped, prodded, goaded with sticks', have 'their tails pulled, twisted and in some cases broken', and 'some had their eyes poked and gouged or were roped and dragged along the ground' (Jones & Davies, 2016, p. 55). RSPCA analysis of the prevalence of pain causing actions being experienced by these animals identified that they were each 'subjected to an average of 22.5 goads, 2.00 tail pulls/twists, 5.75 tail bends and 8.00 eye gouges' (Jones, 2011, p. 19). In describing the expressions of fear and stress by the animals shown in 'A Bloody Business', Bidda Jones, head of science and strategy at RSPCA Australia, described the vocalization of one particular steer: 'the tongue is coming out, so clearly distressed. You can see from his eyes that he's distressed. These are all behaviours that are indicative of fear, anxiety, distress' (as cited in ABC Four Corners, 2011). In the *60 Minutes Australia* episode 'Sheep, Ships and Videotape' (Sacre, 2018), animals are shown suffering severe heat stress and literally cooking to death 'covered in waste and desperately gasping for air and water in extreme heat' (Dalton, 2018).

These exposés thus set out narratives of individualized terror and suffering, displaying the experiences of animals in ways that we cannot help but empathize with and be horrified by. These exposés – and the *Make It Possible* campaign – strive to remind us all that 'A live animal is not a sack of potatoes. He breathes. He thinks. He can suffer' (Animals International, n.d.). They all strongly suggest that permitting such suffering is inexcusable regardless of its profitability. Indeed, in the words of Voiceless Patron and former High Court Justice, Michael Kirby (2011), the narratives filmed and presented to us of the suffering of animals in live export make clear that the 'paramount consideration must now be the ethical one. The live export trade as currently carried out is indefensible. It must stop'. As such, both kinds of campaigns use narrative techniques to call on viewers to recognize and empathize with livestock animal sentience and their suffering. These are calls to recognize the similarities between animals' experiences in such conditions and what our own experience would be in similar conditions, and in that process to recognize ourselves as essentially compassionate and caring but ignorant of what is happening to these animals. Some of these narrative techniques are as follows. First both kinds of campaigns stress the realism of their contexts – this is seen as fundamental for establishing a need for change able to be recognized and understood by a broad audience. As was noted above, although *Make It Possible* does use a range of animation techniques, this work is applied to real footage from Australian intensive farms. This is Ritchie's point about it being critical that the campaign video reflects 'the current situation for most animals raised for food in Australia today' (as cited in van Gurp, 2012). It is footage that correlates with other exposés we might have seen about the situation of animals in intensive farming contexts. In the case of the live export footage, this is documented reality presented through the format of investigative journalism. It is a reality that is further established in the terms of veterinary and scientific expertise. Thus the footage used in 'A Bloody Business' was analysed by the RSPCA's chief scientist with regards to scientifically accepted indicators of animal suffering (Jones, 2011). Indeed, both the RSPCA and Voiceless consistently reference the expertise of scientists, animal welfare and legal experts, the industry itself, and government reports in their responses to the live export exposés (see, e.g., Voiceless, 2019).

Along with highlighting the realism of the need to recognize and protect livestock animals against cruelty and welfare abuse, these narratives also explicitly draw on emotional modes of appeal to foster empathy, care, and support for change. There are three techniques that are in common use across both kinds of campaign. These are (a) the use of emotionally resonant images, (b) the use of emotionally resonant descriptions, and (c) the framing of narratives in explicitly moral terms. Of these, the first marks recognition of the point that images enhance the transmission of 'intended meanings' (Aiello & Parry, 2020, p. 4; Rose, 2016) whilst also being a 'catalyst to set off a chain reaction of mass emotion' (Mitchell, 2013, p. 96). Indeed, 'emotionally arousing images' are highly effective in drawing a response from viewers and facilitating participation in some form of collective action (Kharroub & Bas, 2016, p. 1977), to the point that activist groups use visuals in their campaigning as a 'tactical strategy' (Carty & Onyett, 2006, p. 237). Visuals, furthermore, make things public; they generate individual and public emotion as well as communicating everyday public concerns (Aiello & Parry, 2020; Hariman & Lucaites, 2016). They are an integral part of 'the relational processes through which particular relations of social power' can be reinscribed as issues of personal as well as 'political concern and concrete transformation' (McLagan & McKee, 2012, pp. 9–10). In these campaigns, then, video and photographic images show close-ups of animal eyes and faces in living distress – the sow unable to reach her piglets, the weeping bull, the bull bleeding from the eye, the heat distressed sheep with protruding tongue, sheep panting for air. Such images call for our recognition of our likeness with these animals and for our response to them.

Similar work is done with narrative description. Thus, in the narratives of both campaigns, *vulnerable* animals are described as being subjected to *unimaginable cruelty*. Live export, as another Animals Australia campaign initiative has made very explicit, is named 'a crime against animals'; marking a trip that 'should never be taken' on 'death ships' (Animals Australia, n.d.-a). Animal experience is further described in terms of 'sheer physical terror' (Coghlan, 2012), as enduring 'gross horrible abuse' (see Doyle, 2011) and 'atrocious cruelty' (Textor, 2011). Subjected to a range of 'horrifying acts of brutality', many of these animals 'die slow and hideous deaths' (see Doyle, 2011). In the *Make It Possible campaign*, descriptions focus on the confinement of animals, that they are living in an industrial not a natural environment. They are described as having been denied 'the simple joys in life that we take for granted, freedom, sunshine, fresh air and exercise' (Animals Australia, 2012), and as unable to stretch or to carry out species-natural behaviour such as caring for their own young. Such descriptions of subjective responses are recognized as being emotionally resonant. In one study of Australians' attitudes to live export, those who had seen recent footage from another live export exposé were asked how they felt about it. Their commonest response was 'sadness' (22.5 per cent), with another 20 per cent saying that they mostly felt empathy towards the animals, 15 per cent expressing 'anger', and 13 per cent describing the situation as 'sickening' (Sinclair et al., 2018). This is very similar to responses to the *Make It Possible* campaign, with content analysis of sets of the public responses posted in Animals Australia's websites showing the top seven *feelings* used by respondents to be: sickness, horror, disgust, anger, sadness, shock, and being brought to tears (Rodan & Mummery, 2016).

Finally, campaigns and exposés focused on animal welfare typically frame themselves in moral terms. These terms tend to be drawn from a focus on sentience as a capacity we share with animals, and the associated argument that overriding this capacity in the pursuit of profit is unethical and unjustifiable. These campaigns thus tell moral stories with the explicit aim of reframing audience attitudes (Mummery & Rodan, 2019b; Munro, 2015; Singer, 2018). As such, the narrative of sentience informing these various campaigns and exposés encourages viewers to recognize these animals as being *like* them in important ways, as well as to recognize themselves as being essentially compassionate and caring but ignorant until now of the real situation of animals in these industries. Viewers are thus positioned as not knowing of the terrible price paid by animals in these industries; as not knowing that these industries represent an 'animal welfare disaster of a magnitude this planet has never known' (Animals Australia, 2012). Importantly, as emotionally resonant moral narratives, these campaigns and exposés are also calls for change, aligning with the view that action and mobilization 'arises from affective stakes – caring for someone, outrage about an injustice, or aspiration for a changed world' (Shotwell, 2011, p. 100). Insofar as concern for animal welfare is 'framed in terms of the suffering of individual animals' (Jones & Davies, 2016, p. 58), it is no surprise that 'the public's ire' can be raised 'by the moral shock of seeing animal suffering' (Munro, 2015, p. 10).

A Call to Action and a New Vision

> If it is accepted that animals are sentient and possess certain capabilities, it is logical to believe that these capabilities should be safeguarded through the adoption of rights and freedoms. This sentiment is expressed in the quantitative results, revealing high levels of agreement on rights and freedoms for animals, particularly relating to freedom from pain and cruelty. Specifically, this included the right not to be subjected to unnecessary pain and suffering; freedom from thirst and hunger; pain, injury and disease; fear and distress and from discomfort by providing appropriate environment, shelter and comfortable resting area.
>
> – Futureye, 2018

The narrative of sentience common to animal welfare campaigns thus calls on viewers to recognize their likenesses to (nonhuman) animals, to care about animal experience, and to act so that animal suffering (at least) can be minimized. Such campaigns present the idea that industries permeated by high levels of animal suffering should be unjustifiable to any truly moral Australian (Mummery & Rodan, 2019b; Munro, 2015). With their focus on telling the animal story in terms of sentience and individual experience rather than use value, such campaigns and campaign materials are designed to generate individual and public emotions of shock, horror and disgust, sympathy and empathy, as well as the mobilizing emotions of anger and outrage. And, importantly, there are signs that such attitudes are starting to take hold. Certainly, such attitudes tend to be very visible in the aftermath of activist exposés or other media investigations into animal cruelty. For instance, in the aftermath of 'A Bloody Business' members of the Australian Parliament reported being swamped by calls (Chen, 2016; Jones, 2011); Twitter was inundated with opinion tweets of 'shock and disgust' (Textor, 2011) and a 200,000 strong petition against live export was tabled in the Australian Senate (Animals Australia, 2011). Such peaks in public attention are expected, but public support for the better protection of animal sentience does appear to be growing and becoming more mainstream (Buddle et al., 2018; Coleman, 2018; Coleman et al., 2015, 2017; Futureye, 2018; Parbery & Wilkinson, 2012; RSPCA, 2018). Indeed, a 2018 survey (of 1,521 Australians) showed that 95 per cent of those surveyed view farm animal welfare to be of concern, with 91 per cent wanting at least some reform to address this, and 40 per cent seeing the need for significant reform (Futureye, 2018).

With public concern for livestock welfare gaining traction in the Australian context, public mobilization on behalf of the sentience of livestock animals is effecting some changes in what is normalized in industry, company, and consumer practices. With increasing numbers of Australians engaging in protest activities against live export, for instance (see, e.g., Figure 4; RSPCA, 2018), a number of legislative proposals to change and/or end the live export industry have arisen (see, e.g., Petrie, 2016, 2018; C. Petrie, personal communication, October 17, 2018).

Figure 4. RSPCA, Animals Australia, leading politicians, and thousands of concerned Australians gathered in Australian capital cities to rally for a kinder future free from the cruelty of live export. Image from Ban Live Export National Rally in 2011 in Sydney, New South Wales. *Source:* Photo courtesy of James Morgan and Nikki To.

Policy changes so far might be limited in scope, slow to take hold, and often much less than what is being requested, but the live export industry has itself recognized that there is now 'a very high level of political sensitivity to animal welfare incidents associated with livestock exports' (Keogh et al., 2016, p. 93). As Simon Crean from the Australian Livestock Exporters Council has put this, the welfare of live export animals has become 'front of mind with the community' (as cited in Bourke, 2018). Not only are there 'continuing campaigns in major capital cities by animal welfare activists, seeking to have livestock exports banned' – and both Animals Australia and RSPCA Australia are major players here – but there is recognition that these campaigns have contributed to the industry clearly losing its 'social licence to operate' on several occasions (Keogh et al., 2016, pp. 93, 98). Failure to fulfil the community expectations inherent to social license can lead to 'increased litigation, increased regulations, and increasing consumer demands' (Coleman, 2018, p. 15), each hampering the success of the industry and – perhaps – requiring industry change. Such attitudes have seen Australia's live export industry itself announce some industry-imposed suspensions and sanctions on instances of welfare breaches (Bourke, 2018).

This growth in awareness of the (counter-)narrative of sentience has also seen some changes in the practices of animal agriculture, with consumer pressure contributing not only to industry change but major food chains in Australia changing some of their sourcing and food practices. For example, activist campaigns regarding the containment of sows in farrowing crates (Animals Australia, 2016b) led to the Australian pork industry voluntarily deciding to phase out sow stalls entirely by 2017. Further to this, Coles, a major chain of supermarkets in Australia, subsequently announced that Coles Brand fresh pork products will come from sow stall-free farms. This practice was subsequently extended to all pork products including bacon and ham (Coles, 2016). The consumer advocacy group Choice has also claimed that consumer desire to back better animal welfare standards has contributed to free range eggs becoming the fastest growing egg sector in Australia, with growth occurring at round eight times that of cage eggs (Clemons & Day, 2017). Thus, while cage eggs represented nearly 70 per cent of egg sales in 2009, in 2017 they stood for less than 50 per cent of sales

(Locke, 2017). Furthermore, Aldi, Coles, Heinz, Hungry Jacks, McDonald's, Nestle, Subway and Woolworths have all committed to phasing out cage eggs from their supply chains, along with multiple Foodworks and IGAs, and other food and foodservice distributors (Gantz, 2019; Mummery & Rodan, 2019a; RSPCA, 2019g).[7] These are decisions many of these companies have explicitly tied to their acknowledgement of consumer pressure. As Woolworths' head of sustainability, Armineh Mardirossian, stated in 2013 of their commitment to phasing out cartons of cage eggs from all brands and to only use cage-free egg ingredients for Woolworths branded products by the end of 2018:

> We've seen a significant increase over the past five years in the choices that [our] customers are making in terms of buying barn and free range, and the caged eggs have seen a decline over the years [...] The customer is in the best position to decide how they're going to spend their money, and that dictates everything else. (as cited in ABC Rural, 2013)

It is finally worth noting that the 2018 Futureye report has indicated that 65 per cent of Australian respondents have expressed they are willing to pay more in their shopping to ensure better animal welfare standards (Futureye, 2018).

Although such changes are admittedly uneven and often seem dependent on activist campaigning and media attention to instances of animal welfare abuse (Coleman, 2018), what is nonetheless clear is that public attitudes are becoming sensitized to the counter-narrative of animal sentience (Sinclair et al., 2018). That is, there is an increasing community belief that, given their sentience, animals are entitled to the better protection of their welfare, a view in close alignment with activist campaigning (Futureye, 2018). Research findings are further showing an increasing level of public distrust of both the industry and government when it comes to the protection of the welfare of livestock animals (Futureye, 2018). In addition, there is a growing awareness in the livestock industries that animal welfare is a high priority public issue and that changes in community values will need to be addressed in proactive ways (Coleman, 2018; Fleming et al., 2020; Keogh et al., 2016). As Lynne Bradshaw (2019), the chairwoman of RSPCA Western Australia, has put this: 'Australians want real animal welfare improvements. Locking the farm gate and refusing to change because "we've always done it this way" won't fly any more'. Given these points, and analysis that public concern for the protection of animal sentience in Australia will continue to grow (Futureye, 2018), the questions are what may evolve from this change and what might be able to be imagined differently with regards to human-animal relations.

With regards to these questions, the Futureye report has engaged a 'social maturation' curve approach. Basically, a social maturation approach considers (a) how present a particular issue or value is in the public consciousness, (b) which key bodies are talking about it, and (c) what society is demanding (Futureye, 2018, p. 19). It works on the assumption that community values are formed and can change over time. A social maturation approach further suggests that changes in community values tend to progress sequentially through six steps (McGrail et al., 2013). These steps are as follows. First is *observation* and occurs when a problematic issue or pattern is first identified. *Emergence* comes second and refers to when theories about that issue or pattern are advanced and validated or falsified. Third is *popularization*. This phase sees media coverage about the issue shift to mainstream, and the formation of issue-specific organizations. *Challenge* describes the fourth step. This is the phase of 'societal engagement', advocacy, and pushback; there is greater politicization of the issue or pattern, and growing research interest. *Governance* comes fifth and refers to when 'policy is developed and contested', and government and voluntary regulation established. Finally, a new *normativity* is established. This is the phase where 'socialisation and mainstreaming of the issue occur', and 'new values, behaviours, and practices are [...] accepted as new norms' (Futureye, 2018, p. 19).

In accordance with this sequence, the Futureye report finds that the Australian public uptake of the narrative of animal sentience would fit within the *challenge* phase, although the 'growing awareness of farm animal welfare issues over the past eight years' is indicative of an 'acceleration toward the governance phase' (2018, p. 23). On this basis, the Futureye report makes several predictions. These are as follows. First is that with broad recognition of animal sentience, definitions and considerations of animal welfare will shift from the minimizing of animal pain and suffering to the maximizing of animal wellbeing. The second is that there will be a mainstreaming of activist groups working alongside and with industry. Thirdly, big business visionaries will come to show leadership on issues of livestock animal welfare, with those failing to act coming under increased scrutiny 'as the public takes action against laggards'. Finally, regulators will come under intense scrutiny to prove 'there is no conflict of interest' between their policy and regulatory arms (p. 25).

Such predictions are clearly in alignment with the growing public concern for animal welfare, including calls to improve the welfare – and wellbeing – of livestock animals. At the same time, they are also conservative predictions in that they do not obviously question the fundamental anthropocentric narrative justifying human use of animals. In this vision, perhaps one would say that the narratives of sentience and of economic value come to be equally weighted. There is no doubt that such a vision would bring about immensely improved welfare conditions for livestock animals. Under the framework of equally weighted narratives, it would seem evident that industry practices regarding the *production* of animal products would need to change substantially. The life and death experiences of livestock animals would all have to be formed in alignment to maximized welfare and wellbeing conditions that respect and promote species-specific behaviour in appropriate environments. It is worth noting that such changes are very much in alignment with those argued for by such animal welfare organizations as RSPCA Australia. That is, the RSPCA's mission is to ensure that all animals – including livestock animals – experience not simply the five freedoms of animal welfare, but experience a life accorded respect across the five domains. Adopted by RSPCA Australia in 1993, the five freedoms stand as the first widely accepted evidence-based framework to capture the various dimensions of animal welfare in one model. More specifically, the five freedoms are comprised of: (a) freedom from hunger and thirst; (b) freedom from discomfort; (c) freedom from pain, injury, or disease; (d) freedom to express normal behaviour; and (e) freedom from fear and distress (RSPCA, 2019k).

Recognition of animal sentience requires consideration of more than just physical (dis)comfort, however, and it was this insight that drove the development of a model able to explicitly consider animal wellbeing as well as welfare. This still means accounting for nutrition, environment, health, and behaviour, but also requires an explicit consideration of *mental state*. Together these comprise the five domains. Although the five freedoms and five domains do overlap extensively, the five domains model makes the mental state of an animal as important as their physical state. It rests on the point that for every physical experience in an animal's life, there will be an accompanying emotion or subjective experience that also impacts that animal's welfare. In accordance with this view, ensuring that animals experience *a life worth living* means that they must not just be free from fear and distress but should have the opportunity to have positive experiences, such as anticipation, satisfaction and satiation (RSPCA, 2019j). Put another way, the five domains foreground the importance of actively promoting animal wellbeing as well as preventing poor animal welfare. RSPCA Australia describes this strong understanding of animal welfare in the following terms:

> RSPCA Australia considers that the welfare of an animal includes both physical and mental states. Ensuring good animal welfare goes beyond preventing pain, suffering or distress and minimising negative experiences, to ensuring animals can express their natural behaviour in an enriching environment, feel safe, have healthy positive experiences and a good quality of life. Thus providing good animal welfare means providing animals with all the necessary elements to ensure their physical and mental health and a sense of positive individual wellbeing. (RSPCA, 2019b)

The question, however, is whether we can imagine further than this kind of outline of animal welfare. That is, could the narrative of animal sentience actually come to override our anthropocentric economic narrative? What might such a weighting mean for human-animal relations? There is a broader but extremely important issue here which concerns the kinds of human-animal relationality and moral consideration that the recognition of animal sentience can enable. On the one hand is the kind of relationality and consideration built into improving the protection of animal welfare. This is the orientation behind the campaigning discussed earlier in the chapter, the work of the RSPCA, and the predictions set out in the Futureye report. Here, the recognition of animal sentience drives attempts to significantly improve animal welfare – seeing this as a human duty, and one that has been unjustifiably ignored in the pursuit of human (economic) interests – but the broader *use* value narrative regarding animals remains undisturbed. This kind of welfarist perspective is perhaps best described as a kind of protectivism, but one which only recognizes and protects some animal interests. It marks what could be called a curtailed speciesism. Such a view broadly matches what could be called the Australian attitude to animals: 'in both the general community and amongst professionals working with animals, there is a majority view that animal welfare is a significant issue but that use of animals is acceptable provided that it does not lead to unnecessary suffering' (Coleman, 2008). Under such a view, suffering should therefore count and be minimized no matter the species, and suffering can be understood to encompass psychosocial experience as well as just physical experience. However, as we have noted, the welfarist approach maintains a hierarchy regarding the kinds of moral consideration afforded different animal species.

While this welfarist approach continues as the primary counter-narrative to the anthropocentric animal use narrative in Australia at this point in time, the recognition of sentience can inform other ideas of human-animal relationality. The recognition of sentience can, for instance, also lead to opposition to the exploitation and slaughter of nonhuman animals in any and all forms (Francione, 2008). This is a position that radically calls into question the legitimacy of all animal use value narratives, contending instead that animals possess moral rights to live their own lives. From this viewpoint, variation in substantive animal-based agricultural industries and practices can never be enough (Jena, 2017). What is rather at stake are the lives of nonhuman animals, and human activities that threaten the group and individual sanctity of animals can only illustrate the practice of speciesism. Such a position recognizes first that there are two kinds of animals – human and nonhuman – and second that both, as sentient beings, have rights and interests that should be protected as a moral duty. This is typically called an animal rights perspective (see Francione, 2008; Francione, 2010a), and it contends that unjust or immoral practices – such as ignoring the rights and interests of a sentient being, and treating them as just a (human) resource – can never be justified by an appeal to any economic considerations, no matter how pressing or important they might seem (Francione, 2010a; Rowlands, 2002). In the words of one animal rights advocate, 'animals should have the right not to be treated as the resources of humans and [...] animal exploitation should be abolished' (Francione, 2010a, p. 34).

As should be evident, the ongoing political and economic value given to agriculture in Australia – including animal-based agriculture – has meant that rights arguments opposing the *use* of animals have not gained the levels of public support that are typically held by arguments for improved animal welfare. The lack of support for such rights-based arguments and activism is also evidenced through the turn to *Ag-gag* laws and other legislative attempts to silence and even criminalize activist critique of Australian agricultural practices and industries. Nonetheless, as we have stressed, our imaginings throughout these chapters are drawn by the possibility of developing ideas and practices of human-animal relations that are no longer fundamentally informed by anthropocentrism and exploitation. Such ideas and practices would rather recognize and value alternative relations with animals – relations that could inform the transformation of Australia's current anthropocentric social imaginary to one that is more supportive of animal sentience (and other principles supportive of interspecies care). As noted, these are difficult questions to consider and proposals to develop – let alone implement – given how deeply entangled Australian ways of living are (and have been) with anthropocentric and speciesist narratives. Additionally, at least according to the Futureye report, Australia is still in the *challenge* phase of social issue maturation with regards to the recognition and protection of animal sentience. This means that suggestions here will continue for some time to be at risk of being framed – and dismissed – as extremist and unreasonable. Such dismissals are particularly rife with reference to questions directed to the strong conventional Australian expectations of livestock animals being primarily understood as food and other resources. It is these issues that lie at the heart of the next chapter.

Notes

1. It is worth noting that since the 1980s the National Farmers' Federation (NFF) has been 'one of the most vocal proponents of a deregulated economy and a free enterprise agriculture' (Lawrence, 1987, p. 79), and has continued to pursue neoliberal policy objectives (Botterill, 2005). What is important to note about these objectives is that the neoliberal regime has commodified both animals and the environment (Francione & Garner, 2010; Kirjner, 2015; Sorenson, 2009; Torres, 2007). As we will discuss in the context of the turn in Australia to *Ag-gag* laws, this has meant that attempts to argue for the inherent value of either nonhuman species or the environment – and therefore for fundamentally different treatment – can only be seen as unreasonable or threatening by neoliberal bodies (Munro, 2004; Sorenson, 2009).

2. Animal enrichment, also known as environmental or behavioural enrichment, is the process of providing non-free-living animals with some form of stimulation to encourage natural behaviours, stimulate curiosity, and provide animals with choice and control in their environment (Van Metter et al. 2008). Such stimulation might involve the regular provision of dynamic environments, cognitive challenges (often food-based), and social opportunities. An enriched environment should promote a range of normal behaviours that animals find rewarding as well as allowing animals to positively respond to potential stressors. While practices to do with improving animal enrichment have gained considerable traction in places like zoos, wildlife parks and animal rehabilitation centres, as well as laboratory settings, studies on livestock behaviour have brought to light the need for environmental enrichment in livestock species as well (Riber et al., 2017).

3. The Australian live animal export industry is a major contributor to the Australian economy, in 2018/9 worth AUD 1.79 billion to the Australian economy (LiveCorp, 2019). In 2018/19, Australia exported via sea and air a total of 2.32 million head of livestock. This total was composed of 1.26 million cattle worth AUD 1.64 billion, 989,000 sheep worth AUD 142 million, and 18,650 goats worth AUD 7.2 million. However, the value of the live export industry goes beyond the price of the sales alone, with almost 10,000 people employed in the live export of cattle industry across Australia, including livestock producers, transporters, and exporters (LiveCorp, 2019). Many of these jobs are located in rural and regional areas across the country, meaning that many communities directly and indirectly benefit from this trade (Fleming et al., 2020). As of January 2021, the industry contributes more than $1.8 billion to the annual gross domestic product (GDP) and employs approximately 13,000 people across Australian urban and rural territories. Live animal export is also listed as one of the leading employment opportunities for First Nations people in northern Australia (The Livestock Collective, 2021).

4   It is also worth noting that views of animal welfare and the significance of different welfare indicators can vary considerably between (human) stakeholder groups and over time. For instance, in 2004, the Livestock Export Program identified seven key indicators of animal welfare on-board ships. These were mortality, clinical disease, respiration rate, wet bulb temperature, space allowance, change in body weight, and ammonia levels. Of these, death was seen as the primary welfare indicator (see Pines et al., 2007). Stakeholder research in 2015 saw injury/wounds, ability to stand, disease and ventilation identified as the most important welfare indicators. Note that of the 2004 identified welfare indicators, three are resource-based measures (wet bulb temperature, space allowance, and ammonia levels), while only one is resource-based in 2015 (ventilation) (Fleming et al., 2020). This may demonstrate increasing recognition about the importance of animal-based measures for the consideration of welfare. A similar conclusion was reached by the European Food Safety Authority which released a document in 2012 stating that 'animal-based measures are the most appropriate indicators of animal welfare' because 'animal-based measures aim to measure the actual welfare of the animal and thus include the effects of different input factors' (European Food Safety Authority, 2012). The 2015 research also showed differences between stakeholder support for different welfare indicators. Specifically, while members of the general public and animal welfare advocates saw physiological welfare indicators such as heart rate and body temperature as important, these were seen as less important and not very practical by live export industry workers, perhaps – as the authors speculate – because the collection of such data is labour intensive (Fleming et al., 2020).

5   Voiceless is an independent, non-profit, charitable organization that particularly works toward promoting legal reform towards the protection of animals. Its aims include: empowering and mainstreaming the animal protection movement; shining a spotlight on cruel industry practices; advancing legal protection for animals; raising public awareness and; increasing animal protection in science and technology (Voiceless, n.d.).

6   It is worth noting that these exposés of animal suffering range from examining what happens to Australian live exported animals at their destinations as well as during their transportation to destination ports. Although all of the exposés unquestionably present animal welfare violations when measured against animal welfare science and the international laws on animal welfare set by the World Organisation for Animal Health (see, e.g., Jones, 2011; Osborne & van der Zee, 2020), there has also been some questioning as to the organizational politics of investigating and foregrounding offshore animal welfare abuses. The concerns in this context have been that placing the emphasis on 'foreignness' portrays animal cruelty in a manner that links it to particular social and racialized groups (see, e.g., Dalziell & Wadiwel, 2016; Janssen, 2014; Zoethout, 2013), and that public outrage and condemnation are being curated so that they are directed at foreign animal slaughter

practices rather than those closer to home (Evans, 2018). Whilst we acknowledge such concerns, we would note that Australia's live export industry has had a long history of criticism and of failed attempts at reform (Mummery & Rodan, 2019b; Villanueva, 2012), and that live export practices globally have long been under scrutiny from a range of sources for welfare violations and cruelty (see, e.g., Osborne & van der Zee, 2020). We would also note that live export is but one of the focuses of Australian animal activist and advocacy organizations, with these organizations also driving investigations and exposés of a wide range of domestic instances of animal cruelty: animal cruelty in Australian abattoirs, live baiting in the greyhound industry, duck shooting, the conditions in intensive animal agriculture, puppy farming and so on (see, e.g., Animals Australia, n.d.-b). As Animals Australia explains its investigative approach, 'There is only one way to expose hidden animal abuse...' – 'by shining a light on cruelty' (Animals Australia, n.d.-c).

7   In line with caged hen eggs gradually losing popularity on supermarket shelves in Australia and a range of companies transitioning to using free-range eggs in their products, the new Australian Animal Welfare Standards and Guidelines released for poultry have recommended that standard battery cages for poultry be phased out across Australia by 2036 (Bainbridge & Branley, 2021). Egg producers will have the option of transitioning to larger furnished cages (larger versions of battery cages with enrichments such as scratch pads, perches, and nest areas for the hens), or may decide to move straight to cage-free systems, such as barn-laid and free-range eggs (Parker & Bromberg, 2021). While the shift is supported by animal welfare organizations, it is also under criticism for the lengthy time being offered for transition and for its continued commitment to caged egg production (see, e.g., Bainbridge & Branley, 2021).

CHAPTER 3

Transitioning Australia's Meat-Eating Culture

As the previous chapter has shown, a counter-narrative of sentience is starting to gain some traction with regards to changing Australian attitudes towards livestock animals, particularly as concerns their welfare standards and living (and dying) conditions. It is no surprise that this counter-narrative is informing not simply increased levels of activism and advocacy on behalf of livestock animals – as has been considered in the previous chapter with reference to Australia's agricultural and live export industries – but some changes in everyday Australian culture. Although we contend that such changes can be identified across many areas in everyday life, one important area for consideration here is that of Australians' habits with regards to food and eating. After all, what we purchase and eat has immense ramifications for the kinds of lives and wellbeing experienced by livestock animals – and, as such, the counter-narrative of sentience in this context provides a reminder that eating needs itself to be recognized as an inherently industrial, political, and ethical act. This being so – and given the rising public awareness of issues concerning livestock welfare in Australia – there can be seen a development of some alternative food values and norms supporting different eating practices. These include commitments to full veganism through to vegetarianism, along with a range of other models of *reducetarianism* and *flexitarianism* – such as commitments to meat-free Mondays, to being *vegan before six*, to only eating animal products that are produced through farming practices informed by high welfare standards (organic, free range, sow stall free, to name a few). Also of interest is the development of meat and meat-like products from methods other than through the slaughter of animals.

This chapter thus extends our discussion of the counter-narrative of sentience and its increasing traction in Australia. In particular, it marks a shift from outlining the growing awareness of the need to recognize sentience in livestock animals, to reflecting upon what might change in our own personal eating and consumption practices if animal sentience is recognized. If the quality of animal experiences and lives matters, then under what conditions can we feel justified in consuming them? As noted, however, rather than focusing directly on what needs to change in the lives of livestock animals – although, as the previous chapter outlines, there is a lot of work to be done on this issue – this chapter turns its focus to the kinds of changes Australians are already making, how they are taking hold in Australian society, and some of the possible future trajectories of such changes. This chapter thus comprises four main sections. First is a brief reminder of the main points of the counter-narrative of sentience in the context of livestock animals, given that it is this narrative and our response to it that is an important contributor to the development of new values and norms in eating. Second, we map how the narrative of sentience also gives rise to arguments for animal rights and consider how these arguments also play a role in the development of new norms in eating. Third is a broad outline of some of the forms of eating these new norms are generating. Finally, we consider who is narrating calls for change? How is the expertise and influence of various narrators supporting their calls to care and change towards new ideas and habits? As will become clear through our analyses of the various suggestions for change that we consider contribute to the development of an alternative imaginary in human-animal relations, the gaining and maintaining of traction for calls for change is key. As such, we conclude this chapter with a consideration of the capacities and work of those who do not simply call for change but can be understood to be social or cultural *change-makers*. These are people who can be called *cultural intermediaries*, meaning those who can be seen as playing a significant role in cultural production and in legitimating change (Bourdieu, 1979/1984). As we will suggest, many of the narrators of the new norms considered in this chapter can be identified in this way given their capacities to not just promote the recognition of eating as an industrial, political, and ethical act, but to influence subsequent changes to people's everyday eating and consumption practices.

The Narrative of Sentience and the Question of Care

Blowing the whistle on the use and abuse of animals through industrialized farming in Australia, the animal rights documentary *Dominion* (Delforce, 2018) is another clear illustration of the focus of the counter-narrative of sentience. Broadly, and like both *Somewhere* and the *Make It Possible* campaigns from Animals Australia (2020c, 2012), *Dominion* repudiates the pastoral rural myth of the Australian farm – of open plains, undulating paddocks, and carefree animals grazing under the sky. By means of images gained from drones and frontline investigators, *Dominion's* narrators highlight how livestock animals are instead born into industrial or *factory farms* as the means to our economic ends. Brought into existence on a mass scale purely to provide cheap materials for food, clothing and other products, these animals' own wellbeing and interests are disregarded. In one example of this instrumentalization of animals into resources, the main narrator of *Dominion*, Joaquin Phoenix, reports on the plight of cows and their calves in the average Australian dairy farm (see also Hughes, 2020). Phoenix explains that in order to produce milk for humans, cows are separated from their calves only within hours of their birth. Calves are fed milk substitute so as not to diminish even a little the human use of the cow's milk. Male calves are sent to the slaughterhouse around 5 days of age, while female calves are kept to become the next milking cows. These *lucky* calves join the industrial model of the dairy cow. They are made to give birth every year from around 2 years of age, forcibly separated from every one of their calves, and slaughtered as *unproductive* around 7 years of age if not younger (the natural life span of a cow is up to twenty years). As the documentary shows, cows and their calves call for each other after being separated – viewers see their visceral distress and hear their cries. Through this footage and his narrative, Phoenix reminds us that, just like humans, cows possess a strong maternal bond – they strongly desire to mother their calves, to feed them – but we ignore this desire for our own interests. In his Oscar acceptance speech at the 92nd annual Academy Awards (for his role in the film *Joker*, February 9, 2020), Phoenix (2020) again called us all out over the cruelty of what we have come to take for granted: 'We feel entitled to artificially inseminate a cow and when she gives birth, we steal her baby. And then we take her milk that's intended for her calf and we put it in our coffee and our cereal'.

Just like the campaigns and live export exposés considered in the previous chapter, *Dominion* (Delforce, 2018) makes visible to us that sentience in livestock animals is just as real as our own human experience of it. Livestock animals experience their lives and, as *Dominion* shows us, cows and calves suffer in the short, constrained lives we allow them. In accordance with the narrative of sentience, this is suffering we can and should identify with and feel empathy for. It is the narrative of sentience, then, that enables Phoenix's call – and that of Bidda Jones (2011) in her analysis of the footage screened in 'A Bloody Business' – that we need to consider and protect livestock animals' welfare and wellbeing much more than we currently do so. It also leads to the RSPCA's highlighting of the importance of both the five freedoms and the five domains for all animals, and the various campaigns by organizations like Animals Australia calling on us all to protest the cruelties to animals that are normalized within industrialized farming practices. In other words, the narrative of sentience contains a call for us all to learn to care. Typically referring to ideas of providing for, looking out for, or protecting someone, the idea of care works to generate attitudes such as worry, concern, attention, solicitude, and protection. Care is furthermore always situated and concrete – we always care for someone or something. That is, care is characterized by 'its focus on the concrete and particular, its emphasis on the maintenance and extension of connection, and by its concern for responsiveness and the satisfaction of needs' (Luke, 2007, p. 125). Care is not abstract but specific, and it is also response and action oriented. If we truly come to care about someone or something, we will take action to protect that someone or something (Shotwell, 2011).

Animal activists and advocates thus all argue that to recognize sentience in animals is also a recognition of the need to care for their wellbeing and to act on behalf of that care. As Matthew Evans from the *Gourmet Farmer* television program (Hilton, 2010 – present) has pointed out, because a lot of us have lost awareness of 'the consequences of our eating' and other consumption practices – most particularly with regards to their 'impact on animals' – this has permitted our uncritical acceptance of industrialized practices that prioritize economic value well over animal welfare and wellbeing (Evans, 2019). This is a point that Peter Singer also notes, stressing its further ramifications:

> We [...] grow up eating animals and I think that has a marked effect on the attitudes that we take to animals afterwards. It makes us think of animals as objects for our use, rather than beings with lives of their own, and that is where all the problems start. (Russell & Singer, 1997, p. 44)

In both Phoenix and Singer's views, therefore, once we accept the sentience of livestock animals and recognize the various ways their lives are bound up in our consumption practices, we should (a) come to care about the treatment they receive – given that this treatment is carried out for our benefit – and (b) change what we have taken for granted and what we do in our consumption. The narratives of sentience and care, in other words, by way of transporting us through empathy for suffering to acceptance of the need for change, are considered by a range of people to open into narratives of alternative farming and consumption practices. They are also considered by some to result in narratives of nonhuman animal rights. Although it is the first set of narratives that is the primary focus of this chapter – and which will be examined in the sections below – it is useful to briefly recap the main points of narratives of animal rights, as these ideas are influential in some of the narratives of alternative consumption practices considered later in the chapter.

Understanding Animal Rights

As noted in the previous chapter, counter-narratives of animal rights also start from the recognition of animal sentience. As has been outlined, sentience describes the condition in which a being has its own interests and can prefer a life of wellbeing rather than of suffering. The recognition of sentience in animals also stresses that it is this capacity that should matter more than any other with regards to weighing interests against each other – such as the interest of an animal to live a life of wellbeing versus human interests in consuming low-cost animal products. This is the point that the capacity to suffer – or, more broadly, to have subjective experiences – should always be the first consideration with regards to any and all proposed treatments. (Remember that Peter Singer suggests that to only recognize sentience in humans is to be *speciesist*.) These ideas have provided fertile grounds for the development of animal rights narratives. Tom Regan, for instance, stresses the point that our shared sentience means that 'we are each of us' – humans and nonhuman animals – 'the experiencing subject of a life, a conscious creature having an individual welfare that has importance to us whatever our usefulness to others' (1986, p. 186). All *experiencing subjects of a life*, Regan continues, should thus be considered as individuals with worth *in their own right* and should therefore never be considered purely as resources for the benefit of others. And, as all such subjects have an interest in living a *good* (to them) life, being an experiencing subject of a life is enough to require that those interests should be respected and protected. As he explains,

> What's wrong – fundamentally wrong – with the way animals are treated isn't the details that vary from case to case […] The fundamental wrong is the system that allows us to view animals as *our resources*, here for *us* – to be eaten, or surgically manipulated, or exploited for sports or money. (Regan, 1986, p. 179, original emphasis)

Regan's argument, then, is that the most basic right of all sentient beings is the right never to be treated merely as a means to the ends of others – or, put another way, the right to be consistently treated with respect.[1] Under such an argument, it would never be enough to better regulate industries that use animals so that animal welfare is better protected – the focus of the RSPCA, for instance. The assumption is rather that all such uses are always wrong and should be abolished (Cochrane, 2012; Francione, 2008, 2010a; Regan, 1986). Put simply, animals should be liberated from being framed – and used – in terms of human interests, as such interests under an unbounded anthropocentrism have resulted in near total disregard of animal sentience (see, e.g., Figure 5).

Figure 5. Vegan activist groups such as Vegan Australia strive to remind Australians that there is no substantive difference between the animals we choose as companions and those we treat as resources for our consumption – both are sentient. This image is from a 'Choose Vegan' campaign carried out by Vegan Australia in Melbourne, Victoria, in 2017. *Source:* Photo courtesy of Vegan Australia.

As is typical of all rights arguments, however, there is debate as to what rights precisely should be recognized with regards to animals. The real point of debate regarding ideas of animal rights, however, concerns whether nonhuman animals have a right to not be used by us at all (Cochrane, 2012). This is a particularly important question for this chapter, given that animal rights narratives tend to end in calls for veganism and the complete abolition of human use of animals – a point we will consider in the next section. The issue is also a continuing question for this book, and different responses to it will turn up in our explorations of the narratives our society has told and is telling about our relations with animals.

From Reducetarians and Flexitarians to Vegans

Despite some change in recent decades, Australia is still predominately a meat-eating culture (OECD, 2015).[2] Meat consumption and the consumption of other animal products are clearly still sanctioned and normalized, and animal agriculture and animal lives are still primarily viewed under an economic lens. The everyday, taken for grantedness of the consumption of meat and other animal products is evidenced through meat consumption statistics, the ubiquity of advertising linking meat and dairy consumption to human health and vitality, and ongoing negative stereotyping of alternative diets such as veganism (Cole, 2008; Masterman-Smith et al., 2014).[3] A further factor is the denigration of the work of animal advocacy and activist groups if conventional ideas of animal agriculture and diet are challenged (Rodan & Mummery, 2019). At the same time, however, Australia's National Farmers' Federation (NFF) noted in 2018 that veganism and other dietary habits that minimize or avoid the consumption of meat and other animal products are rapidly gaining in prevalence and status throughout Australian society (KPMG, 2018; see also Admassu et al., 2020). Indeed, growing visibility of vegan diets, food, and meals on social media and within popular culture has seen an 'exponential growth' in vegan consumers (Estok, 2021, p. 335). Google Trends reveals, for instance, that search interest in the word *vegan* has increased dramatically in Australia from 2004 to 2021. Analysis of Google Adwords data has also found that vegan-related searches in Australia surged by 51 per cent year-on-year in 2020 (Tasmanian Times, 2021). Searches for vegan restaurants almost tripled in four years, searches for vegan recipes recorded an all-time high in 2020 (aided perhaps) by pandemic lockdowns, and demand for non-food related vegan items is also strong (Tasmanian Times, 2021). An international report by Euromonitor International (2016, as cited in Wan, 2018) also highlights these developments when it describes Australia as the 'third-fastest growing market in the world' with regards to veganism and vegetarianism. In addition, Roy Morgan (2019) found over 12 per cent of Australians are 'following vegetarian or almost vegetarian diets' (Curtain & Grafenauer, 2019, p. 2). Interestingly, research has also shown that the percentage of Australians who identified as vegan or vegetarian was similar regardless of political persuasion, incomes, education, place of birth, race, or religion (Sutton, 2019). Further reports also show that since 2012 more Australians are trending towards eating less meat or adopting some version of a flexitarian diet (Curtain & Grafenauer, 2019). Similarly, market analysis has reported that in Australia 'per capita plant-based milk consumption is anticipated to increase at an annualized 16.9 per cent' (Reeves, 2019). It is this growing commitment to alternative diets and eating and what these might look like that is the focus of this section.

The first point here is that a majority of these commitments are made on ethical grounds, although the precise nature of these ethical grounds can vary widely (including possible considerations of human, animal and environmental factors), as can the resulting individual commitment. As one recent convert to a diet with less meat explains, 'If you're trying to fix all of the inequities of this world through your own personal consumption decisions you'll go mad, because pretty much everything has exploitation and cruelty' (as cited in Wahlquist, 2019b). However, as this same respondent stresses,

> when there are decisions that you can make, where I can make this slight change in the way we live, at no cost or ill-consequence to myself, and it does make an improvement to the world around us, then that's not a particularly difficult decision to take. (as cited in Wahlquist, 2019b)

Such points remind us that the kinds of dietary changes we are talking about can be broadly linked to ideas of ethical consumption which propose that consumers might make their product and consumption choices in accordance with their ethical values (Oh & Yoon, 2014). As noted, such values might prioritize human factors (e.g. health, the safety of workers, and fair work practices), and/or those concerning animal welfare, the environment and sustainability (Searchinger et al., 2018; Shaw & Shiu, 2002; Springmann et al., 2018). Some common examples in the Australian context of such decision-making would include the choice to buy Australian or locally grown or made products, the choice to buy organic or Fair-Trade products, or of course the choice noted in market research to follow dietary habits that minimize or avoid the consumption of meat and other animal products.

It is this latter choice that is being increasingly reported in market analysis and print and online media under the aegis of choosing a *flexitarian* diet and/or acting in accordance with the *reducetarian* movement. Each of these promotes 'dietary changes towards healthier, more plant-based diets' as a means of reducing the negative impacts of the current global agricultural enterprise (Springmann et al., 2018; Willett et al., 2019). More specifically, a flexitarian diet puts the focus on healthy eating and environmental sustainability by reducing meat and dairy consumption (Henderson, 2017; McGrath, 2018; Rosenbloom, 2016). Similarly, the reducetarian movement, launched in 2015 as a non-profit organization, claims to help 'unite vegans, vegetarians, flexitarians and anyone else interested in ending factory farming' (as cited in Tait, 2017) by directing people to eat less meat and dairy and fewer eggs. Such commitments are described as being able to not just improve people's own health (Bittman, 2013), but better support animal welfare and the environment (Kateman, 2017). The reducetarian movement further urges consumers to mount a 'countermarketing effort' in the face of the meat industry's promotion of meat as the only affordable protein-rich food choice (Godin, 2017, p. 20). Both strategies, then, focus on changing the behaviour of meat-eaters by encouraging them to reduce their consumption of meat – hence slogans of 'meatless Mondays' (Lerner, 2017, pp. 141–2; Luna, 2017) and of being *vegan before six*.

As is evident, these are not *all or nothing* messages. They instead promote the idea that even if we each make only small reductions or changes in our diets, such changes, together, can 'collectively result in a significant difference in the world' (Schawbel, 2017). This was also the message included in the *Make It Possible* campaign by Animals Australia which, in asking people to 'imagine a world without factory farming', encouraged them to publicly pledge to either (a) boycott factory-farmed animal produce, (b) reduce consumption of animal products, or (c) become meat-free (see Figure 6). The argument here being that taking any of these steps would help make the world a better place for livestock animals and the environment, and facilitate better human health.

Figure 6. *Make It Possible* campaign poster from Animals Australia. *Source:* Photo by author.

While the main intention in flexitarian or reducetarian diets is to eat more plant-based foods and to maintain a commitment to eating less beef, chicken, pork and seafood (Kateman, 2017), different kinds of reducetarian diets suggest changing meat intake in type as well as in volume. Thus, while one focus is to reduce the amount of commercially produced meat being eaten, another is to substitute the consumption of some livestock meats (e.g. beef, lamb, and pork) with others with smaller environmental footprints, most commonly chicken or fish (Cushing, 2019). Indeed, such preferences are gaining in traction with fish and chicken now the animals 'most frequently killed for food' globally (Adams, 2020, p. 51; see also Sanders, 2020). For those willing to make deeper changes, the meat of free-living animals, especially those which are killed accidentally on the roads or deliberately for environmental management purposes, is another option, as is engaging a *nose-to-tail* philosophy of eating[4] according to which no parts of an animal killed for food go to waste.[5] One proposal here, and specific to Australia, is *kangatarianism*, a term describing an omnivorous diet in which the only meat consumed would be that of the kangaroo (Cushing, 2019). The *kangatarian* argument, for instance, is that:

> Kangaroo require less water than cattle, sheep or pigs, and no land is cleared to grow feed for them or give them space to graze. Their paws cause less erosion and compaction of soil than do the hooves of common livestock. They eat less fodder than ruminants and their digestive processes result in lower emissions of the powerful greenhouse gas methane and less solid waste. (Cushing, 2019)

Further points often made are that because kangaroos would not need to be *farmed* as such, their general welfare would remain higher than that of animals produced for and into factory farming,[6] and that kangaroo meat is healthier for human consumption than that of conventional livestock animals. It is also noted that if kangaroos were deliberately supported alongside other livestock – instead of being killed to preserve resources for sheep and cattle – this would help facilitate a reduction in the numbers of cattle and sheep, lessening the environmental harm they do (Ellicott, 2018; Wilson, 2018).

Vegetarianism is another reducetarian dietary possibility, according to which all meat products are eliminated from the diet, although eggs and some dairy products may continue to be consumed. Calls for alternative diets can also, however, take the stronger line outlined in the previous section of protecting what some would call the right of animals to not be exploited by humans at all. Such views are most typically framed in terms of veganism, which is generally defined as:

> a philosophy and way of living which seeks to exclude – as far as is possible and practicable – all forms of exploitation of, and cruelty to, animals for food, clothing or any other purpose; and by extension, promotes the development and use of animal-free alternatives for the benefit of humans, animals and the environment. (Vegan Society, 2008)

While veganism may also be a decision informed by health, religious or environmental factors (see, e.g., Estok, 2021; Wright, 2021), the important point is that in dietary terms it always denotes the practice of dispensing with all products derived wholly or partly from animals. Veganism requires that no animal products are eaten or otherwise consumed (Vegan Society, 2008). As we have noted, when informed by ethical grounds – a key one being the rejection of all forms of human exploitation of animals – veganism is the only flexitarian or reducetarian choice that aligns with arguments for animal rights. Thus, from a vegan perspective, other reducetarian options might help diminish some levels of animal suffering, but they do not fundamentally and consistently challenge anthropocentric and speciesist assumptions.

Interestingly, the second object of vegan living noted by the Vegan Society – the promotion of the development and use of animal-free alternatives for the benefit of humans, animals, and the environment – is key not only for vegans but the broader reducetarian movement. Indeed, two major food technologies – plant-based meat and cultured or cell-based meat – are being promoted as promising directions for a global agriculture that would no longer be invested in animal suffering. While plant-based meat substitutes have long been available (van der Weele et al., 2019), new technologies have been developed to produce a 'significantly better tasting' product which possesses an aesthetic appeal, smell, texture, flavour and mouth-feel closer to conventional meat (Joshi & Kumar, 2015; Keefe, 2018). In Australia, the impact of such developments has meant that consumers can now purchase plant-based burger meals from well-known burger chains, and plant-based mince, burger, sausage and other meat-style options are sold in Australian supermarket chains. Recent market analysis reports that '6 in 10 Aussies' have sampled or shown interest in trying 'the new generation of plant-based meat products' (Food Frontier & Life Health Foods, 2019, p. 5). Indeed, in the fiscal year of 2020, Australia's plant-based meat sector increased its grocery retail sales by '46% over the previous year' and doubled their products on supermarket grocery shelves (Food Frontier, 2020, p. 2). Such points suggest that plant-based meat has the potential to help conventional meat-eaters reduce their meat-eating by transitioning to at least some food items that still look and taste much like meat, and all without having to significantly change their cooking and eating habits.

Although cellular agriculture and cell-based meat[7] – also termed cultured meat, cell-based meat, cellular meat, synthetic meat, clean meat, lab-grown meat, bio-meat, artificial meat, factory-grown meat or in vitro-meat – is at the early stages of its development (Keefe, 2018; Post, 2013), it too has potential to be a game changer in bringing about reducetarian as well as vegan aims. Unlike plant-based meat, cultured or cell-based meat is 'made out of flesh' (Sebo, 2018, p. 165), but is grown from animal muscle cells in a laboratory (Ackland, 2019). It is the production of 'meat', but not from 'slaughtered animals' (James, 2019). Memphis Meats (n.d.), for example, explain their process of making their cell-based meats as:

> Our cells follow their natural process to form muscle and connective tissue, just like they would when growing on an animal. This happens in a vessel we call a 'cultivator' … The entire process from cell to meat takes between 4–6 weeks. Once the meat is ready, we simply harvest it from the cultivator and it's ready to enjoy!

Although such products have yet to transition into marketable and affordable products, once on the market they too have the potential to provide another means for meat-eaters to continue with their dietary and cooking preferences whilst avoiding subsidizing the cruelties of industrialized animal agriculture and slaughter (Bryant & Barnett, 2018). It has also been argued that such cellular agriculture products could play a role in reducing greenhouse gas emissions and protecting the environment (Verbeke et al., 2015b; Wilks & Phillips, 2017).

Of course, thus far, the cultured meat field predominantly offers 'promissory narratives' to agribusiness, governments and consumers that cellular agriculture will be able to meet the challenge of providing affordable,[8] safe, cruelty-free food, improved animal welfare and environmental benefits (Stephens et al., 2018, p. 161). In addition, even though cultured meat promises meat-eaters dietary and cooking preferences of cruelty-free meat, there are strong critiques and doubts expressed about the industry's claims to consumer, animal welfare and environmental benefit. Firstly, doubts have been expressed about the social acceptance and level of demand by consumers for cultured meat. Many studies so far cite consumer concerns about (a) the 'unnaturalness' of the product and procedures (Welin, 2013, p. 29; Gertenbach et al., 2021; Verbeke et al., 2015a); (b) high 'yuck' and 'disgust' factors (Bogueva & Marinova, 2020, pp. 5, 9; Verbeke et al., 2015a); and (c) the possibility of health dangers and food safety issues around agro-food technologies (Mayhall, 2019, pp. 162–3; Verbeke et al., 2015a). Secondly, animal welfare claims have been disputed owing to the use of foetal bovine serum as a growth medium, meaning that the cultured meat industry still involves the use of animal 'living tissue' (Leroy & Praet, 2017, p. 81) and still incorporates a commitment to animal slaughter (Chauvet, 2018; Cole &

Morgan, 2013; Fernández, 2021; Fernandes et al., 2021; Metcalf, 2013). Thus, while the cultured meat industry might have the potential to substantially reduce levels of animal slaughter, it also continues the 'fetishization of meat' that animal rights activists in particular are at pains to challenge (Cole & Morgan 2013, p. 212). Finally, doubts have been expressed about the claim that cultured meat products will manifest better environmental outcomes (Broad, 2020; Fernandes et al., 2021; Mayhall, 2019). Here critics note that it may in fact 'require more energy' and 'fossil fuels' to manufacture lab-grown meat-products than the production of meat through conventional models of animal agriculture (Mayhall, 2019, p. 162). Such concerns may well be merited, but the newness of the industry may also mean further innovations address them, making cultured meat a potentially key player in reducing animal suffering without necessitating substantial change to people's eating habits.

Making the Change

Because our everyday personal habits – including our food purchasing and eating habits – are so very ingrained, they can be difficult to challenge, let alone change. Such change thus needs to be motivated – *we* need to be motivated to make and maintain such change. Although there are many theories of motivation that could be considered here, with regards to these issues of changing personal patterns and habits to meet the challenges set by new ethically inflected ideas, the insights from social movement theory concerning motivation to action are useful. Snow and Benford (1988), for instance, identify three main tasks for social movements with regards to motivating action: diagnostic, prognostic, and motivational framing. Under this model, diagnostic framing identifies some event or condition as problematic. Considering the narratives of *Dominion* or the *Make It Possible* campaign, then, the diagnosis is that intensive livestock farming is wrong because of its inherent cruelty to livestock animals. Linked to the diagnostic task, the prognostic task is to outline a potential solution – for Phoenix, Animals Australia, and the reducetarian movement, this means mobilizing individuals to change their everyday behaviour so as to influence both the livestock and the retail industries. The third task, motivational framing, involves the actual shift into action. Thus, for activists as well as for social movements more broadly, the important step is that from prognostic to motivational work. Jasper and Poulsen, examining the animal rights movement, argue that one common strategy has been the use of *moral shock*, meaning 'when an event or situation raises such a sense of outrage in people that they become inclined toward political action' (1995, p. 498; see also Jasper & Nelkin, 1992). They also suggest that moral shocks are most effective when they are 'embodied in, or translatable into, powerful condensing symbols' that are able to 'neatly capture – both cognitively and emotionally – a range of meanings, and convey a frame, a master frame, or theme' (Jasper & Poulsen, 1995, p. 498).

As both animal rights and welfare movements have realized, representations of animals are highly effective condensing symbols able to convey a master frame of cruelty and suffering to produce the moral shock required to engage and motivate people towards change (Lowe, 2006; Nabi, 2009; Wrenn, 2013). Visual representations are particularly important due to their capacity to cut through verbiage and draw immediate attention to key issues (Aiello & Parry, 2020). Images can furthermore be used as forms of documentary evidence to not only position the viewer as a 'moral witness' to that situation (Hariman & Lucaites, 2016, p. 189) – a key strategy in documentary style exposés such as 'A Bloody Business' – but to also generate the viewer's response-*ability*. Put another way, seeing in the form of witnessing – whether in the contexts of mainstream media, social media, activist websites and public posters – can be extremely effective in evoking not just moral shock but empathy, and hence preparedness to call for and to make change. *Dominion*, the *Make It Possible* campaign, and the exposés of live export discussed in the previous chapter all use images of animal suffering in this light. Indeed, proponents of the narrative of sentience all rely on the use of images to bring mainstream Australians to care about the experience of livestock animals.

At the same time, it has been recognized that having our everyday habits brought into question through moral shock, can also lead to responses of resistance and defensiveness. Such responses may be due to the cognitive dissonance that can occur when individuals find themselves caught between divergent beliefs, ideas, or values. The obvious instance is the psychological conflict that can occur when individuals have preferences for a meat diet but also feel a 'moral response to animal suffering' (Bastion & Loughnan, 2017, p. 278). Experiencing what has been labelled the 'meat paradox' (p. 278), this conflict can see people attempting to talk down the harm being done to animals through animal agriculture, deny their responsibility and generally attempt to 'buffer their identities' in the face of learning that their preferred models of consumption are a cause of animal suffering (pp. 280–2; see also Adams, 2020). Such denial is made easier when the proponents of change hold a minority view: 'the less common a moral view is, the more likely its proponents will be dismissed as kooks who don't deserve to be taken seriously' (Paytas, 2018). Such conflicts thus mark a defensive denial of a need for change. This is something vegans and other proponents of animal rights have long experience of, as one commentator has noted with regards to campaigns for animal rights promoting vegetarianism and veganism:

> [W]hen one is confronted with personal and aggressive attacks condemning meat consumption, it is one's own behavior being condemned: the enemy is thyself. [...] Thus, it could be that moral shock campaigns are ineffective when promoting vegetarianism, because condemning meat consumption (as opposed to other violations of animal rights) inevitably forces people to confront their own behavior (as opposed to that of others), and they are less likely to join a cause that requires them to make fundamental changes in what is such a deeply ingrained lifestyle. (Mika, 2006, p. 932; see also Joy, 2008)

Jasper (1998) argues similarly that if the emotional reaction to moral shocks involves a need to change personal behaviour, feelings of dread or of being overwhelmed can paralyse mobilization. Moral shocks may also not motivate long-term behavioural change, and even while individuals may have several motivations to purchase and eat ethically there may well still be an attitude-intention-behaviour gap in their practice, with people's everyday practices simply not matching up to their stated commitments (see, e.g., Bray et al., 2011; Burke et al., 2014; Eckhardt et al., 2010).

These are complex issues, and individuals' responses – short-term and long-term – to the potential moral shock of such campaigning and the narrative of sentience may well depend on a variety of individual, psychological and social factors. But this is always going to be the case with regards to attempts to shift social norms and inculcate a different social imaginary. As the social maturation curve analysis used in the 2018 Futureye report makes clear, there is no single societal shift with regards to any social change (see the previous chapter for more details). Social change is rather a messy process where issues go through periods of resistance and challenge before they are accepted into policy and governance models and finally accepted as normal. Given that the 2018 Futureye report diagnoses Australian attitudes towards the recognition of sentience and animal welfare as being still in the *challenge* phase, it is not surprising that individuals show themselves as being at different points regarding called-for transitions into new positions and practices. Rather, therefore, than seeing such shifts as too difficult to achieve, this is where we would note that the achievement of all social change comes out of a multitude of motivating factors. Multiple narratives for change are needed, narrated and promoted by different groups, so that messages are repeated across different domains, can interact with each other, and can engage different audiences. This can mean that individuals will come across these messages in a variety of contexts and from a variety of different narrators. While a kind of scattergun approach, this does allow for the gradual normalization and legitimation of new knowledge – such as that of the narrative of sentience itself and its call to care. This process is to the point of eventually – hopefully – ensuring changes in both individual and collective behaviour and, thus, bringing about a social change.

In the context of the counter-narrative of sentience and the way it is engaged to communicate the need for, as well as the possibilities of, alternative patterns of eating and consumption, there are two main groups of narrators visible within the Australian community who can be seen as *opinion leaders* (Dubois & Gaffney, 2014) or *cultural intermediaries* (Bourdieu, 1979/1984; Smith Maguire & Matthews, 2010). The first of these are of course the activist and advocacy groups and organizations we have already noted. These include peak Australian bodies concerned with animal welfare (Animals Australia and RSPCA Australia), as well as a range of reducetarian, vegan and animal liberation groups. A second group is made up of high-profile chefs who narrate versions of a reducetarian message about healthy animal products and good food needing to come from well-treated animals. In Australia this cohort includes Jamie Oliver, Hugh Fearnley-Whittingstall, Curtis Stone, Paul West, and Tobie Puttock, among others. Another related influence in Australia in this context is the *paddock to plate movement* (a form of locavorism; see Stănescu, 2010) which focuses on environmental sustainability, animal welfare and the importance of consuming local produce. Importantly, the point is that the various members of each of these groups are considered to hold expertise by at least some parts of the broader Australian community, meaning that their narratives of change are taken seriously by those who value their points of view. Because such groups and their individual members are socially embedded, with their messages visible within the public domain, their narratives of change can gain followers and traction, and be broadly shared (Dubois & Gaffney, 2014; Katz & Lazarsfeld, 2006; Shafer & Taddicken, 2015). There are also, of course, other narrators who continue to call for the maintenance of conventional habits and practices of eating. These include the NFF, along with Meat & Livestock Australia (MLA), Australian Dairy Farmers, and various other national and state industry bodies that continue to promote mainstream ideas with regards to animal industries and the *proper* norms of Australian eating.

Activists and Advocacy Groups

We have already discussed the work of some of these activist and advocacy groups in making a narrative of sentience visible within the Australian community, both with regards to promoting activism on behalf of livestock animals and as concerns changing Australian eating patterns towards those that are more respectful of animal welfare, if not animal rights. Animals Australia's (2020c, 2012) campaigns *Somewhere* and its predecessor *Make It Possible* stress, for instance, that livestock farming – particularly factory farming – is a major cause of animal cruelty. As they have made visible through these and other campaigns and exposés, all of Australia's factory-farmed animals experience high levels of suffering during their lives, and all livestock animals are treated as resources and potential commodities rather than as sentient individuals who themselves would desire to live a life free from suffering. Indeed, both *Somewhere* and *Make It Possible* ask us all to imagine what livestock animals would dream of from their sterile confinement, and suggest kindness, freedom from suffering, and to live in a natural rather than artificial environment supporting species natural behaviours, including mothering. As these campaigns make clear, conventional industrial treatments of livestock animals have meant that considerations of animal welfare consistently come second to human economic interests, a ranking that is justified via our longstanding anthropocentrism and speciesism. It is on the basis of this longstanding lack of consideration for livestock animal sentience and wellbeing that these campaigns have called all Australians to change their eating and consumption habits so as to stop enabling such farming practices.

Both *Somewhere* and *Make It Possible*, for instance, stress the importance of choosing to become meat-free or at least reduce consumption of animal products (the earlier *Make It Possible* campaign also suggests the action of boycotting factory-farmed animal produce). Similar reducetarian messages have also been the basis for a range of other campaigns by the organization. These have targeted both everyday consumers as well as the supermarket and fast-food industries: see, e.g., *That Aint No Way to Treat a Lady*, *A Cage Egg is a Bad Egg*, and *Please McDonald's*. Adding to their impact and their capacity to achieve public support (see Chen, 2016; Rodan & Mummery, 2014, 2016, 2018), campaigns have been endorsed and narrated by a range of high-profile Australians. For instance, Pat Rafter, Mick Molloy, Missy Higgins, Claire Hooper, and Dave Hughes, among others, endorsed the *Make It Possible* message. Arj Barker, Carl Barron, Anthony Lehmann, and Peter Rowsthorn narrated *That Aint No Way to Treat a Lady*.[9] For ABC investigative journalist Sarah Ferguson, Animals Australia's strategy of targeting mainstream consumers with reducetarian messages which are endorsed by high profile Australians has brought about a paradigm shift, moving animal welfare issues 'from the fringe to the mainstream' (as cited in Barrowclough, 2015).

Campaign work and the mobilization of Australian consumers have further been instrumental in facilitating changes in both the supermarket and fast-food industries with regards to product choice and use (see, e.g., Mummery & Rodan, 2019a). Indeed, research suggests that because campaigns by animal welfare organizations can create new food norms, they in fact open up a 'new market opportunity for industry buy-in' (Carey et al., 2020, p. 296). Campaigns and associated organizations can thus be 'co-opted or "corporatized" for industry and market consumption' (p. 287). Coles, in phasing out the use of sow stalls and battery hen cages for home brand products in early 2013 – a year earlier than initially scheduled – explicitly referred, for instance, to its ongoing dialogues with Animals Australia and other animal welfare organizations, including the RSPCA, regarding issues of animal welfare and sustainability (Sampson, 2013, as cited in Rodan & Mummery, 2014). Woolworths and Aldi, as well as independent Australian supermarket chains, have also all committed to phasing out cage eggs. Although these companies cite consumer pressure as the basis for their decisions, it is worth noting that Animals Australia has been working with other advocacy organizations to mobilize consumers with regards to pressing for change. International and Australian manufacturers that supply major Australian supermarkets have also committed to phasing out their use of cage eggs, with some directly citing Animals Australia as influential in the development of their cage-free egg policies (see, e.g., Berry, 2019). A range of fast-food companies, some of which were targeted by specific campaigns by Animals Australia – the *Please McDonald's* campaign, for instance, which was narrated by Australian children – have also committed to phasing out their use of cage eggs. Although it is impossible to ascertain precisely how much impact Animals Australia has had in various companies' moves away from cage eggs, it was for its capacity to induce change that Animals Australia was recognized as a 'standout charity' by the U.S.-based Animal Charity Evaluators, the first time an Australian based organization has been so recognized (Animal Charity Evaluators, 2017; see also Animals Australia, 2016a). It is finally worth noting that the NFF too, along with a range of its affiliated organizations, recognize the impact of Animals Australia in generating change. Indeed, the NFF has directly identified Animals Australia as a threat to the practices of conventional industrialized animal agriculture in Australia (Farm Online, 2013; Rodan & Mummery, 2014).

To further strengthen its work in advocating for the recognition of sentience and the take-up of alternative – reducetarian as well as vegetarian or vegan – eating habits, Animals Australia shares plant-based recipes and has put out its own vegan cookbook – *Taste for Life* (Animals Australia, 2017) – along with a free vegetarian starter kit. Other advocacy organizations are following suit. Edgar's Mission, a Victorian not-for-profit sanctuary for animals rescued from factory farms, released *Cooking with Kindness* (Ahern & Edgar's Mission, 2017), a compilation of recipes from some of the best vegan-friendly cafes and restaurants from around Australia. Voiceless, another strong proponent of the narrative of animal sentience, also promotes plant-based, dairy-free – that is, vegan – eating and shares recipes through its blog. Such aims are further supported by the development and introduction of alternative *meat* products into supermarkets, allowing consumers to cook with and eat plant-based *meats* without any sense of needing to learn a different cooking style. Indeed, the growth of the industry and its support by mainstream supermarkets is being described by Animals Australia as having the potential to bring new consumer cohorts to try plant-based meat alternatives and reducetarian diets, as well as supporting consumers to make the change to fully meat-free diets (Animals Australia, 2020c).

In its turn, RSPCA Australia stresses the importance of stronger considerations of animal welfare and reducetarian eating habits through its implementation of the RSPCA Approved Farming Scheme. Under this scheme, farms can be certified as meeting the RSPCA's standards – specified welfare conditions that support the physiological and psychological needs of livestock animals – meaning that subsequent animal products can be identified as having been produced under humane conditions.

> The aim of the RSPCA Approved Farming Scheme is ultimately to improve the conditions for farm animals. The RSPCA believes that farm animals must be treated in a way that meets their physiological and psychological needs. As well as having appropriate food, shelter and veterinary care, they must have the freedom to express natural behaviours.

> Many common practices in animal farming do not meet the animals' needs. However, these practices are not illegal. By raising public awareness and ensuring that consumers have access to higher welfare alternatives, the RSPCA aims to create demand for these higher welfare products. As consumer demand increases, producers will have a greater incentive to adopt humane farming practices. The RSPCA Approved Farming Scheme forms part of this strategy. (RSPCA, 2019i)

As this scheme is explained, what stands out is the focus on improving the welfare conditions of livestock animals in Australia, whilst also working to educate and mobilize Australian consumers on the welfare needs of these animals. The RSPCA's Humane Food Manager, Hope Bertram, explains that the number of RSPCA approved farms is growing because consumers continue to drive the demand for higher welfare: 'Australians are becoming more and more interested in the treatment and conditions for farm animals, and if they eat meat or eggs, they want to make humane choices' (as cited in RSPCA, 2019d). Alongside this focus, the RSPCA also offers an additional service to everyday consumers. Called *Choose Wisely*, this online directory is the RSPCA's initiative to make it easier to choose humane food when eating out at restaurants and cafes (RSPCA, n.d.). Although the RSPCA presently only certifies eggs, chicken, or pork as humane – that is, as produced with the welfare of the animal as a high priority – such an initiative clearly supports the broad principles of reducetarian and flexitarian, as well as vegetarian eating. Recognized as possessing a substantial amount of mainstream recognition and legitimacy (Chen, 2016), the RSPCA is an important narrator with regards to not just calling for improvements in welfare conditions for livestock animals but further for promoting reducetarian or flexitarian decision-making.

The Australian Marine Conservation Society (AMCS) provides similar advice to consumers with respect to choosing sustainable and responsibly sourced fish and seafood in Australia. Independent of government and industry and dedicated to protecting Australia's ocean ecosystems and wildlife, the AMCS advocates for real, evidence-based solutions based on the best available science for the complex problems facing our oceans (AMCS, 2021a). One of these problems is, of course, that commercial fishing practices (including the uses of trawlers, gillnets and longlines) take an enormous toll on our oceans due to (a) the continued overfishing of many species; (b) the devastation some of these methods (such as trawling) can wreak on marine ecologies and habitat; and (c) the poor record these methods continue to have with regards to accidentally killing other species such as dolphins, dugongs, sea lions, turtles, seabirds, and even whales (AMCS, 2021c). In addition, the farming of fish – aquaculture – is also problematic, with fish farms contributing to the pollution of marine ecologies, and further risking other marine species (Animals Australia, 2021b; Flanagan, 2021). The first point here is that because farmed fish such as tuna and salmon are carnivores, they need a suitable diet. This has resulted in farmed fish actually being fed commercially caught wild fish – with around 2 kilograms of wild caught fish needed to produce just 1 kilogram of farmed salmon, and between 10 and 20 kilograms of wild caught fish needed to produce just one kilogram of tuna (Animals Australia, 2021a). Secondly, common practices used to protect fish farm stocks against predation by seals, for example, have been found to be cruel and harmful to seals (Zwartz, 2018). In addition, farmed fish are prone to diseases and parasites from overcrowding which can in turn be easily transmitted to wild fish populations (Animals Australia, 2021b). With the aim of helping consumers make more ethical decisions about what they eat, the AMCS' *GoodFish: Australia's Sustainable Seafood Guide* (2019b) – available as a website and an app – includes sustainability ratings for more than 170 wild caught and farmed fish choices, covering over 90 per cent of the seafood eaten in Australia. More specifically, *GoodFish* assesses fisheries and aquaculture operators on a range of practices, such as the stock status of the species, the methods used to catch or farm them, and impacts on other marine wildlife and habitats (see AMCS, 2019a, 2019d). These assessments result in a three colour-coded classification: green-listed *Better Choice* options, amber *Eat Less* species, and those currently on the red, *Say No* list. The guide also contains information on the chefs and leading restaurants in Australia who have committed to no longer serving unsustainable, red-listed seafood. This is a reminder that, as the *GoodFish* Program Manager Sascha Rust has said,

> Chefs are real arbiters of our seafood choices and the best chefs are closely connected to the supply chains from the ocean to the plate. That so many of them are now using Australia's Sustainable Seafood Guide to make sure their customers get the

best seafood choices is not just a testament to the guide, but to them as guardians of the future of food. (AMCS, 2019b)

High-Profile Chefs and the Paddock to Plate Movement

At a time when so-called *lifestyle* choices can be seen as politically and ethically inspired, and are enacted throughout a broad range of private, public and institutional arenas (de Moor, 2016; Wahlen & Laamanen, 2015), lifestyle media provide a key site through which eating can be considered via 'ethical' frameworks (Arcari, 2019, p. 169). As was noted above, high-profile chefs can also play a substantial role in endorsing certain lifestyle choices concerning eating over others:[10]

> Celebrity chefs have taken on a significant role in influencing food cultures, consumption practices and public policy. As a group of powerful cultural and political intermediaries, celebrity chefs have used their public profile to address causes related to food ethics and sustainability, and to shape consumer 'choice' [...]. (Phillipov & Gale, 2018, p. 400)

High-profile chefs such as Jamie Oliver, Hugh Fearnley-Whittingstall, and Curtis Stone have thus been identified as playing a significant role in Australia with regards to the shaping and promotion of ethical consumption habits and broadly reducetarian inflected changes in food and eating norms (see also Ankeny et al., 2019; Phillipov, 2017; Phillipov & Gale, 2018). Other chefs also recognized in Australia for explicitly promoting such norms to mainstream audiences include Paul West and Tobie Puttock.[11] Critical of the practices common to industrialized and intensive farming and of the impacts such practice have on both animal welfare and the quality of animal products, these chefs actively engage their high profiles to promote sustainable and humane farming practices as well as the importance of high animal welfare in the production of good and healthy food. More specifically, they use their status, culinary knowledge and, typically, their televisual skills and media networks to (a) explicitly campaign against industrialized agricultural practices; (b) raise people's awareness about the sources and treatments of what they eat; and

(c) influence people to choose higher welfare and more locally produced foods over industry-produced and low welfare foods. Consistently, then, these chefs call on viewers and readers to care about the foods they consume and to avoid overconsuming what Michael Pollan (2008, p. 1) has described as 'foodlike substances' – that is, highly processed and industrially produced food products – and food products produced through inhumane agricultural systems. A number of these chefs further counsel consumers on how they can contribute to ensuring the welfare of livestock animals – for example, through buying RSPCA approved meat and eggs. Some further promote specific models of eating, with Oliver, for example, describing himself as flexitarian, and also using his media networks to call for everyone to reduce their meat-intake, as well as to try vegetarian and vegan food and cooking (Oliver, 2019; Pointing, 2019). Indeed, as he stated in an interview in 2015, 'It's a no-brainer, a plant-based diet is the future' (as cited in Vegan First, n.d.).

What is very evident is that such chefs are using several strategies to share and promote their messages throughout the broader community. One focus – shared with campaigns like *Somewhere* and *Make It Possible*, and the documentary *Dominion* – has been towards making the inhumane ways in which many livestock animals are treated much more visible to consumers. Some notable examples of this have been *Jamie's Fowl Dinners* (presented by Oliver, aired on UK Channel 4) and *Hugh's Chicken Run* (presented by Fearnley-Whittingstall, aired on UK Channel 4). Both shows have been credited with changing consumer attitudes towards certain intensive farming practices, and both chefs are credited with influencing supermarkets in both the United Kingdom and Australia to stock higher welfare brands of animal products, such as those recognized by the RSPCA (Phillipov & Gale, 2018). Some major Australian supermarkets have also partnered with high-profile chefs to promote their message of recognizing and responding to consumer concerns about livestock welfare: Coles partnering with Stone, and Woolworths with Oliver. Other chefs have partnered with animal welfare organizations such as the RSPCA (Puttock and Stone), and others model the high animal welfare standards and environmental sustainability possible through hobby farming (Fearnley-Whittingstall and West).

Similar messages are shared and promoted through Australia's paddock to plate movement which focuses on environmental sustainability, animal welfare and the sourcing of food from local purveyors (see, e.g., Campbell, 2008; Greengarten, 2010; Waters, 2018). More specifically, the paddock to plate movement stresses the importance of understanding the progression of food items from being *on the hoof* or *in the ground* to being on a plate. Campaigns that foreground the importance of moving away from industrialized farming practices and the overprocessing of food thus include the From Paddock to Plate (FP2P) Schools Program that aims to encourage students to reflect and think about 'food, sustainability, environment, business and agriculture beyond the classroom' (FitzRoy, n.d.), and Stephanie Alexander's Kitchen Garden Program which aims to influence 'children's food choices' and their 'attitudes towards environmental sustainability' (Gibbs, n.d.). Television programs such as *Gourmet Farmer* with Matthew Evans (aired on SBS; Hilton, 2010 to present), *Paddock to Plate* with Matt Moran (aired on Foxtel, Lifestyle Channel; Parker, 2013–14), and *River Cottage Australia* with West and Fearnley-Whittingstall (aired on SBS 2020; Boylan, 2013–16) further promote the idea of agriculture needing to become a 'sustainable enterprise' again, and one which better considers environmental sustainability, animal welfare, and the importance of eating seasonal and locally produced food products.[12]

Importantly, then, Oliver, Fearnley-Whittingstall, Stone, Paddock and West, alongside Alexander, Evans, Moran, and the broader paddock to plate movement, are all championing new food norms and reducetarian habits through their media presence. This is achieved as television presenters (as termed in Bennett, 2010; Bonner, 2011), as writers of cookbooks, contributors to magazines and newspapers, and as general promoters of reducetarian and flexitarian models of eating. Considering the high level of interest in lifestyle media and the ways lifestyle choices can be used to demonstrate one's identity and commitment, it is worthwhile remembering that several of these chefs reach audiences in the millions. This is especially so with Oliver, who is a considered to be a global media celebrity, and Fearnley-Whittingstall who is also well known in Australia. Given that much of the focus of these narrators of alternative models of eating is on issues of food quality – which they then directly tie to high animal welfare levels and their impact on human health and wellbeing – these are messages that can be heard and listened to by broad audiences. It is also worth noting that several of these narrators also directly challenge attempts to present such a focus as the province only of the well off.

Maintaining Existing Norms

Reducetarian messages are not unopposed in Australia. As noted, the NFF and affiliated organizations and individuals continue to challenge attempts to call for changes in Australian agriculture and eating norms, variously describing such calls as *un-Australian*, against the national interest, and as ignorant of the realities of farming. Unsurprisingly, given meat-eating is still an Australian norm, such messages tend to be spread most widely in response to particular peaks in advocacy activity or other developments. For instance, the distribution and impact of the *Make It Possible* and associated campaigns concerned with livestock welfare by Animals Australia – including those challenging live animal export practices – led to extensive denigration of such animal advocacy as pedalling a pro-vegan agenda with the intent of 'ending the meat production and meat consumption in this country' (Bettles, 2013b; see also Clark, 2013; Farm Online, 2013; Nason, 2013). Protest activities organized by animal activists also provoke spikes in the denigration of vegans and animal activists. Direct protest actions have thus led to activists being described as 'blatantly anti-farming' (Nason, 2013), 'green-collared criminals' (Sullivan, 2019), and as engaging in activities 'akin to terrorism' (ABC News, 2013; Greer, 2013) that place 'the community at risk' (ABC News, 2019; see also Coughlan, 2019). The introduction of alternative meat products such as plant-based meat into mainstream supermarkets has also prompted the circulation of media messages disparaging reducetarian and vegan ideals. In this instance, the products have themselves been the target – being described as fake, not authentic, with labelling misleading to consumers (see, e.g., McCarthy & Henderson, 2018; Wahlquist, 2018). Proponents of such products and of alternative diets are also targets, described for instance as a 'bleeding edge' minority 'at risk of losing touch with reality' (Marshall, 2019).

These are messages aiming to maintain existing meat and animal product eating norms and protect existing models of animal agriculture. Under this framing, Australian farms and farmers are being targeted by and need protection from hostile extremists and radicals. Meat-eating norms also tend to be endorsed by affiliated industries, with such industries in Australia touting the authenticity and naturalness of their products (as compared against, for instance, their plant-based alternatives), and promoting the message of meat and dairy being essential parts of a healthy diet. Such messaging is exemplified in the annual MLA Australia Day lamb advertising campaigns which, for multiple consecutive years since 2004, have explicitly promoted meat-eating as natural, healthy, and integral to being Australian, and presented vegetarians and vegans as being self-evidently ridiculous and faddish. Indeed, across these campaigns, Sam Kekovich – dubbed the lamb ambassador or *lambassador* – has labelled vegetarians and vegans as 'hairy legged sandal wearing lentil eaters', 'soap-avoiding', 'pot-smoking' and 'un-Australian' (as cited in Bembridge, 2016). Although such advertisements are presented – and defended – as satirical, their repeated screening towards and on Australia Day highlights deeply held views about the normalcy of animal agriculture and meat-eating, and the purported fringe status of alternative diets. Graeme Yardy from MLA (as cited in Workman, 2021) embodies this norm when commenting about their 2021 Australia Day lamb campaign *Make Lamb not Walls*: 'Lamb has always been the meat that brings people together'. At the same time, as noted previously, these same industries have also recognized that dietary changes are gaining in traction among Australians. So, while Australian producers can work to ensure that consumers recognize the *naturalness* of animal products, they also need to remain adaptable in the face of these new eating and dietary patterns. This will arguably need to involve a stronger response to concerns about animal welfare and sentience.

Gaining Traction

What these various engagements and promotions of the narrative of sentience in the context of changing ingrained habits around food and eating make clear is the importance of influential narrators. Certainly, as noted above, different narrators of the narrative of sentience and its associated calls for care and change in the treatment of livestock animals will speak to more or less effect with different audiences. Those already concerned about animal welfare, for instance, may actively support activist sponsored reducetarian messages, although they may also find some campaigns to be too minimal in their animal welfare and wellbeing demands. Existing vegans or animal rights proponents, for example, may view welfare-oriented campaigns as ineffective for addressing the anthropocentrism and animal cruelty inherent in industrialized animal agriculture. On the other hand, mainstream meat-eaters may find the messages delivered by high-profile chefs to be more convincing regarding the importance of considering animal welfare in food selection than if delivered by animal activists, who may be dismissed as extremist or faddish. This is an important point because, as noted earlier, Australia is still predominately a meat-eating culture and, as we have noted, vegan-specific messaging has a history of being dismissed as extremist, and unsupportive of Australia's economy and conventional way of life (Rodan & Mummery, 2019). Cognizant of these tensions, some animal welfare organizations such as Animals Australia are explicitly working to reach and influence mainstream Australians and are, as such, presenting broad reducetarian messages rather than vegan or even vegetarian ones. Such messaging addresses the point again that even small personal changes can have big impacts for animal welfare.

What is important to note here is that the various narrators that have been outlined in this chapter – animal advocacy and welfare organizations, animal activists, high-profile chefs, even the NFF and MLA and other affiliated organizations – are all striving to become and be change-makers, what we have previously described as *cultural intermediaries*. In other words, through persistently spreading their messages through traditional and social media platforms, websites and Apps, each of these groups and their various spokespeople see themselves as being in the business of constructing value, culture, and ways of living. They might be supporting the values and culture of the existing anthropocentric social imaginary – work performed by the NFF and MLA and other affiliated organizations, for example. Conversely, they might be proposing new values and counter-narratives that would contribute to an alternative imaginary that better recognizes and respects the sentience and interests of other animals. This is the work of animal activist and advocacy organizations as well as by several high-profile chefs and other associated reducetarian movements. Put another way, cultural intermediaries strive to 'impact upon notions of *what*, and thereby *who*, is legitimate, desirable and worthy, and thus by definition what and who is not' (Smith Maguire & Matthews, 2012, p. 552). In this instance, of course, this is the very question as to whether livestock animals deserve to have their sentience and interests respected – whether they should be treated as individuals worthy of respect in their own right rather than simply as resources.

The effectiveness of this work is of course dependent on whether purported cultural intermediaries are recognized as possessing professional expertise and authority within key fields. Thus, high-profile chefs are recognized as experts in their fields, and trade on this in promoting – and in some cases even branding – their specific messages about food, diet, and eating. Their activities and expertise around food are reported on in traditional media, re-circulated in social media, and reiterated in their social media channels. Such chefs thus become food leaders influencing public discourse – whether audiences agree with them or not – and offering a counter-narrative to conventional views and models of meat-eating. In this same manner Australia's peak animal welfare bodies and spokespeople also attempt to stress their professional authority and/or mainstream legitimacy with regards to delivering messages about animal welfare, the narrative of sentience, and the importance of reducetarian eating practices. These are key capacities. Although not all audiences will accept messaging from Animals Australia, certain of its spokespeople are recognized as possessing cultural authority. For instance, Animals Australia's Director of Strategy, Lyn White, was honoured as a Member of the Order of Australia in 2014 and was the recipient of the Australian Humanist of the Year award in 2019. And, of course, it is the RSPCA Australia brand that is relied upon with regards to its certification of certain Australian farms as humane and certain animal products as humanely produced.[13] It is these capacities, then, that underpin and make possible the motivational work of these groups in calling for change, and that perhaps facilitate the imagining and introduction of alternative eating norms that go some way to responding to the call to improve animal welfare and recognize animal sentience. As the 2018 Futureye report reminds us, while there is rarely any single defining societal shift with regards to any social change, what these cohorts of animal advocates and high-profile chefs all make clear – as proponents of reducetarian messages – is that different eating practices are possible and not hard to implement. They also all tend to stress the point that small local changes all count and will all make a difference. And, as has been noted, the difference already being made with regards to Australian eating patterns is one that has been identified as requiring substantial adaptability from Australia's animal agricultural industries.

Despite these points it is worth noting that the majority of current reducetarian messaging and practice in Australia does not substantially question the ways anthropocentric narratives play out with regards to livestock. That is, while it certainly has the capacity to draw attention to a need to better consider animal welfare (along with broader issues of environmentalism and sustainability), comparatively little of Australia's reducetarian messaging calls for an end to the use of animals as resources. While such a demand certainly lies at the heart of veganism, it is unsurprising that the broader reducetarian movement in Australia does not make this demand central. This too is an issue of traction insofar as the reducetarian message is very much aimed at the members of a mainstream meat-eating culture. And yet, because the reducetarian message nonetheless does demand recognition of our eating as always being an industrial, political, and ethical act, this message does have the potential to inspire some initial steps towards a broader and deeper change. This could involve the collective exploration of the kinds of relations that might be possible between humans and animals if the anthropocentric assumptions justifying human use of animals are destabilized, perhaps to the point of transitioning Australia's meat-eating culture and also of seeing in not just new norms for everyday eating and consumption but an alternative imaginary around our relations with all animals, even the ones we once saw only as livestock.

Notes

1   It is important to note that this view that a subjectively aware being should never be treated merely as a means to the ends of others and should have the right to be consistently treated with respect is also integral to much of the theory of human rights. Although human subjective awareness has typically been awarded special status under the framework of human exceptionalism – and thus historical conceptions and the award and protection of rights have been restricted to humans – this is being challenged. There have been, for instance, various legal attempts to have at least some of the protections of human rights also awarded to some non-human animals (dolphins, great apes) (see, e.g., Kurki, 2021; Great Ape Project, n.d.; Nonhuman Rights Project, 2021). More broadly, there are increasing calls for human rights to be reconceptualized as sentient rights (see, e.g., Cochrane, 2013; Woodhouse, 2019a, 2019b). This is to argue that human rights are not qualitatively distinct from the basic entitlements of other sentient creatures, and that attempts to differentiate human rights – and respect for human rights – by appealing to something distinctive about humanity, their unique political function or their universality ultimately fail (Cochrane, 2013).

2   Australia's longstanding appetite for meat-eating has been argued to be largely derived through the intersection of White settler attitudes towards meat – which were developed in a home country where it was scarce – and its relative affordability in Australia. With meat continuing to hold powerful social and cultural connotations, Australia remains among the world's most enthusiastic carnivorous nations (Ting, 2013).

3   See Masterman-Smith, Ragusa and Crampton's (2014) content analysis of articles about veganism published in 2007 and 2012 in Australian newspapers. They found two predominant and interrelated discourses: 'a discourse of irrationality directed at vegan motivations' along with 'a discourse of impossibility directed at vegan practices' (p. 7). For further analysis of discourses and stigmatizing stereotypes about veganism, see Cole (2008) and Williams, Archer and O'Mahony (2021).

4   Nose-to-tail meat eating is defined as 'an approach to consuming meat that sees us enjoy as much of the animal as possible – organ meats, fat, unpopular and uncommon cuts', reminding us that 'farmers raise whole animals, not pork chops'

(Ethical Farmers, 2020). In practice, nose-to-tail meat eating might entail purchasing and using a share of an animal (e.g. 1/8 cow, 1/2 lamb, 1/4 pig) (2020), or it might mean choosing 'less popular cuts of meat like neck cuts, liver, ears [...] After all if an animal has to die for us, we should respect it enough to eat all of it' (Sustainable Table, n.d., p. 8).

5   In its guide to ethical meat eating, the Sustainable Table provides meat-eaters with six tips for becoming a more mindful meat-eater. These tips are as follows: (a) make meat a treat, (b) support a nose-to-tail philosophy, (c) choose more eco-friendly meats (and fish), (d) choose grass-fed or pasture-raised organic meat, dairy and eggs, (e) buy ethically farmed meat from small organic or biodynamic farms, and (f) shop at accredited farmers' markets (Sustainable Table, n.d., pp. 8–9). The Sustainable Table generally makes the argument that while 'there are strong [...] arguments for collectively turning to vegetarianism, it's an unrealistic and potentially alienating message to push. [...] This booklet is set to challenge and empower you to make a difference, one meal at a time' (n.d., p. 2). Such a message has some similarities with that spread by Animals Australia through its *Make It Possible* campaign.

6   Similar arguments are made for the consumption of other wild caught meats, such as rabbit or hare (see Sustainable Table, n.d.).

7   To date the preferred terms for cell-based meat are 'cultured meat' or 'clean meat' (Stephens et al., 2018, p. 163; Fernandes et al., 2021).

8   The issue of affordability is important. A recently published study that looked at public opinion on plant-based diets has found widespread support for the ethical and environmental benefits of veganism and vegetarianism among meat-eaters, but identified price, along with taste and convenience, as key barriers to taking up such a diet (Bryant, 2019). These findings suggest that if cultured meats can meet their promises of affordability whilst also meeting requirements of taste and convenience, many meat-eaters should feel comfortable in transitioning away from meat produced through conventional animal-based agriculture.

9   Pat Rafter is a former world number one tennis player. Mick Molloy is an comedian, writer, producer, actor, television and radio presenter. Missy Higgins is a singer/songwriter winner of nine ARIA music awards. Claire Hooper, Dave Hughes and Anthony Lehmann (Lehmo) are all well known stand-up comedians, and television and radio presenters. Arj Barker is an American stand-up comedian and actor well-known in Australia. Carl Barron is a theatre and television comedian, and Peter Rowsthorn is a stand-up comedian, writer and actor in the highly successful Australian television comedy *Kath & Kim*.

10 Australia has a booming food culture and Australians evidence a commitment to high-profile chefs and television cooking shows. Viewership of high-profile chef cooking shows in Australia remains high, for example, with *MasterChef* remaining one of the highest-rated shows on domestic free-to-air television (News Chant, 2021). Cooking shows have furthermore been shown to influence home cooking

practices – the so-called '*MasterChef* effect' (AAP, 2016; Yu, 2016). As another commentator suggests, 'TV chefs are rock stars in Australia' (Hamilton, 2014).

11 There are many additional high-profile chefs featuring in Australia, of course, with several gaining television and media renown from their *Masterchef* exposure. There are also a variety of other television cooking shows screening in Australia led by high-profile chefs – for instance, Asian Australian chefs Kylie Kwong, Adam Liaw, Poh Ling Yeow and Luke Nguyen. However, whilst these chefs and shows certainly promote a variety of eating styles, they do not use their media presence to actively and continually promote reducetarian food norms or food norms oriented to the improvement of animal welfare. Such chefs might therefore incorporate some vegetarian and vegan recipes and meals in their repertoire, but their priority is not on actively challenging the practices of industrialized animal agriculture.

12 Of course, there has been some critique of this kind of *locavorism*. For an in-depth discussion, refer to Stănescu (2010).

13 Despite this achievement Australian media reports have also exposed some of the discrepancies between the ideal and practice of RSPCA's certification of Approved Farming Standards of free-range, bred-free and improved animal welfare (see, e.g., Denholm, 2021; The Age, 2004).

CHAPTER 4

Making Friends and Kin with Companion Animals: Skippy to Red Dog

In the previous chapters we have considered the work by such animal welfare organizations as Animals Australia and RSPCA Australia towards building empathy and concern in mainstream Australians for those animals – typically livestock – that have had a history of treatment by humans that is primarily informed by their economic value under anthropocentric interests. As these chapters have also noted, the majority of Australians have little direct involvement with the actual lives – and deaths – of livestock animals, and these animals have consequently had relatively little visibility to us as sentient beings in need of our care. This is despite the fact that these animals live and die for us in enormous numbers each year, and animal products make up a significant part of most Australians' diets as well as contributing to multiple other aspects of everyday consumption. Making the experiences of these animals visible to – or, at least, imaginable by – mainstream Australians has thus become an important aim for animal welfare and advocacy organizations. Typically, the tactic has been to foster our awareness of them as sentient beings who experience their lives in ways similar to how we do. This is hoped to foster our empathy and our outrage for what these animals have been made to experience under the auspices of our anthropocentrism. Animal welfare and advocacy organizations thus call on us to imagine how these animals could be envisioned and treated with more respect – from implementing substantially better welfare standards for them, through to fundamentally revisioning our everyday consumption practices. Rather than continuing to trace such imaginings, however, this chapter turns its focus from these issues to the animals that are already visible and important in the lives of the majority of Australians – our companion animals.

This is an important shift in focus. While our relations with livestock animals definitely do need to be rethought and reset via the triggering of our empathic imagination and care – fostered through recognition of our shared sentience – a majority of Australians would *already recognize* themselves as being in relationship with companion animals. Indeed, such relationships are common across many of our shared stories. These include stories of friendship and kinship, of trust and respect across species lines, and of the close connection of lives. As such, in this chapter we examine a range of Australian imaginative narratives that feature close companionate and caring relationships with companion animals and that have been broadly available to Australian publics through such popular screen media as films and television series. These include stories about Skippy ('the bush kangaroo'), Mr Percival, Red Dog, Larry ('the Wonderpup'), among others. Each of these stories can be seen as modelling a range of caring and companionate relationships between humans and animals – relationships that many Australians would already recognize and accept. Most importantly, these stories model relationships and interactions that cannot be fully described by the economic narrative and values of anthropocentrism. Such stories are important because, as we have noted, the common horizons needed for any social imaginary – existing or potential – are built not simply through the revision of concepts and theories but through the mainstream sharing of narratives, images, and myths. As we have shown in Chapters 2 and 3, then, animal activist campaigns actively develop and promote stories stressing our *likeness* to animals in that we all experience our lives and prefer a life of wellbeing. These are stories typically tied to the narratives of sentience and care, and that have – at least in Australia – been used to generate calls for improved animal welfare. However, the stories we already share of our friendships and our care-based interactions with companion animals can also be read as counter-narratives to anthropocentrism, and thus as also containing important insights with regards to enabling the reimagining of human-animal relations beyond the norms set by anthropocentrism.

This chapter thus comprises three main sections. The first of these is a brief contextual description of both the broad status of companion animals within the Australian context and the basis for our selection and analysis of screen media stories in this chapter. The second section comprises our examination of these stories. Identifying and exploring the two main forms of relationship visible within these stories – friendship and kinship – this section also notes several other issues that would seem important for broader attempts to revision human-animal relations. The final section encompasses our concluding reflections regarding these stories and their potential for contributing to an alternative Australian social imaginary that is less driven by anthropocentricism. Of particular interest here is to consider what such stories might offer in the context of reimagining human relations with other animals, whilst also noting some of the points of contestation that have arisen with regards to these stories and models of human interaction with companion animals.

Contexts

Australians and Companion Animals

Australians favour companion animals – *pets*, in contemporary parlance – with about 61 per cent of Australian households sharing their lives with a member of at least one companion species (Animal Medicines Australia, 2016, 2019). Research suggests that the most popular companion species is the dog (40 per cent), followed by 'cats (27%), fish (11%), birds (9%), small mammals (3%) and reptiles (2%)' (Animal Medicines Australia, 2016, p. 6, 2019). Other households report living with members of other species, including 'horses, goats, cows, alpacas and hermit crabs' (2016, p. 6, 2019). Many Australian households also live with more than one pet (22 per cent), and even if a number of Australian households are described in this report as not having a pet, about 90 per cent of households report having lived with one or more pets at some point (2016, p. 6). Furthermore, and still according to this report, about 88 per cent of Australian pet owners describe their experience with companion animals as positive. Indeed, only 3 per cent of the Australian households surveyed named what they saw as some of the main disadvantages of living with companion animals. These included economic costs,[1] limitations for travel, added cleaning requirements, inability to be spontaneous and the emotional pain of mourning their deaths (Animal Medicines Australia, 2016, 2019).

What these points suggest is that Australians as a whole are comfortable in (at least some) interspecies relationships, and that some of these relationships can matter to us. Thus people may share stories of their *fur babies*, and describe them as psychological-kin and equal members of the family (Kao, 2018; Topolski et al., 2013). For some, companion animals are 'family members' who are and should be cared for through 'everyday acts of devotion to make their lives better' (King, 2021, p. 13). Or – on a darker note – people in violent relationships may find their companion animals targeted for abuse as well, because of their recognized value (Coorey & Coorey-Ewings, 2018).[2] With regards to the companion animals we care for and live with, it is by now a well-established research finding that people who enjoy the day-to-day companionship of animals suffer less loneliness and anxiety in their lives and can gain a range of health related benefits including improvement in overall health and psychological well-being (Robinson, 2020; Smith, 2012). Further, we relax with our companion animals. We enjoy their spontaneous and unconditional affection and physical closeness. We can *be ourselves* in their presence, given that they have no socially acquired expectations of us. To put this otherwise, they offer us psychological and emotional release. Indeed, our living with companion animals brings:

> an element of slapstick and anarchy into the cool, smart, self-absorbed world of business and public affairs. They make us miss work; they muss up the perfect clothes, perfect hair, that are needed to assure our 'professionalism', our presentability, in this public world; they strew shit and dirt around the manicured gardens, and leave paw marks through the tidy houses, that announce our hard-won social status. They gently lead us back from the obsessive quest which is definitive of the modern ethos [...] (Mathews, 1997, p. 8)

As well as enhancing our capacities to care across species lines, living with companion animals can enable our recognition of 'how odd or arbitrary our human priorities might appear from a non-human perspective' (p. 5). Both such reorientations are clearly extremely valuable with regards to the possibility of shifting away from the values and assumptions typically foregrounded in anthropocentrism, human exceptionalism and speciesism, and which – as we have discussed in the previous chapters – continue to be presented as the norm in many Australian narratives regarding human-animal relations. To reiterate, these are: that human exploitation of animals is justifiable; that human development interests justify animal harm and animal deaths; that all (nonhuman) animals should be regulated for their impact on the (human) community and relevant ecosystems; and, that little to no attention needs to be given to animal interests unless there is some legal obligation to do so. If care for animal welfare triggered from a recognition of animal sentience offers one path away from taking these assumptions as normative, our hope is that reflection on the models of companionate interaction we take for granted with our companion animals offers another.

Stories and Screen Media

In the Introduction, we noted that while the members of a society might live under the influence of a diverse range of imaginaries and associated narratives, what is called the social imaginary provides the framework for the broader normative regulation of that society (Castoriadis, 1975/1987; Taylor, 2004). We have also noted that a social imaginary is always both symbolically and materially produced and maintained, and that a key part of this production and maintenance is carried out through the distribution and sharing of stories which are considered to be 'visibly rooted in everyday life' (Anderson, 2006, pp. 35–6). Given, then, the position of companion animals in the everyday lives of Australians, it seems clear that Australian stories of our relations with our companion animals – whether these are consumed privately, shared with family, friends or colleagues, or shared across our screen media – have the potential to strengthen the idea that we can share our lives with animals in accordance with non-economic values. Such stories can also remind us that sharing our lives with animals in interspecies relationships has its own benefits – a key recognition for the possibility of a relationship that is not driven by purely anthropocentric assumptions. A further important point here is that stories about animals have been shown to 'have a positive impact' with regards to how readers (or viewers) can come to 'value the well-being of other species' (Malecki et al., 2019, p. 123). This is to remember the importance of storytelling more generally – that a focus on stories and storytelling can improve our empathy and our capacities to understand others (Oatley et al., 2016; Witherell & Noddings, 1991).

Given the ubiquity of animals in Australian everyday life, it is unsurprising that animals and human-animal relations feature in a significant portion of Australian stories and storytelling. First of all, Australia's First Nations people feature animals as a key 'part of our dreaming [...] their stories are embedded in our landscape' (Smallacombe, 2020). Indeed, relationships with animals play an important role throughout Australian First Nations kinship systems, communities and knowledge systems, with animals – and the *Dreaming* and other ancestor stories that involve them – providing important foundations for both individual and collective identities for First Nations people (Glynn-McDonald & Sinclair, 2021). More specifically, these systems and stories reinforce identity, social cohesion and social control by providing a deep sense of belonging, not only to other people and community but also to the land, the animals, plants, songs, rituals, art, stories and the Law as laid down in the *Dreaming*.[3] Animals have also played significant roles in constructions of Australian identity and culture post Anglo-Celtic settler occupation (see, e.g., Markwell & Cushing, 2014). Given the influence of agrarianism or country-mindedness for ideas of national character post Anglo-Celtic settler occupation (Aitkin, 1985; Berry et al., 2016),[4] it is unsurprising that associated narratives star not only country, bush or outback human residents but the *jumbuck* or sheep (Waltzing Matilda), *brumbies* or feral horses (*The Man from Snowy River*), or other working animals such as dogs (e.g. the dog on the tuckerbox, Red Dog). Other animals significant in post-settlement stories that have become influential with regards to ideas of national character include Simpson's donkey and the Waler horse known as Bill the Bastard (both memorialized for their roles in rescuing ANZAC soldiers in World War One), and Phar Lap (an exemplary *rags to riches* story memorialized through the feature film *Phar Lap: Hero to a Nation*). Australian folklore is also heavily populated by animals, with stories of drop bears (an imaginary predatory, carnivorous version of the koala), bunyips and yowies, hoop snakes, as well as the various legends of the Tantanoola Tiger, the Grampians Puma, and the Gippsland Panther (see, e.g., Waldron & Townsend, 2012). Animals are also key national symbols adorning the country's currency and coat of arms (Berlin, 2019; Markwell & Cushing, 2014). In addition, as noted earlier, many of the stories of post-settlement Australia promoted through Australian tourism industries star animal populations: cuddling a koala, taking a selfie with a quokka at Rottnest Island, watching the penguins come ashore at Philip Island, swimming with whale sharks at Ningaloo, snorkeling or diving in the Great Barrier Reef, whale watching trips, seeing birdlife massing at Lake Eyre when it is in flood, for example.

In the instance of this chapter, with its focus on examining Australian companionate human-animal relations and interactions, the decision was made to focus on stories starring such relationships as told and publicly shared through Australian screen media.[5] Screen media are particularly influential in storytelling. Not only are they broadly accessible and shared, but they are able through their visuality to easily and effectively reflect multiple aspects of our lives to us and build connectedness, empathy and concern (Aiello & Parry, 2020; Elkins, 2008; Rose, 2016). And, indeed, one important role of screen media in Australia has been to create cultural stories that engage with familiar Australian locations, settings and narratives – that allow Australians to recognize themselves. It has been identified that, when watching specifically Australian screen media, Australian audiences generally want to experience a 'sense of "home"' and 'a way of life' that they relate to and recognize (Screen Australia & Ipsos Australia, 2013, p. 2; also see Olsberg.SPI, 2016, pp. 13–14).[6] The further point here is that as broadcast media, screen media programs – particularly those released through public broadcasting platforms[7] – can connect millions of Australian viewers and audiences (Bignell, 2004; Turow, 2017), reaching viewers of all ages in nondidactic, 'mimetic' ways that both engage and inform them (Dant, 2012, pp. 119–20). In addition, due to their wide accessibility and distribution, screen media have been shown to also enable social connectedness through generating private and public conversations with family, friends and colleagues, in education classrooms and workplaces, and on social media (Deloitte Access Economics, 2016; Turow, 2017). To put this another way, screen media are highly effective in enabling spaces for viewers to reflect, individually and collectively, on a variety of social ideas (Aiello & Parry, 2020; Hariman & Lucaites, 2016). Together these points mean that screen media are a prime medium for the circulation of Australian cultural narratives and national tropes, and have indeed been recognized as helping foster norms of shared identities and places (Dant, 2012; Deloitte Access Economics, 2016; Screen Australia & Ipsos Australia, 2013). A further point here is that, as we have noted earlier, for many people, animals are predominantly visible through representational and mediated forms – and the primary human-animal experience is via television, the most watched medium (Mills, 2017).[8]

Given our ongoing focus on human-animal relations and specifically on companionate human-animal interactions in this chapter, the decision was made to make the featuring of such interactions one of our primary selection criteria – the second being that selected programs needed to be Australian produced and to foreground recognizable Australian content.[9] Regarding this first selection criterion, companionate human-animal interaction also needed to feature significantly in a program to warrant its selection – simply having animals or human-animal companionate interactions occasionally present within a program was not enough for its selection.[10] Using this selection criterion we also excluded screen media films and programs that primarily featured animal-animal interactions (such as *Bluey*) or played out conflictual human versus animal or *man as hunter* narratives (this latter typical of Australian fishing adventure programs, for instance). Educational and observational documentaries (e.g. David Attenborough type programs) were also excluded given their focus on *presenting* animals and animal behaviour as if independent of human intervention. Adding the second selection criterion of programs needing to be Australian produced with recognizable Australian content led to the selection of nine feature films, three documentary films and four television programs, that date from the first decades of Australian television to the present. These are, in alphabetical order: *Baxter and Me* (Leahy, 2016), *Dot and the Kangaroo* (Gross, 1977), *Emu Runner* (Thomas, 2018), *Healing* (Monahan, 2014), *Koko: The Red Dog Story* (McCann & Pearce, 2019), *Larry the Wonderpup*

(Crook, 2018), *Oddball* (McDonald, 2015), *Penguin Bloom* (Ivin, 2020), *Red Dog* (Stenders, 2011), *Red Dog: True Blue* (Stenders, 2016), *Send in the Dogs Australia* (Quail, 2011–), *Skippy the Bush Kangaroo* (Robinson & Hill, 1967–68), *Storm Boy* (Safran, 1976), *The Camel Boy* (Gross, 1984). See Table 1 for details of each of these.

Table 1: Selected Screen Media

Title	Year(s)	Key Details
Baxter and Me	2016	Director: Gillian Leahy Medium: Documentary film Awards: • Best Documentary Script (Australian Writers Guild 2016)
Dot and the Kangaroo	1977	Director: Yoram Gross (OAM) Medium: Feature film (animation and live background) Awards: • Best Feature Film, Children's International Jury (Tehran International Film Festival 1977) • Sammy Award for the Best Animated Film (Australian Film and Television Awards 1978)
Emu Runner	2018	Director: Imogen Thomas (with Indigenous Script Consultant Frayne Barker) Medium: Feature film Awards: • Best Australian Independent Film Peer Award (Gold Coast Film Festival 2019)
Healing	2014	Director: Craig Monahan Medium: Feature film Awards: • Flinders Series Award (Australian Director's Guild 2015) • Best Film Audience Award (Rencontres Internationales du Cinéma des Antipodes 2014) • Director's Choice Award for Best Foreign Film (Sedona International Film Festival 2015) • Best Picture in Cannes 21st Cinephiles
Koko: The Red Dog Story	2019	Directors: Aaron McCann & Dominic Pearce Medium: Documentary film

Larry the Wonderpup	2018–20	Directors: Ben Shackleford, Luke Jurevicius Medium: Television (Family program), live action and animation, 2 seasons
Oddball	2015	Director: Stuart McDonald Medium: Feature film Awards: • Best Family Film Audience Award (Mill Valley Film Festival Audience 2015) • Best Kids Film Hollander Prize (Traverse City Film Festival 2016) • Films4Families Youth Jury Award (Seattle International Film Festival 2016)
Penguin Bloom	2020	Director: Glendyn Ivin Medium: Feature film
Red Dog	2011	Director: Kriv Stenders Medium: Feature film Awards: • Best Narrative Feature (Heartland Film Festival Grand Prize Award 2011) • Best Feature Film (Inside Film Awards 2011) • Best Film (Australian Academy of Cinema and Television Arts Awards 2012) • Australian Film Institute Members' Choice Award (Australian Academy of Cinema and Television Arts Awards 2012) • Best Feature Film Audience Award (Rencontres Internationales du Cinema des Antipodes Award 2012) • Grand Jury Award (White Sands International Film Festival 2012) • Favourite Film of the Decade Audience Choice (Australian Academy of Cinema and Television Arts Awards 2020)
Red Dog: True Blue	2016	Director: Kriv Stenders Medium: Feature film Awards: • Grand Prize Narrative Feature (Heartland International Film Festival 2017) • Young People's Jury Award (TIFF Kids International Film Festival 2017)

Send in the Dogs Australia	2011	Director: Claire Meech Medium: Television Documentary, 2 seasons
Skippy the Bush Kangaroo	1967–68	Producers: Lee Robinson & Dennis Hill Medium: Television, 3 seasons, 91 episodes Awards: • Best Export Production (Logie Award 1969)
Storm Boy	1976	Director: Henri Safran Medium: Feature film Awards: • Best Film (Australian Film Institute 1976) • Jury Prize (Australian Film Institute 1977) • Awgie Award (Australian Writers Guild 1977)
The Camel Boy	1984	Director: Yoram Gross (OAM) Medium: Feature film (animated)

Note: All the films are available on DVD and many others, at different times, are aired on Netflix, AmazonPrime, Foxtel and SBS ONDemand, ABCiview. Some films can also be purchased from Internet streaming sites.

As will be seen, each of the selected films and programs feature Australian stories of care-based companionate human-animal relations and interactions. In addition, the programs we have selected make a variety of human-animal relationships visible. Not only do they outline some of the dimensions of human-animal companionate relationships with conventional companion animals (dogs being the standard example: Red Dog, Larry, Oddball and Baxter), but they also explore the possibilities of companionate relationships with animals not normally recognized as companions: kangaroos (Skippy), birds (Mr Percival, Penguin, and Yasmine), camels (Binta, Aziza, and Ali), for example. Most importantly, these various programs illustrate models of relationship in which human interests do not appear to be the sole or primary motivator of action or value, rather showing modes of interaction wherein the needs and interests of nonhuman others are considered alongside human interests. This illustration of valuing more than human interests is a key point for our inclusion and consideration of these programs in this chapter insofar as these representations of human-animal companionate interaction suggest how we might relate to nonhuman others in non-economic and non-hierarchical ways. In particular, the selected programs reveal two different forms of human-animal companionate relationships – forms of friendship and kinship. The different attributes and possibilities of these two forms of companionate interaction and relationship are explored next and are illustrated through reference to selected screen media. Although we acknowledge that such illustrations exceed strict realism in many of these programs, this is not problematic per se. The representations of companionate human-animal relationships found in these screen media – even if they are imagined, partial and, on occasion, inconsistent[11] – nonetheless effectively illustrate how we humans *might* learn to pay better attention to animal interests and build interspecies relationships, as well as modelling a possible regulation of our tendencies to automatically give value to our anthropocentrism in human-animal relations.

Being Friends with Other Animals

To some people, it is simply obvious that friendship with an animal can exist. And, indeed, as noted above, storytelling about the capacity for friendship between humans and animals – in both adult and children's literature as well as screen media – has a long history (see Cosslett, 2016; Ratelle, 2014). It also seems necessary to say that friendship with an animal companion could conceivably be the strongest relationship in the lives of some humans, whether temporarily or permanently. Of course, whether animals, especially companion animals, are considered able to count as friends depends on the conception of friendship in play as well as on the conception of and relationship with animals. For instance, friendship is typically thought to need mutuality and reciprocity. This means that both sides of the friendship need to regard the other with affection as well as trust – there needs to be mutual care. A friend, after all, is someone you would *choose* to spend time with, and you further recognize that the other's interests matter alongside your own. Friendship is also considered to be based in irreplaceability and uniqueness (Telfer, 1991). Namely, friends are non-substitutable. We might, of course, have several friends, but every friend is considered unique and valued in themselves, because of themselves.

A friendship is also not made simply through requiring – and finding – someone with specific attributes, or by simply *liking* a person's various attributes (whatever they may be). That is, we always befriend an individual not a type, and although we may be friendly in our approach to another, an actual friendship needs to be built from and recognized by both sides of the relationship. This is also the reminder that friendship does not arise simply out of proximity, although the choosing to spend time together and engage in joint pursuits is one of the indicators of friendship (Telfer, 1991). Here the point is that while we would all certainly share our lives with different individuals, living in proximity with another does not by itself guarantee the building of a friendship. The same applies to sharing our lives and living spaces with animals. Although some animals are designated as companion animals by virtue of their specific attributes (e.g. being a member of a particular species and breed), simply living or being with such an animal does not mean that animal will be considered a friend, or that the animal so concerned may act in ways that suggest they regard the human with any affection. That is, what matters for friendship is that time spent together is by choice as well as by preference. Together these various points suggest that the properties of friendship include, for instance, companionship, trust, loyalty, commitment, affection, acceptance, sympathy, concern for the other's welfare, as well as time spent together and the maintenance of a friendship bond despite separation (Brent et al., 2014).

It would seem clear that, for at least some Australians, such conditions would fit with how they view their relationships with their companion animals. At the same time, it is also important to note that there is debate as to whether humans and animals can be friends in the same sense that humans can be friends with each other. And here some of those who suppose that animals cannot be our friends might do so because they doubt that animals are able to make such a bond (Clark, 2008). This is another iteration of the typical anthropocentric narrative that finds it easy to say that humans always matter more than animals, and that if animals are to be valued at all, then this is entirely for their usefulness to us. Such objections might further note that to even frame some human-animal relationships as friendships is to *anthropomorphize* animals and animal behaviour (meaning to unreasonably identify human characteristics, emotions, or behaviour in nonhumans). And yet the point that human-animal friendships may differ from human-human friendships is not necessarily a problem insofar as human friendships can also vary in type. As a result, while it would certainly seem reasonable to accept that some ideas and expectations of friendships may involve particular kinds of interactions that animals would not appear to be able to participate in, there is still a plausible case for thinking that (at least some) humans and animals might be able to be considered as being friends (Hens, 2009; Townley, 2010). This is because most of the properties (as outlined earlier) that a relationship should have in order to be characterized as friendship also seem identifiable in at least some human-animal relations.

It is this possibility of human-animal friendship that we consider is played out across many of our selected films and programs, especially *Skippy the Bush Kangaroo*, *Storm Boy*, and *Larry the Wonderpup*. Each of these, for instance, features what can clearly be described as a human-animal friendship: between Sonny and Skippy (*Skippy*), Mike and Mr Percival (*Storm Boy*), and Sasha and Larry (*Larry the Wonderpup*). In each case, these stories show both members of the relationship – human and animal – demonstrating not only their affection for the other but their preferences for the other's company. More specifically, each of these human and animal characters are shown as displaying other-regarding behaviour in ways considered necessary for and illustrative of friendship. These friendships, of course, are presented through these stories with varying levels of realism – Skippy, for example, far exceeds the dimensions of realistic kangaroo behaviour and vocalizations, and *Larry the Wonderpup* imaginatively presents what Larry himself might be thinking and feeling in his interactions with his human friend, Sasha, and other humans and animals. Regardless, each film and program is effective in illustrating how human-animal friendship might play out. Such possibilities make visible additional dimensions of human-animal relations that might be able to help reshape the anthropocentric expectations and practices regarding animals that have been legitimated through Australia's social imaginary.

Sonny and Skippy

A children's television show, *Skippy the Bush Kangaroo* tells the adventures of 10-year-old Sonny Hammond, younger son of Matt Hammond, the Head Ranger of the fictitious Waratah National Park, and Skippy, an orphaned eastern grey kangaroo, in the National Park in the late 1960s. The stories – unfolded over three series and ninety-one episodes (made from 1966–69 and broadcast 1968–70) – revolved around events in the park, including its animals (including koalas, wombats, emus, kookaburras, lyre birds, among others), the dangers arising from natural hazards, and the sometimes villainous actions of visitors to the park (e.g. poachers and illegal zoo collectors). Sold in 128 countries, watched by 300 million people worldwide, and translated into twenty-five languages (Oliver, 2009), as well as generating a variety of spin-offs and merchandise, *Skippy* introduced Australia and its wildlife to the world. A full-length feature film titled *Skippy and the Intruders* was also released to theatres in 1969. Not just a children's adventure series, the show carried strong environmental themes about protecting the bush and its inhabitants, and was one of the first Australian programs on Australian television featuring First Nations people and other recognizably Australian characters, Australian native animals, and Australian landscapes.[12] Most importantly for our focus, however, the series featured the possibility of interspecies friendship, with Sonny and Skippy's friendship providing a foundation for the show and enabling their working together effectively to save the day in the face of a multitude of different threats – albeit with Skippy often acting in extremely unkangaroo-like ways (Idato, 2009).

Sonny and Skippy's friendship is as such a key focus across the episodes, with the series' theme song further describing Skippy as 'a friend ever true'. Each episode opens with Sonny and Skippy's reunion with Sonny whistling into the bush and Skippy and Sonny respectively running (or hopping) to find each other. Importantly, as is a requirement for friendship, this response is presented as uncoerced. Skippy, as the series is at pains to stress in multiple episodes, freely roams the park; she is free to go as she chooses, meaning that it is always her choice to respond to Sonny's whistle. Skippy is also shown as actively seeking Sonny out to alert him – and the others – as to illegal or dangerous activity in the park. Thus, in episode 1, Skippy is shown looking into Sonny's bedroom, and Sonny asks: 'Now Skippy what is it? [...] You want me to come with you Skip [...] Wait till I get something on'. Other episodes show her warning Sonny about the actions of two fishermen (who are spearing seals) (episode 3), finding a woman collapsed in the bush (episode 6), finding a baby (episode 16), finding sheep rustlers (episode 19), finding horse thieves (episode 46), finding a military deserter (episode 48), finding men stealing wildflowers (episode 91). Skippy and Sonny are also repeatedly shown as working together as a team to achieve a goal (episodes 46, 52, 60, 81). Sonny explicitly describes Skippy as his 'friend' (episodes 6, 78), his 'best friend' (episode 15), says that he loves her (episode 13), and makes her his rugby team mascot (episode 88). Both Sonny and Skippy also express their friendship and their affection for each other through their actions. Not only do they show their mutual affection through nuzzling and hugging, but they are both shown as trying to protect each other from threats (episodes 7, 11, 13, 48, 53, 57). Skippy alerts others when Sonny is in danger (episode 48), while Sonny reassures Skippy when they are both in danger (episode 62). Both finally show despondency when the other is critically ill or away (episodes 4, 14, 45, 89).

Although Sonny and Skippy's adventures are clearly not realistic, with Skippy being represented as behaving very atypically throughout the episodes – kangaroos are not known for their capacities to play the piano, operate radios or machinery, unlock car doors, play monopoly, kick rugby balls or tie and untie knots, for example, and they do not communicate or hold extended conversations in the way of Skippy's trademark *tch tch tch* – this is not an issue. It is rather the idea that Sonny and Skippy, as a human and a kangaroo, can share parts of their lives in other-regarding ways – that they can build a friendship – that is important. Even if the stories in *Skippy* are not realistic in their exact depiction of either this friendship or kangaroo behaviour, they are still important for their illustration of non-anthropocentric ways of interacting with nonhuman others.[13] The stories of Sonny and Skippy, then, not only illustrate the idea that humans and animals can build friendships, but they raise the possibility that human-animal interaction in the form of friendship does not need to be limited to members of already recognized companion animal species.

Mike and Mr Percival

Another non-standard human-animal friendship, this time between 11-year-old Mike, known as Storm Boy, and Mr Percival, a pelican, is outlined in the 1964 novella by Colin Thiele, and in the film of the same name (Safran, 1976). Set in the Coorong wetland in South Australia, and also carrying strong environmental themes and a refusal to disavow First Nations people, *Storm Boy* tells of Mike finding three orphaned pelican chicks (their mother was shot) and deciding to rescue and raise them, naming them: Mr Proud, Mr Ponder and Mr Percival. In so doing, and raising the three to adulthood, Mike builds a bond with each of them: handfeeding them, encouraging them to learn to fish and to fly. Although all three pelicans are released back into their habitat when grown and deemed able to care for themselves – because 'wild things should be free', as Mike's father reminds Mike – Mr Percival returns to a despondent Mike: 'he's come back, Mr Percival's come back'.

From this point of the return by Mr Percival, the story traces the friendship developing between Mike and Mr Percival. As with Sonny and Skippy, the visual representation of this interspecies friendship is through the expression of physical affection and through the sharing of activities. Close-ups show Mr Percival rubbing his neck onto Mike's face and Mike returning this physical affection by rubbing his forehead onto Mr Percival's neck and then nuzzling him. The film also shows Mike sailing his raft while Mr Percival glides alongside; both running along the beach together; both fishing together; playing hide and seek together. Mike further shows his father how Mr Percival can play fetch: when Mike throws the ball, Mr Percival catches it in his beak, then walks back to Mike with the ball. Mr Percival is also Mike's confidante, the friend to whom Mike turns to express his anxieties about going to school. In his turn, and as was the case with Skippy and Sonny, Mr Percival displays his friendship by his physical presence at and participation in Mike's activities. As is typical of friends, Mike and Mr Percival are shown as choosing to spend their time together and, by remaining in each other's company, show their enjoyment of this time and activity. As with the Sonny and Skippy friendship, Mike and Mr Percival's friendship is a vivid reminder that human-animal friendships could be possible outside of the standard companion animal relationship. A key point here is choice. Indeed, both Skippy and Mr Percival are shown as being present in the friendship by choice – they are both wild animals, they have agency over their actions. Further, they do not need human companionship (as illustrated in the *Storm Boy* narrative by having neither Mr Proud nor Mr Ponder return) and they live outside of the pet *ownership* model. This stress on choice means that perhaps the relationships between Mike and Mr Percival, and Sonny and Skippy, illustrate stronger depictions of friendship than those often claimed with reference to conventional companion animals. Such animals may, after all, have little choice about spending time with a specific human.

Sasha and Larry

This is not, of course, to suggest that human-animal friendships involving companion animals are not possible. After all, the chances for the majority of Australians to feel they have developed any human-animal friendship do rest with the companion animals they might choose to share their lives with. This is the reminder that the development of any friendship takes time in that it entails the building of a relational bond between individuals, and few Australians live in such relational contact with non-companion animals as to have the opportunity to build such a bond. Furthermore, as noted earlier, the idea that at least some people develop friendship bonds with their companion animals is well-entrenched and, of course, illustrated by the stereotypical framing of dogs as being our *best friend*. It is just such a relationship that is explored in *Larry the Wonderpup*, a live action and animated television series of thirty-four episodes following the adventures of Larry, a small terrier dog rescued from the streets, and his 'little owner' Sasha who explores the wonders of the real world and the realms of the imagination with him. Other characters include Sasha's mother, Sue 'Mummy with the red shoes and the red face', Sasha's friend Mateo and his dog Norman, Arthur a terrier dog who lives next door to Larry and is Larry's 'best dog friend', and Grimesly, the cat, who lives next door to Larry. All the nonhuman animals are given voices and opinions that express some seemingly species-natural views, and the show is presented from a variety of perspectives, including Larry's – to the point of the use of 'dog-cam', allowing viewers to see Larry's life from his point of view. This consideration of perspective is a key theme throughout the series, with each episode including animated storytelling and songs which outline both human and animal perspectives and interests. Songs from Sasha's (and Mateo's) perspective are thus about how much she loves and wants to be with Larry (You Make My Day, Ain't No Lie, Better, Dig Deep); the fun Larry and she have together even on rainy days (Fun Fun Fun, Rainy Days); and the naughty things puppies do such as roll in the mud, eat shoes, school projects, plants and hide socks (No Good for You, Oh No and Mudslide). Such songs are an expression of the love and care a human child has for their companion animal. Conversely the songs presented from the dog's point of view are about how much the dog loves to be with their human to play in the park, go for a walk, play ball, get a belly scratch, and be off the lead doing his own thing (My Best Friend, Itch to Scratch, Doggy Dance).

As with Sonny and Skippy, and Mike and Mr Percival, the friendship between Sasha and Larry (and Mateo and Norman) is depicted through their actions and their being together. It is also explicitly voiced by both characters in song form and through their interactions during their adventures. Thus Sasha gives Larry lots of cuddles and pats throughout the episodes, further saying that he is the 'bestest friend 100 per cent of the time' (episode 10), the 'bestest pet in the world ... [a] girl's best friend' (episode 20). Larry in turn comments that Sasha and he 'do everything together that's amazing' (episode 5), and that 'Sasha always gives the best hugs' (episode 18). *Larry the Wonderpup* thus reminds viewers that friendship between human and companion animal can enrich the lives of both, but that both sets of perspectives and interests need to matter and be considered for any relationship to be considered a friendship. The program also makes clear that the only true measure for a human-animal friendship must be the consistency of other-regarding actions. This is the point that intentions cannot be truly told to or known by the other in such a relationship other than via action with and toward the other. Thus, even in a show that deliberately imagines and presents both of Sasha's and Larry's perspectives regarding the world and their adventures, they are also both shown as being unable to fully identify or understand the motivations of the other. For instance, Sasha says to Larry: 'I never know what is going through your head' and Larry replies: 'But I never know what's going through her head' (episode 1). This, of course, is the reminder that even without mind-reading capabilities, both Sasha and Larry can develop, practise, and live their friendship, as can all of us in all of our friendships, whether they are with humans or animals.

These are not the only human-animal friendships that are presented throughout the selected stories. There is also that of Mick and Blue in *Red Dog: True Blue*, that between Penguin the magpie and Sam in *Penguin Bloom*, and the two beginning friendships between human boys (Ali and Peter) and young camels (Binta and Ali) depicted in *The Camel Boy* (one of the few screen stories presenting human-animal companionship in Australian settler history from a migrant perspective). There are also multiple other friendly or companionate relationships between humans and animals brought into focus through these and other stories, even if these might not demonstrate all of the criteria typically deemed essential for full friendship. For instance, there is the relationship between Leahy and her dog Baxter which, as described in the documentary *Baxter and Me*, shows them interacting, eating and dancing together, and as sleeping side-by-side. Like Sasha, Leahy might not feel absolutely certain as to Baxter's motivations – are they love or food security? – but she is nonetheless clear that she loves Baxter and that she feels 'blessed' that they spend time together. There are also the human and police dog companionate relationships explored in *Send in the Dogs Australia*, a television documentary series showcasing two of Australia's largest State Police Dog units. With each episode depicting trained police dogs and their police handlers carrying out their work together, the series foregrounds the strong bonds of trust and cooperation that need to be built and maintained between the handlers and their dogs for this work to occur. Thus Senior Constable Gray, who is shown with his police dog Diesel in Melbourne searching for a runaway driver, comments that 'as he [Diesel] develops, I develop and the bond just grows stronger and stronger, I love his determination and I love the fact he has got determination' (series 1, episode 1). Senior Constable Cowling in Brisbane strokes and pats police dog Zac, saying 'he's a wonderful dog, he's my best mate' (series 1, episode 2). Similarly, Senior Constable Ellison with his police dog Edge comments that 'I just gotta trust my dog', and that 'I cheer at him the whole time', 'I let him know he is the man' (series 1, episode 2).

There are several final points to make about the depictions of human-animal friendship that can be found, for instance, in *Skippy*, *Storm Boy* and *Larry the Wonderpup*. The first is to note that it is no surprise that these friendships are built between children and young animals. If uncoerced and other-regarding action is a key measure in any friendship and the fundamental measure of a human-animal friendship – and action, of course, is a key element in visual storytelling – then the activity and the enjoyment of shared and repeated play (and of shared adventures) demonstrates this relationship well in visual terms. Thus each of these programs uses the display of human-animal play as a major demonstration of friendship, and each of the pairs – Sonny and Skippy, Mike and Mr Percival, and Sasha and Larry – is shown in play together, sharing their enjoyment. Shared play time and adventures also provide the time, space, and proximity for the relationship building considered essential for friendship. Similarly, the framework of shared cooperative work as in the case of police dogs working with their handlers also enables human-animal relationship building. In a different instance, in *Penguin Bloom*, the magpie Penguin builds a strong companionate relationship with the Bloom family. This begins from when she is found as an injured nestling and cared for and raised by them during a time when the family was also struggling to cope with the repercussions of the mother, Sam, having received a serious spinal cord injury from an accident. With the film narrative covering Penguin's growth to adulthood and learning to fly and Sam's working to regain some independence in her life, it is also clear that it is by spending extensive time together that Penguin and Sam are able to build a deep relationship of care and friendship for each other. Together these points remind us that all friendships take time to build, and need time to grow, so that they can withstand separation. Friendship, after all, is typically thought to encompass not just spending time together but the pleasure of being reunited after separation. This is particularly the case in human-animal relationships where it may be the choice to return to the other that is seen as being the test, demonstration, and non-verbal communication of friendship. Thus the choices made by

Skippy to return to both the park and her friendship with Sonny after being captured and removed from the park on several occasions, and by Mr Percival to return to Mike despite having been released back into the wild, have particular weight in these narratives of friendship.

Recognizing Kinship

In many ways, friendship is a gold standard relationship, and as such has quite stringent requirements. That is, friendship is an intimate, other-regarding relationship of mutual affection, loyalty, empathy, and trust, marked also by mutual pleasure in the sharing of time, space, and activity. Once built, and if maintained, friendship should also be a relatively stable relationship, able to weather separation as well as other challenges. It is not the same as friendliness, although the display of friendliness on the part of an individual may well be an initial important step in the building of any friendship. In its turn, understood literally, friendliness is a demonstration of some of the qualities necessary for friendship. Someone who is friendly is thus devoid of hostility, open to and respectful of others, aware and considerate of others' interests. Put another way, friendliness – and, indeed, friendship – would seem to require a recognition of what may be called kinship. This is the recognition that the other is akin to you – a possible friend if circumstances allow, but at the least another being worthy of respectful treatment. Like friendship, friendliness and kinship are not driven by assumptions of one's superiority to the other. Indeed, it seems evident that anyone who remains highly invested in and driven by their own sense of superiority will struggle to be other-regarding to the extent necessary for recognizing kin or developing any truly strong friendships. Indeed, even though such individuals might call some others friends, it can be questioned whether the resultant models of interaction are, strictly speaking, those of friendship.

However, rather than continuing consideration of the necessities and possibilities of human-animal friendship, the focus in this section is on kinship – on what it means to recognize kin, and what this might look like in the context of human-animal relations. Here it is worth noting that while the ideas and models of human-animal kinship can vary – a point to be discussed shortly – they do share the point that kinship does not need to be understood as actually entailing a genealogical relationship. They vary, however, with regards to the kind of human-animal relationship that different models consider essential for there to be a kin relationship. For instance, one understanding of human-animal kinship takes the idea that human relationships with companion animals would approximate the close connections ideally holding between members of a human *family* (Charles, 2015; Charles & Davies, 2008; Gabb, 2011). So conceived, companion animals would be considered psychological-kin and as such beloved members of the family (Cudworth, 2011; Kao, 2018; Topolski et al., 2013). Interestingly, in some illustration of this, the Suncorp Home Index Report of 2019 (a national survey of more than 1,600 Australians) found that more than a third (38 per cent) of the Australians surveyed admitted to loving their companion animals – *pets* in the report's parlance – more than their human family members, and over half (51 per cent) admitted to placing more importance on a potential home's pet-friendliness than its geographical proximity to family and friends (Duffy, 2019; Suncorp, 2020).

Donna Haraway sets out another story regarding human-animal kinship. In her view, while the idea of kinship should definitely be understood as a non-genealogical connection, kinship does not need to follow the model of (some) companion animals becoming seen as part of *the family*. Haraway, rather, sees kinship as a kind of situational inheritance that comes from our living in conjunction and connection with others, and that whether they are human or animal is beside the point. Under this conception, kinship does not depend upon whomever we like or dislike – a point key, of course, to friendship – but is simply the result of our being closely involved with and in the lives of others (Čičigoj, 2019). Thus, whilst we clearly can recognize kinship relations of the kind that illustrate a kind of family connection, just through our living in the world we are also connected with – and thus potentially kin with – a multitude of other beings, human and animal. This is not to say that we should simply call all possible connections as kin. Rather kinship refers to those who interact in ways that have real consequences. Thus, as Haraway puts this,

> By kin I mean those who have an enduring mutual, obligatory, non-optional, you-can't-just-cast-that-away-when-it-gets-inconvenient, enduring relatedness that carries consequences. I have a cousin, the cousin has me; I have a dog, a dog has me. (as cited in Paulson, 2019)

Kinship, in other words, is certainly co-situatedness, affinity and choice – all necessary for viewing others as family, for example – but also the idea that relationships bring obligations. And, indeed, Haraway stresses that the recognition of human-animal kinship means seeing the nonhuman as having their own demands and interests and recognizing that these should be respected. (In human terms, this recognition of kinship might reflect the basic premise of the recognition of human rights – that all humans deserve respect and consideration of their needs and interests by virtue of their being human.) Under these conditions, the ideal for acting with regards to kin, as Haraway says in reference to companion species, is 'to have regard for, to see differently, to esteem, to look back, to hold in regard, to hold in seeing, to be touched by another's regard, to heed, to take care of' (2008, p. 164). Promoting a recognition of kinship between humans and animals has indeed been key to work by animal activists who – in the context of live export, for example, or with regards to reforming intensive models of animal agriculture – effectively argue that our shared sentience with animals automatically makes us kin to them, and that we need to learn to act regarding them with this kinship in mind.

Although Haraway has argued for coming to recognize kin in a very broad way – 'all earthlings are kin' (2016, p. 103), she suggests – it is also the case that this is not our usual practice, and perhaps not where most of us would start in our recognition of human-animal kinship. Indeed, Haraway (2016) stresses that recognizing kin outside of the beings we already find easy to count as family – beloved members of companion species, for example – is perhaps the hardest and most urgent challenge we humans face today. Both narrow and broad ideas regarding kinship with animals are nonetheless useful for helping dismantle the structures of anthropocentrism and human exceptionalism that drive and constrain our existing social imaginary. As she puts this, making kin with 'other sorts of "we", other sorts of "selves", and unexpected kinds of […] nonhuman critters […] is crucial for imagining and crafting […] flourishing worlds, now and still to come' (Haraway, 2018, p. 102). And kin, she also notes, do this crafting of new worlds together, with each other: 'when you touch someone and someone touches you, one question that emerges is, Okay, who are we? Who do we make each other?' (as cited in Paulson, 2019). These are particularly important questions given our interest in the development of a less anthropocentric imaginary.

A third set of ideas concerning human-animal kinship is visible in Australian First Nations' beliefs. As we have noted, these beliefs concerning kinship between humans and animals rest on broader ideas of kinship. As Aboriginal novelist Amberlin Kwaymullina (2005) argues:

> Rock, tree, river, hill, animal, human – all were formed of the same substance by the Ancestors who continue to live in land, water, sky. Country is filled with relations speaking language and following Law, no matter whether the shape of that relation is human, rock, crow, wattle. Country is loved, needed, and cared for, and country loves, needs, and cares for her peoples in turn. Country is family, culture, identity. Country is self.

The point here is that, according to an Indigenous point of view, separation – between humans and animals, between humans and *Country*, and so on – is artificial. Rather than separated into different stories and given different values as in anthropocentrism, 'human, rock, crow, wattle' are all kin, all connected in *Dreaming* – all part of the same story (Milroy & Milroy, 2008). Kinship is thus the recognition of the interconnection of the different but equal threads that form every shape of life in *Country*: 'Some – like human, kangaroo, paperbark – are known to western science as "alive"; others, like rock, would be called "non-living". But rock is there, just the same. Human is there, too [...] neither the most nor the least important thread' (Kwaymullina, 2005). Instead of stressing differences such as *alive* versus *non-living*, human versus animal, and assuming an anthropocentric hierarchy of species, this idea of kinship is the reminder that every thread is equal with others. And, as we have previously noted, such kinship is further formalized in *Dreaming* and Law with specific responsibilities set out for preserving one's *Dreaming* kin.

Importantly, unlike the possibilities of human-animal friendship explored in the section above, these ideas of kinship between humans and animals are less of a demand for mutuality and specificity. Rather the expectation is that kinship describes the recognition of 'situated partial connection[s]' of curiosity, trust, and respect (Haraway, 2003, p. 140). In addition, while both friendship and kinship are recognitions of relatedness, kinship does not assume mutual enjoyment and pleasure in the bond. Kinship also does not require reciprocity of forms of relatedness. Nonetheless, as with friendship, kinship is a recognition of connection and care, of responsibility. We care about those we recognize as our kin, and we (should) act in accordance with this care. Put another way, kinship means we are 'touched by another's regard' (Haraway, 2008, p. 164) and, further, that this regard marks a co-shaping of both our relationship with that other and our understanding of the world (see, e.g., Figure 7).

Figure 7. Many Australian birds such as these wild sulphur crested cockatoos in Sherbrooke, Victoria, are intensely curious, intelligent, social and prepared to interact with humans. Such transient interactions could be considered as a building block for the extension of kinship outside of companion species. *Source:* Photo by Laya Clode on Unsplash.

It is thus to find such points of human-animal kinship – whether these are partial, transient, situated relationships of human-animal curiosity, trust and respect or illustrate the co-shaping of new relationships and worlds – that we have also examined selected screen media. Although not as common as stories of human-animal friendship, the recognition of kin and the making of new kin worlds is visible within some of our selected films and programs, especially in the films *Emu Runner*, *Dot and the Kangaroo*, and *Red Dog*. As we will show, kin recognition and kin making can take various forms, but all such possibilities add to the narratives regarding new models of human-animal relations that may be able to destabilize assumptions and practices normalized throughout Australia's post-settlement anthropocentric social imaginary.

Gem and the Emu

Emu Runner is a story about 9-year-old First Nations Ngemba girl, Gem (Gemma) Daniels, who lives in the remote Australian town of Brewarrina, in New South Wales. Gem is a champion runner at school who measures her speed against the emu, her totem animal passed down on her mother's side. Gem's mother describes the emu as 'our animal, that's what connects us to this land, our people'. As Gem explains at one point: 'all our different mobs got them something that ties them back to their mother's country'. From this beginning, the story traces Gem's finding solace in the company of a wild emu after her mother dies. As her mother's and her own totem or *Dreaming* animal which ties their family together, as well as a symbol of speed, agility and grace, the film shows Gem forging a relationship with this emu in a way clearly demonstrative of recognizing and making kin. In the case of the story of *Emu Runner*, this is a recognition of kin as family. That is, the emu stands in for Gem's connection with her mother. Its arrival after the mother's death marks an event that cannot be ignored as it can be understood as Gem's mother's spirit connecting with her daughter, as well as Gem receiving her mother's spiritual role as she begins to grow out of childhood. Gem's Uncle Wes also explains to her that male emus have the full responsibility of raising their young, further saying 'Your dad no different'.

If the human recognition of animal kin demands recognition of a relationship as well as an acceptance of care for the other, both are clearly played out in *Emu Runner*. For instance, on the way to school Gem is shown as often sitting in the bush, waiting and watching for the emu. When she speaks to the emu, she asks: 'Where's your mob ...? Are you "hungry"'? Wanting to establish a bond and accepting the importance of care for the other, Gem shares her lunch with the emu, and collects discarded vegetables from the back of the Co-op shop to feed her. In an endeavour to bring the emu – physically and emotionally – closer to her, Gem also practices imitating different emu sounds, and becomes expert in using a small metal tin to produce a female emu sound. In a later scene Gem is shown using the tin box to call the emu and, when it comes, handfeeding her watermelon and flowers. Gem's provision of food thus becomes the basis for both the relationship and the care. Walking through the bush on her way to school, Gem tells her friend: 'She's [the emu] gonna need more food'. This anxiety for the emu sees Gem playing truant from school so she can find more food for the emu: she takes the cat biscuits from her uncle's house, she steals groceries from cars, she steals an older couple's morning tea from their outdoor table, and she steals fruit from a tree of a suburban house. As she feels further connected with the emu, and as the emu starts to approach her more closely, Gem also steals a scarf from a washing line to dance with in front of the emu as well as her friend. When her actions come to light (she is reported to the police for stealing), Gem further comes to understand through her father's and her uncle's teachings that even as kin, the emu needs respect for her own sake – in this instance as a wild animal who has already been taught (by her father) to find food for herself.

With regards to issues of kinship, it is clear that both Gem and the emu are touched by the other's regard. Both alter their behaviour for the sake of the other in a way that suggests curiosity and trust. They choose to see and spend time with the other. Gem's sense of kinship with the emu also manifests itself in her regard for her – seeing her as a connection back with her mother, providing food, and finally coming to understand that the best care is to let the emu feed herself. *Healing* provides similar illustrations of kinship, with its depictions of inmates of the Won Wron Correctional Centre, a low-security prison farm situated in rural Victoria, participating in the rehabilitation of injured birds of prey (for instance, eagles, kestrels, and owls).[14] The film focuses mainly on Iranian-Australian Viktor Khadem, one of the prison inmates, showing how his participation in the project – rehabilitating Yasmine, the wedge-tailed eagle he cares for, from a near death experience – offers him hope for the future. In this film, while the relationships between the inmates and birds are slow to build, needing time for the development of trust, these relationships also reflect remakings of understanding with regards to experiences of confinement (both physical and through loss of connection) and freedom, and of the difficulties of learning to trust. As Haraway might also note, such remaking of understanding does not require the mutual making of an interconnected world (as in the way of friendship), but is rather an illustration that the recognition of connection – at least from the human side – means also considering the world from the perspective of kin. In the case of *Emu Runner*, Gem is certainly shown to be reconsidering her life and world with reference to the perspective and context of the emu, and the inmates in *Healing* are shown as building their relationships with the birds they work with whilst remembering that these birds are and must remain wild.

Dot and the Kangaroo

A human-animal familial kinship relation is also established strongly in *Dot and the Kangaroo*, the first Australian animated film to find commercial success (Buckmaster, 2015).[15] Set in the Blue Mountains and the Jenolan Caves, and combining live-action and animation, the film *Dot and the Kangaroo* is based on Ethel Pedley's (1899) children's book of the same name. Comprising an explicit attempt to awaken young settler Australians' senses of kinship with their new country – and to counter other (anthropocentric) settler narratives that framed the Australian bush as threatening and Australian native animals as pests and threats to agriculture and national prosperity (as well as rejecting the disavowal of Australia's First Nations Peoples)[16] – Pedley's (and Yoram Gross') *Dot and the Kangaroo* emphasizes the unique qualities of the Australian bush and its native animals, and calls for their conservation (Rahbek, 2007; Taylor,

2014; note that we examine more of these anthropocentric threat narratives in the next chapter). Tracing the fear of a child losing its parents alongside the fear of a parent losing its child, the film opens with 5-year-old Dot going into the bush to find special grass for her pet rabbit to eat, falling down a hill and losing her way. In her distress, Dot is approached by a red kangaroo who has lost her joey (baby kangaroo) when chased by hunters, and who – acting as a surrogate mother with Dot as her surrogate child – promises to help Dot find her way home. This becomes a journey of discovery which has the kangaroo introducing Dot to a number of other animals, including the Council of Animals (who decide all forest matters), and teaching her a greater appreciation for nature and for how interconnected her world is with the natural world. The story also stresses how akin humans and animals are.[17] Dot, for instance, says she is thirsty, and Kangaroo responds: 'Of course you are, everyone is thirsty at sundown. I am thirsty too. But the nearest water hole is a long way off so we better get started at once'. Dot starts walking in her bare feet but gets prickles, so Kangaroo offers Dot her pouch so they can hop down to the water hole with some speed. Dot says 'it's lovely in here, thank you', and Kangaroo says 'and I feel as if I had my baby again'.

After various adventures and dangers through which both Dot and the kangaroo help and support each other – including the Council of Animals being reluctant to help Dot because humans have not done much to help animals – the way home is found for Dot and the path shown by another animal character, Willy Wagtail. Once Dot's home is reached, Dot wants Kangaroo to come in but Kangaroo says: 'No I can't come with you. These are your people, your family [...] my home is out there'. Willy Wagtail tells Dot not to be sad: 'your kangaroo must have freedom, must live her own life [...] She's gone home, and her home is the bush. Kangaroo must have her freedom, freedom, freedom'. This is a clear reminder for Dot – and very similar to the reminder Gem is given by her uncle with regards to her emu kin – about the need to not just recognize animal kin, but that this recognition and associated care must always be driven by respect for the needs of that other.

Red Dog in Community

Set in the 1970s in the Pilbara mining town of Dampier, Western Australia, and based on a true story, the feature film *Red Dog* (Stenders, 2011) pays homage to an Australian Kelpie cross. Known as Red Dog, the kelpie became legendary for having wandered the Pilbara searching for John, his deceased human companion (he was even rumoured to have caught a ship to Japan in his search), as well as for his position as a community dog on his self-chosen return to Dampier. In the words of residents, Red Dog was a 'dog for everyone'. As outlined by one resident, Nancy, 'I don't own Red Dog. Nobody does. He's common. He's accepted and cared for by the town. The Community'.[18] Indeed, he is registered as *a common* community dog, and recognized as being free to follow his own interests. In this capacity he travels with the mine workers each day on the bus to the mine site and joins the men in the pub each evening, as well as attending community gatherings such as BBQs on the beach and fishing. He was also made a member of the Trade Union. As the film makes clear, Red Dog is considered a member of the community in his own right, offering companionship to others – albeit on his own terms – and he is mourned as a member of the community when he dies. Indeed, after his death, the question is raised why the town should have a statue of the town's namesake, William Dampier, when all he did in relation to the place was say that there were 'too many flies'. As one of the miners, Jocko, asks: 'Why should we have a statue honouring a poncey, pommie, fly-hating aristocrat? Or for that matter a fat-bloody general or, god help us, a stinking politician?' Jocko instead suggests that they erect a statue to 'somebody that lives and breathes this vastness and desolation. Somebody that's got red dust stuck up their nose, and in their eyes, in their hair and up their arse! Somebody that's like all of us. Men and women who understand the meaning of independence, the importance of a generous heart'. To unanimous approval Jocko suggests that a statue rather be erected in the town in celebration of the dog who knew life, love, loss, and the community. As he puts it, the statue should be of 'somebody that represents the Pilbara in all of us and I say that somebody, dammit, IS A DOG!'

If both *Emu Runner* and *Dot and the Kangaroo* outline possible recognitions of human-animal kinship using models of familial connection, *Red Dog* is a demonstration of the recognition of kinship at the community level. As an individual, Red Dog demands respect for his freedom, but also offers respect and friendliness back to other community members. Indeed, it could be argued that it is Red Dog who is making kin given the way he was able to bring people together, galvanizing community feeling. That this is always on his terms is exemplified again towards the end of the film. Red Dog is sick, dying, and the community has come together to be with him and to, wake-style, share stories about his life and exploits. Rather than staying with his human companions, however, Red Dog walks away from them, unnoticed. He is the community's glue – the one who makes them kin – but he is also always his own dog, making his own choices and needing them to be respected, even at the end of his life.

The Importance of Time and Trust

Whether friendship or kinship, what stands out in the relationships depicted through these various stories of human-animal interactions is their common foregrounding of not just respect for the other but trust, and how this also places an emphasis on the time needed for relationship building. This is unsurprising insofar as trust cannot be willed into existence – that is, we cannot simply choose to trust (Baier, 1986). Rather, trust comes into being as a side effect of our being in relationships with others, from understanding others' behaviour toward us given our familiarity with them (Gambetta, 1990). This is the point that to have trust in someone is to have established and maintained a belief that the object of our trust means us no harm, and that they are sincere in their expressed motives and intentions. This is also a reminder that trust is established through actions and interactions over time – a point that is even more significant in the human-animal relationship when action and interaction are the only measures available for intention. 'It is a stretch to think that animals can bargain for or even understand mutual advantage, but none at all to recognize that they can and do trust' (Silvers & Francis, 2005, p. 71). Establishing and maintaining trust between human and animal involves each learning the right sort of behaviours to exhibit to the other, a process of mutual interactions and of learning from one another – and a process essential for the making of either friends or kin. The importance of this is spelled out in *Dot and the Kangaroo* when Dot and Kangaroo face opposition from the Council of Animals with regards to providing help to Dot. As Council representatives and other animals point out, few humans have shown themselves to animals to be worthy of trust. This is because they disregard the interests of animals in a myriad of ways, including by (a) not bothering to understand the lives of bush animals (and thus believing untruths or 'dreadful stories' about them); (b) not listening to animals (as Kangaroo says to Dot, 'all animals can talk, [it is] just that nobody listens to us'); (c) destroying or taking over bush animal habitat without care; and (d) killing the animals in the bush. And yet, of course, based upon their time together, Dot and Kangaroo are represented as placing their trust in the other as they make a kinship bond, as do all the other human and animal characters discussed as being friends or kin.

These are important issues. Time and trust are the foundations for making friends and kin. They also remind us that friendships and kinships are concrete not abstract – they are always with concrete, situated others who are in proximity with us. That is, whilst we may express an abstract interest in interspecies friendships and kinships, both forms of relationship only take hold and have meaning in the context of being with and spending time with actual others. In turn, as noted above, trust – and friendship and kinship – is reliant on being able to gain some understanding of the other, their needs, and interests. This does not mean that a strong understanding must be attained – and, indeed, this is a message in several of the human-animal stories we have considered. Thus both Sasha and Larry express their perplexities with the other (and Larry expresses his even greater lack of understanding of 'Mummy with the red shoes and red face'), Leahy notes she cannot ever know what Baxter really thinks of her, and Gem recognizes the emu as kin on the basis of only some understanding of her needs and interests. Conversely, if one is unprepared to gain any such understanding, no such relationships can be built. This view is clearly represented by some of the visitors to Waratah National Park, Skippy's home, who express their distaste of and lack of interest in animals. These include the art critic Miss Sonia Pearson who declares, when Skippy enters the room, that she 'can't abide animals in any shape or form, I detest them' (episode 15), and the fashion photographer tasked to do a fashion shoot in the park who, when informed of the expected locations of animals in the park, retorts: 'Animals, dear boy, if I want animals I'll go to the zoo' (episode 82). Pearson and the photographer express views that are fundamentally anthropocentric – animals, in their view, should be present only with reference to human interests. To learn to think otherwise, we need to let ourselves be curious about these others who are with us in our world and recognize that they too can be friends and kin. This is particularly the case with respect to those many animals that live outside the boundaries we typically set for companion animals.

Final Considerations

There are, of course, further complexities in this kind of storytelling. First, as we have noted, these stories of friendship and kinship are not strictly representative of real human-animal relationships, given that the terms and context of all human-animal relationships are undoubtedly determined by human beings (Cudworth, 2011).[19] (For instance, it could be argued that all of these screen media stories are reliant on the reframing of an animal's natural behaviours 'within human-centred narratives', both 'onscreen and offscreen'; see Molloy, 2011, p, 63.) Hence, these stories need to be recognized as imaginative, perhaps idealized, representations of human-animal friendship, and as based on the assumption that such relationships are possible despite all the anthropocentric power relations that would burden them. As we have noted in previous chapters, these power relations are by no means imaginary and have led to perceptions of human-animal interactions and relations being almost irredeemably marked by violence (see, e.g., Nibert, 2013; Wadiwel, 2015) and almost inconceivable beyond anthropocentric assumptions regarding use-value, exploitation and the animal-industrial complex (see, e.g., Wrenn, 2017). And, yet, we argue alongside these stories, to not imagine that other relationships with animals are possible – whether these are of friendship, kinship or some other model of other-regarding relationship – is also to affirm anthropocentrism as a norm. It is further worth mention that some of these issues are in fact explicitly considered in the screen media stories we have examined. Thus, in *Baxter and Me*, Leahy explores some of the ethical problems and issues of power that her *ownership* of Baxter raises, and *Skippy*, *Storm Boy*, *Red Dog*, *Dot and the Kangaroo*, *Emu Runner* and *Healing* all examine human expectations regarding the lives of free-living animals and question what forms of human interaction with such animals might be acceptable from an animal's perspective.[20]

A second point relates to the limitations of friendship (and even kinship) as a model for becoming more attentive and other-regarding. This is the point that no friendships – whether between humans or humans and animals – are generalizable. That is, as we have noted, friendships are always between particular individuals. Sonny is friends with Skippy, not with every kangaroo in the park and not with every animal he meets. Mike is friends with Mr Percival; his relationships with Mr Proud and Mr Ponder do not become friendships. Leahy reflects on her relationship with Baxter specifically; she does not see this relationship as extending to every dog. While kinship models may enable a broader reach, in practice these relationships too start with specific others – and the screen media stories we have considered also start from this specificity: Gem with the emu, the Won Wron inmates with the specific birds of prey they care for. The point here is that while we suggest that these various stories of human-animal friendship and kinship offer some avenues towards becoming less anthropocentric, these avenues are also inscribed with ambivalence. That is, while such forms of interaction may support 'a broadening of concern with issues of animal welfare', they may also simply illustrate a 'reinscribing of species barriers' in other places (Cudworth, 2011, p. 169). Thus, Mike is shown to be friends with Mr Percival, but sees nothing wrong with fishing.[21] While such instances may be read as marking instances of inconsistency and the reinscription of speciesism, it must be remembered that all of these stories have been produced from within a strong context of anthropocentrism and with the (screen media) objective of depicting identifiably Australian characters and views. Against the norms of anthropocentrism, these stories thus offer imperfect representations of human-animal companionate interaction that connect with but still challenge existing Australian perspectives, rather than perfected ideals of interspecies cohabitation and multispecies communities.

Notes

1. Research shows that Australians are spending over 20 billion annually on their companion animals, often in response to new fast growing cultural trends (Bakan, 2021; Kestenbaum, 2018). Trends include consumers purchasing premium pet food and treats through to individualized diets; customized or designer pet products; pet grooming, care, transportation and hotels; an uber service SpotOn.Pet that transports pets; end-of-life services such as palliative care, cemeteries and grief counselling; DOGTV (television tailored for dogs); Dig the dating app for people who have or want a dog; kitty litter boxes with 'innovative features'; and other products that measure health disorders, provide 'personalized medications', or are designed to increase wellness (Kestenbaum, 2018).
2. This is the reminder too that 'violence towards domesticated animals is routinized, systematic and legitimated' (Cudworth, 2015, p. 14), often due to the physical environments and human-pet relationships in which companion animals are commodified, controlled and owned (Sutton, 2021).
3. For example, in First Nations kinship systems, every person has a *Dreaming* or totem. This might be an animal, but may also be lands, waters, or geographic features. These *Dreamings* or totems in turn create a network of physical and spiritual connections – connections we will discuss later in the chapter as kinship – between people and the world, to the point that every individual is accountable to their *Dreaming* and must ensure its protection for future generations. Thus, First Nations people who have an animal *Dreaming* hold a unique connection to that animal and the responsibility to look after that animal and maintain their connection. For example, if someone holds a kangaroo *Dreaming*, they would not hunt kangaroo but would rather work to protect them (Glynn-McDonald & Sinclair, 2021; Rose et al., 2003).
4. As we have noted previously, many Australian narratives post Anglo-Celtic settler occupation have stressed the significance of agriculture as well as agrarianism or country-mindedness (Aitkin, 1985). In accordance with Australia's agrarianism

(see Aitkin, 1985; Berry et al., 2016; Ward, 1966; Wear, 1991), the characteristic Australian is from the bush, and the core elements of the national character come from the struggles of country people to thrive in the Australian bush environment. City people, conversely, are much the same the world over and offer little to the development of national character. Informing both the bush balladists of *The Bulletin* in the 1890s (e.g. Henry Lawson, Banjo Paterson, Edward Dyson, among others) and the painters of the Heidelberg school (e.g. Tom Roberts, Arthur Streeton, Frederic McCubbin, among many others), agrarianism became a key foundation for the Australian legend, further explaining the collective experience of generations of Australians in a way that even now continues to make sense (Berry et al., 2016; Wear, 1991). It is worth noting that the stories considered in this chapter from the television series *Skippy* and the Red Dog films also strongly reflect this agrarian legacy of bush living.

5   While we use screen media here to stand for any media that is produced for or distributed via the screen – for instance, the cinematic screen, the television screen, the computer screen, as well as the small screens accessed on a smartphones and other handheld devices – our focus in this chapter is specifically on the television and cinematic screens.

6   While seeing one's own cultural stories and identities on screen is a significant 'source of [the] cultural value' of screen media (Piper, 2016, p. 171), it should also be noted that much of Australia's screen media has represented – and, indeed, continues to represent – White settler-Australian cultural points of view. This is the point that there continues to be limited representations of the actual diversity of the Australian population (Arvanitakis et al., 2020; Screen Australia, 2016b; Screen Australia & Ipsos Australia, 2013) – a discrepancy visible in the screen media considered in this chapter. There have been some gains with the Australian Broadcasting Corporation (ABC), Special Broadcasting Service (SBS) and National Indigenous Television (NITV) all committed to better representing a diverse Australia, and some reality television genres (particularly cooking and singing competitions) appear successful in better representing Australia's cultural diversity (Khorana, 2020; Lattouf, 2016; Pobjie, 2016). With regards to our focus in this chapter on human-animal interaction and friendship, it is interesting to note that while most of these interactions do feature human characters possessing what can be presumed to be White settler-Australian origins (Sonny, Mike, Leahy, Dot, for instance), First Nations characters are also represented (in *Skippy*, *Storm Boy*, *Dot and the Kangaroo*, *Emu Runner*, *Healing*), as well as characters from other cultural backgrounds (in *Camel Boy*, *Healing*, *Red Dog*).

7   It is worth noting that subscription networks such as Netflix, amazon, *Prime* and STAN not only rarely offer the national and local 'embedded everydayness' (Piper, 2016, p. 181) typical of Australian cultural narratives, but also that as subscription

8   networks they remain 'complimentary to traditional linear broadcast and platforms' (Lewis, 2014).

8   This is the point that television allows people to encounter 'a wider variety of animals in a wider variety of contexts' than they would in their everyday lives (Mills, 2017, p. 20). In addition, as a medium, television offers a 'range of representational strategies and conventions' allowing human audiences to connect with animals (pp. 2, 26).

9   Note that our focus on foregrounding extended companionate human-animal interaction and recognizable Australian content is not translatable to an attempt to be fully representative of Australian cultural diversity. As we have noted previously, much of Australia's screen media has represented – and, indeed, continues to represent – White settler-Australian cultural points of view, with limited representations of the actual diversity of the Australian population (Khorana, 2020; Rodrigues, 2020; Screen Australia, 2016b; Screen Australia & Ipsos Australia, 2013). While this situation is slowly changing, its legacy means that selected programs do vary in their representations of Australian cultural diversity. As such, it is unsurprising that many of the selected programs primarily reflect White settler-Australian culture albeit with some representations of migrant and/or First Nations cultures – with the latter being particularly integral to some programs, *Skippy* and *Storm Boy* for instance, where First Nations views of animals come to inform some of the companionate interactions displayed. First Nations cultural views regarding human-animal relations are conversely highlighted in *Emu Runner*, albeit against a broader Anglo-Celtic settler-Australian culture. While screen media programming foregrounding more representative patterns of cultural diversity is gradually increasing in Australia, it is perhaps not surprising that its focus is primarily on the diversity of human cultural interactions and reflections rather than on interspecies companionate interactions and relationships.

10 This requirement for programs to feature extended human-animal companionate interaction ruled out screen media that reflect the presence of animals within Australian urban and rural settings, but do so without making human-animal companionship a primary focus. This includes such programs as *Bondi Vet* (2009 – present), *Bush Mechanics* (1998–2001), *Crocadoo* (1996–98), *Snake Tales* (2009–12), *Little J & Big Cuz* (2017 – present). Even though the latter puts a focus on education regarding native animals, the main animal character, Old Dog, is not presented as a close companionate animal.

11 That is, in foregrounding one companionate human-animal relation – such as between Sonny and Skippy, for example – the same program might also illustrate non-companionate human-animal interactions, such as fishing.

12 While regular television transmissions commenced in Australia in 1956, until 1963, 83 per cent of screened programming was sourced from the US and 97 per cent of drama was imported – thus during this period Australian television featured

American and British landscapes and characters (Cunningham, 2000, p. 17; Flew & Harrington, 2010, p.157; see Senate Select Committee on the Encouragement of Australian Productions for Television, 1963). *Skippy the Bush Kangaroo* was thus one of the first television shows showcasing Australian landscapes, animals, characters and accents, and indeed – as developed by Australian scriptwriters for, initially, an international audience – specifically aimed to 'give the world [an image of] what Australia might be like' (Oliver, 2009; also see Olsberg.SPI, 2016). *Skippy*'s cultural value is also drawn from being one of the first Australian television programs to represent First Nations people in ways that were not (for the time) overly romanticized or exoticized. Indeed, it is worth noting that Skippy was developed in the same period as Australian voters recorded the largest ever 'Yes' vote (over 90 per cent) in a 1967 referendum to grant First Nations Australians the right to vote and the same citizen rights as other Australians. *Skippy's* cultural value is also recognized as having extended into economic value through its enabling of 'set-jetting', meaning Australian screen media locations being viewed by mass audiences as tourism destinations (Deloitte Access Economics, 2016, p.3; Olsberg.SPI, 2016).

13 Interestingly, it has become clear that kangaroos actually are in fact like conventional companion animals in their capacity to engage in 'intentional communication' and interaction with humans (McElligott et al., 2020, p. 3; see also Cooper, 2017). This suggests that they too possess the capacities to act in companionate ways.

14 *Healing* is inspired by a partnership between the Victorian penal system and the Healesville Wildlife Sanctuary in country Victoria. As depicted in the film, this partnership takes the form of the Raptor Rehabilitation Program, in which inmates learn to care for wounded birds of prey in a minimum-security prison farm. With the aim of exposing prisoners to an environment facilitating care and nurturance, inmates take on all aspects of the care of injured birds, building aviaries, as well as feeding, cleaning, and bringing the birds back to health – and, where possible, seeing them released back into the wild (Department of Justice and Community Safety, 2013).

15 *Dot and the Kangaroo* was the second animated feature film made in Australia, and its commercial success led Producer and Director, Yoram Gross, to make eight sequels. These films, along with others by Gross, were distributed to Australian schools (Buckmaster, 2015). As Paul Byrnes from the National Film and Sound Archive of Australian notes, 'It is no exaggeration to say that a generation of Australian school children grew up on *Dot and the Kangaroo*' (Byrnes, n.d.).

16 Pedley's positioning of Kangaroo-as-narrator allows her to refuse Dot's 'disavow[al]' of Australia's First Nations Peoples as not being 'her fellow human beings' (see Taylor, 2014, p. 177).

17 It is worth noting that Pedley's story sets out quite complicated ideas regarding kinship. Stressing a conservationist perspective, Pedley presents humans (both Anglo-Celtic settler and First Nations) as unfriendly to the bush and to bush

animals – they both kill kangaroos, although only White settlers, Pedley has Kangaroo explain, kill for pleasure. Dot is entreated by Kangaroo to become a better human and a friend to bush animals. Here Pedley sets up several points concerning kinship. She challenges 'the taken-for-granted binary order, which usually pits White settler society as a civilising and improving force against the backward, unhomely and threatening traits of wild Australian nature', rather presenting the 'wild animals in the Australian bush as the civilised and knowledgeable ones and the White settlers as ignorant, mindlessly cruel and in need of educating' (Taylor, 2014, p. 175). At the same time, and counter to the beliefs in White superiority that were then staunchly defended in Australia (see, e.g., Anderson, 2007), Pedley has Kangaroo explain to Dot that 'All Humans are the same underneath [the skin]' (Pedley, 1899/2013, p. 53). Pedley further uses her narrative to present and defend the idea that all creatures have the potential to understand that they are 'the same underneath [the skin]', that all creatures are kin.

18   The importance of this commonality is made much of in the film, with the town presented as both a multicultural community (many of the film characters are presented as migrants from Italy, Poland, Ireland, New Zealand, Greece, Lativia, and America – admittedly somewhat stereotypically) and a celebration of Australian working class values wherein mateship and loyalty transcend difference (Blagrove, 2013).

19   Scholars working in the fields of animal rights, Critical Animal Studies, Sociological Animal Studies and Pet Studies critically examine the variety of complex power relations between human and nonhuman animals made normative by anthropocentrism and speciesism.

20   Whilst we argue that there is potential in such screen media stories to enable the reimagining of at least some forms of human-animal relationship, such stories can also exacerbate use-value considerations regarding nonhuman animals. As Molloy (2011, pp. 10–1) points out, films featuring a particular breed of animal can increase the breeding and purchase of those specific animals. After the film *Red Dog* was released, for example, requests for red kelpies increased 'over 600 per cent' with a majority returned to shelters because their temperament as an active working dog made them unsuitable urban companions (Hill, 2017). Additionally, after the success of *Oddball*, thousands of maremma dogs had to be 'rehomed' as they too were 'not a suburban backyard dog' (Whetham, 2020). The problem with individual animals or breeds becoming an 'animal star' is that stardom leads them to becoming a 'commercial commodity' and this is a major pitfall for thousands of animals outside the entertainment industry (Molloy, 2011, pp. 41, 44).

21   Such discrepancies in other-regarding interaction may align with ideas that some nonhuman animal species are more easily relatable with than others – that some species are seen as more charismatic, as possessing symbolic value or widespread popular appeal (see, e.g., Ducarme et al., 2013). Although there are inconsistencies

regarding the attribution of 'charisma' to animals, animals so recognized tend to have what are (for humans) identifiable faces and large eyes, and are of a size to be recognized – and interacted with – as identifiable individuals.

CHAPTER 5

Neither Food nor Friend: Cohabiting with the Animals Out There

What should be clear so far in our readings of these texts, is that the narratives that a society shares about animals and the different forms of relations with them that are considered normal fundamentally shape its members' ideas and interactions with animals – as living beings, as friends or kin, possible resources or products, and as imagined beings in our storytelling. It is, for instance, Australians' increasing acceptance of the narrative of animal sentience – and that animals have lives and interests that matter to them, and that this should matter to us – that can help us to take on ideas and practices that challenge the conventional economically driven narratives which inveigle us to engage primarily in instrumental ways with animals. As we have also suggested, in our analyses of selected texts, it is our narratives regarding animal capacities for not just sentience but companionship and support that can drive our preparedness to relate to animals as friends and kin, to see them as members of our private communities – our families – who deserve consideration. Both of these kinds of narratives of our interactions with animals can help us imagine and practice broader human-animal relations differently, as well as find other ways of living with and alongside animals that are not driven by anthropocentrism.

Our explorations of these various narratives have so far focused on our most known animals, the animals that are visible in our various communities one way or the other – either through their living presence as our companions or through our recognition that their lives and bodies are the foundation of products ready for human consumption. These are the animals whose presence and visibility fit categorizations as either friend or food. This chapter takes another orientation, choosing rather to examine our relations with and narratives about those animals who escape such categorization – and subsequent consideration – as food or friend. These are the animals who may be either native to Australia or introduced, *wild* or feral, but who are typically framed as living outside of the standard sphere of our actions and consideration. These are the animals the majority of us typically do not feel much – if any – responsibility for. They include those species that might be simply called Australian *wildlife*, or *pests*, or dangerous predators in need of strong, often lethal, management. If the experience of animals in agriculture raises questions with regards to our tendency to automatically think of them in economic and anthropocentric terms – even in the face of our shared sentience – these animals who mostly live outside the scope of our interests prompt within us further questions with regard to how Australians have tended, as a community, to balance human and animal interests. These questions are also important with regards to configuring any new imaginary of human-animal relations. In this chapter we thus examine our relations with kangaroos, *brumbies* (feral horses), *salties* (saltwater crocodiles), sharks and a range of other species which – as represented in public and science reporting and debated in the public sphere – not only live outside our standard sphere of action and consideration, but do not always live in ways that easily align with human interests and human ideas as to the *proper* inhabitation of the Australian landscape. The chapter thus digs further into issues of speciesism and anthropocentrism in the process of investigating some of the assumptions and narratives concerning Australians' relationships with and responsibilities for these animals, and the places of these *other* animals in our past, current and future social imaginaries.

Considering Other Animals

As has already been made clear, under anthropocentric and speciesist imaginings, animals – when they have any visibility in such imaginings – appear ranked in terms of their use values and subsequent interest to us. Under such imaginings, livestock animals do, for instance, hold value, but collectively as a resource rather than as individuals with their own interests. Animals we count as companions also hold value. This could be through their specificity and individuality in being a pet or companion, or their status as a member of a specific breed (typically tied to our ideas of *pure* breeds). Both of these general classes of animal – livestock and companion – are thus given some protection under the anthropocentric imaginary although, as has been noted, this protection may have little or nothing to do with protecting and supporting the interests and possible wellbeing of individual animals. Of other animals, the anthropocentrist will say that it all depends on their utility and perceived value – and this may be at the level of species rather than individuals. Hunters and fishers may well support actions to ensure the continuation of those animal species and populations which are hunted or caught seasonally, whilst still choosing to disregard the interests of individual animals from these populations. *Trophy* hunters who value only the rarity of the animals they hunt or their status as a trophy may have zero consideration for the wellbeing of the population or species. Australian tourism industries based on providing specific experiences with living animal populations – swimming with whale sharks and dolphins at Ningaloo, snorkeling or diving in the Great Barrier Reef, whale watching trips, birdwatching at Lake Eyre, to name a few – will try to preserve those populations and ecosystems. Working animals – dogs trained in various service roles, for instance, and perhaps animals used in racing or other entertainment industries – also have value, and will receive protection for at least as long as they can work or are successful.[1] And if the human-working animal relationship is strong, this protection may continue in the animal's retirement. Such calculations are clear – if an animal or population has value to us, that animal or at least its generic *class* will receive some consideration.

As we have noted, however, consideration organized under the anthropocentric view may not include much consideration of individual welfare or wellbeing. This has been the case for livestock animals in particular, where the individual interests of animals have been easily and indeed typically overlooked in support of human interests. Thus the ongoing removal of very young calves from dairy cows with the cows' mothering interests being overridden for human interests in having cheap dairy products. Thus the containment of animals in intensive settings that ignore the significance of species-natural habitats and behaviours for meeting animal wellbeing. And thus livestock welfare standards and practices that remain oriented towards industry revenue rather than the individual wellbeing of the animals they purport to consider. These issues have been examined in some detail in Chapters 2 and 3.

While companion animals do have more of their welfare interests protected through legal mechanisms in Australia than livestock animals, their individual lives are still expected to align with those of their human owners. Thus an animal that is not considered properly socialized (to humans) or that presents an inconvenience one way or another (again to humans) will be *unwanted*, and may thus be surrendered to a shelter, abandoned or killed. Such practices point again to the prevalence of anthropocentric narratives stressing the 'relative disposability of non-human animals' (O'Sullivan as cited in Berry, 2020; Houlihan, 2019; Wrenn, 2017). What is considered acceptable socialization also gives more priority to human over animal needs and behaviours. Thus companion animals assessed to be traumatized or otherwise unpredictable may well be named risks towards human safety, and euthanized on the basis of this decision. One example here is the case of the South Australian border collies who were rescued in 2019 from an illegal dog breeder by the RSPCA. Although some of these animals were rescued in a full sense, meaning they were rehomed, four of the dogs were deemed unfit for human company and subsequently euthanized (Flint, 2019; RSPCA, 2019a, 2019f). The phrasing used by the RSPCA with regards to this decision is typical of an organization with an animal welfare focus oriented by anthropocentrism: 'Despite […] rehabilitation measures, these 4 dogs continued to show deeply entrenched signs of chronic fear and potentially dangerous unpredictability. As a responsible rehoming organization, we simply cannot and will not release animals to the community that pose a significant risk to public safety' (RSPCA, 2019f).

Given that such practices illustrate the anthropocentric norm of putting human interests first, even when in relation with *valued* animals – and even whilst using an explicit welfarist lens as the RSPCA does – considering the interests of distinctly different animals is even more challenging. That is, if animals have little or no perceived value to human eyes, or simply live outside the sphere of immediate human interests and consideration, then they also tend to receive little welfare protection. To put this another way, being out of the sight of our immediate interests puts these animals, for most of the time, out of mind as well. This is the case for most of Australia's native animals, for example. *Other* to companion and livestock animals, these animals are not often visible to those Australians living in urban environments[2] – although there are exceptions, mostly although not always in the form of bird life and reptiles. The populations and individuals comprising native wildlife inhabit a liminal position in the perceptions of many Australians and in associated senses of care and responsibility for animal welfare and wellbeing. That is to say, we might be aware that there are populations of native animals living *out there* – in the bush or the outback, or just somewhere else – but we do not generally think about these animals or populations as needing our individual care or consideration. (Indeed, even if we do come to think of such populations as in need of care – koalas, e.g. in the aftermath of the Black Summer fires – the provision of such care is typically considered someone else's responsibility.) This is the case even with regards to the native animals we might occasionally see in our urban and suburban regions – birds, koalas, possums, flying foxes, bandicoots, or turtles. They are just there. We might be saddened by the sight of an individual animal killed on the roads or found mauled by a dog or cat, and some of us might go out of our way to provide some support for those native animals that live in our vicinity (planting native species, setting out water for birds, constructing nesting boxes, and calling native animal rescue services to assist injured animals), but we generally do not consider the welfare or wellbeing of these populations or their individual members to be under our remit.[3] It seems that without the kinds of personal connections to individual animals discussed in the previous chapter or identifiable responsibilities for individuals and populations, we do not recognize any responsibilities for them. These are animals and populations that we typically expect to fend for themselves – although, typically anthropocentric in our views, we also prefer that any native animal behaviour does not impinge on us or our interests (an important issue that we will come back to later in this and other chapters).

Responsibility: Questions and Gaps

This general lack of consideration for animal wellbeing is very apparent when considering the situation of Australia's native animal populations under legislation. Analysis carried out for the 2019 Senate Inquiry into Faunal Extinction and the associated review of Australia's federal *Environmental Protection and Biodiversity Conservation Act 1999* (Cth) (*EPBC Act 1999*) shows, for instance, that existing federal and state laws designed to prevent native animal species and populations from being injured, or their overall presence endangered through loss of habitat, are weak, poorly implemented and regulated (Commonwealth of Australia, 2019; Keck, 2020). This was made very clear in both the interim and final reports from the inquiry. Findings have stated that despite legislation to protect biodiversity, Australia is a global leader (see, e.g., Australian Conservation Foundation, 2020; Sanda, 2018) in 'faunal extinction and decreasing biodiversity' (Commonwealth of Australia, 2019, p. 61). In the words of the chair of the 2019 Senate Inquiry into Faunal Extinction, Greens Senator Janet Rice,

> The most important issue to arise from the Inquiry thus far is how totally inadequate our current Federal legislation is in terms of protecting the environment. We're dealing with the drivers of extinction. The EPBC Act is totally inadequate to deal with the situation – the problem is there's no framework for protection. (as cited in Arnold, 2019)

Existing laws and policies are identified as being particularly inadequate with respect to the protection of wildlife habitat – a key issue for stemming these rates of extinction (see Commonwealth of Australia, 2019). As a spokesperson for the Australian Conservation Foundation (ACF) has also noted, existing laws and policies are 'subject to the political whims of ministers who are afforded broad discretion and may be subject to the pull of vested interests' (2018). These vested interests are, of course, human and tend unsurprisingly to align with frameworks prioritizing economic interests and development. The situation is well summed up in the Australian Greens' additional comments to the findings of the interim report:

> The inquiry has heard evidence that the scale and speed of decline for Australian threatened fauna is nothing short of scandalous. We are in the midst of the sixth great mass extinction event.
>
> What is clear from the evidence so far is that there is nothing inevitable about species extinction, it is a choice. With adequate laws and funding, we can ensure that not one more Australian species goes extinct.
>
> But our existing laws and compliance mechanisms are little more than processes to be stepped through by project proponents. They have failed to prevent faunal extinction and species decline. (as cited in Commonwealth of Australia, 2019, p. 65)

This lack of protection is further exacerbated through climate change, with nearly half of Australian species (both animal and plant) already identified as threatened having also been identified as vulnerable to climate change (Lee et al., 2015). Further research suggests that up to half of the plant and animal species endemic to the world's most naturally biodiverse areas – such as southwest Australia – could face local extinction by 2100 due to climate change if carbon emissions continue to rise unchecked (WWF, 2018f). Even if the Paris Climate Agreement 2 °C target is met, these places could lose 25 per cent of their species (Warren et al., 2018). Some of these species may be directly affected by factors such as sea level rise (changing freshwater swamps to saltpan marshes, for instance), snow melt, or rising temperatures,[4] whereas others may lose a prey or other food species (or a pollinator, in the case of plants) that they rely on. Changes in breeding times or migration dates or even rainfall patterns may also lead to an ecosystem no longer being able to sustain its animal and plant populations.[5] All of these points mean that those species that cannot move (or be moved) to more suitable habitats – or which have no suitable habitat left – are at high risk of extinction. Such was the case for the Bramble Cay mosaic-tailed rat, the first mammalian extinction attributed to climate change. Once prolific on an island in the Torres Strait, off the tip of northern Australia, surveys have revealed that due to the island being repeatedly inundated by storms and rising sea levels, the last native animals would have simply drowned (Purtill, 2019).

This is not to say Australians and others do not care about the wellbeing of Australia's wildlife. The findings from a survey carried out by the World Wide Fund for Nature (WWF) Australia in 2018, note for instance that Australians do believe that it is important to protect oceans, wildlife and nature, even if only because these are considered 'critical to maintain the balance of nature' (WWF, 2018b, p. 5). Furthermore, the report shows that 93 per cent of Australians view protecting Australia's endangered animals, forests, and trees to be a priority. Also, according to the report, 89 per cent of respondents agreed that there should be increased investment in restoring wildlife habitats and natural places that have become degraded, as well as the setting of limitations to land-clearing (WWF, 2018a). In total, Australians expressed a 'strong affinity with our forests, beaches, oceans and wildlife, and overwhelmingly support more action to protect Australia's natural habitats' (WWF, 2018a). However, although this expression of a 'strong affinity' both suggests and endorses the possibility of living in ways that better protect and support wildlife and habitat, at the same time it is an abstract relation in the sense that it does not by itself generate concrete forms of relation, responsibility and action.

This is most clearly seen when this 'strong affinity' is compared with Australia's First Nations Peoples' ideas of kinship and relations with *Country*. As the Gay'Wu Group of Women[6] explain these relations:

> When we talk about Country, we are using an Aboriginal English term that refers to specific places, specific Aboriginal peoples' homelands. There are many, many Countries within Australia, many peoples, many languages, many Aboriginal nations. To talk of Country means not just land, but also the waters, the people, the winds, animals, plants, stories, songs and feelings, everything that becomes together to make up place. Country is alive for us, it cares for us, communicates with us, and we are part of it. (2019, p. ix)

As 'the way humans and non-humans co-become' and 'emerge together' (p. xxii), *Country* is 'the centre of society, culture, family, spirituality, the past, present and future' (Jagot, 2017). It is thus the basis for 'infinite cycles of kinship, sharing and responsibility' (Gay'Wu Group of Women, 2019). In the words of a First Nations commentator on the platform IndigenousX, 'The land owns us. [...] we now need to support and sustain it, repair it and save it from any further destruction. [...] we must continue to advocate for it' (Bell, 2020). Not only is *Country* never just symbolically 'imagined or represented' – standard abstract models of thinking – but all of the ways it is 'lived in and lived with' require acceptance of a whole suite of concrete responsibilities relating to the land, its physical features, and the animals and plants that inhabit it (Rose, 1996, p. 7). This is to say that Australia's First Nations Peoples live a social imaginary that has always included responsibility for and active participation in *caring for Country* even as that work has also always involved the *management* and use of Australia's wildlife (Aslin & Bennett, 2000; Collins et al., 1996; Young et al., 1991). Even though the most obvious contrast with such a view is of course the anthropocentric imaginary, Australia's First Nations Peoples' thick, concrete demands of living in and with *Country* do also show up the thinness and abstractedness of expressions of claiming a *strong affinity* with Australia's wildlife and natural habitats whilst not accepting attendant responsibilities. If, then, Australia's post-settlement anthropocentric imaginary shows little to no regard for animals and the environment to the point of generating 'environmental and animal extinction' (Bell, 2020), then expressing a strong affinity is akin to expressing an abstract benevolence. A majority of Australians – and note that the majority of Australians live in urban settings – might thus express their care for wildlife and habitat, but this is more a concern that it continues to exist *out there*. As animal and environmental activists are well aware – and as noted in Chapter 3 – it takes a further push to move concern into action. This, of course, would be the push needed for Australians to really address ongoing losses in biodiversity and faunal extinction.

All of these issues point back to a key question of responsibility. More specifically, it can be argued that it is because the conditions and actualities of these animals' lives and habitats remain *out there*, outside of most Australians' typical spaces of action, interest and consideration, that they remain outside of what most Australians would recognize as our sphere of responsibility.[7] This is the point that our acceptance of responsibility and of a need for action regarding any being or thing typically arises from already having some sense of concrete connection to that being or thing. This is most often a direct causal connection ('I caused this and thus should "fix" it') but might also include a relational connection such as friendship or kinship or, more broadly, care. In the absence of any connection beyond an abstract benevolence, however, the lives and wellbeing of the animals living *out there* can fall through the gaps between (a) our capacities to act and make change, individually or collectively, and (b) any sense of responsibility and accountability that might cause us to so act. This is a reminder that simply being able to respond does not by itself constitute any actual response. As the review of the EPBC Act has made clear, such gaps have been identified as a key determinant for Australia's high rates of species decline and extinction. Actually addressing – slowing or stopping – these rates is, however, a more complicated problem.[8] The next sections in this chapter consider some of the reasons for this as they unpack the range of narratives and counter-narratives in common use regarding animals *out there*.

Issues of Visibility and Action

One strategy often used by animal activists and advocates to overcome this gap between capacity for action and actual action concerns tying recognition of responsibility to awareness. This is the ethical idea that once we know about an issue – and recognize in it an injustice or desperate need – we should feel bound to act. (Note that the assumption is that we can indeed act.) Namely, it is through an increase in awareness that response-ability becomes responsibility. This, of course, points to the problem of visibility. If these animals live outside our usual spheres of action and consideration – and hence outside our understanding – they need to first become visible to us as being in need or as experiencing an injustice due to cruelty or some other disregard of their interests. This, of course, is where public reporting and animal advocacy and activism campaigns come in.

Given this framework, it is unsurprising that Australia's 2019–20 wildfires – the Black Summer fires as they have become colloquially known – have been partially successful in shifting Australians' general and abstract concerns for wildlife into action on their behalf, for at least a period of time. As was widely reported throughout the nation, these megafires had a devastating impact on wildlife, with almost 3 billion native vertebrates (mammals, reptiles, birds and frogs) estimated to having been killed or displaced (WWF, 2020c), and 17 million hectares of forest destroyed, including 6 million hectares of threatened species habitat (see, e.g., Figure 8; Mathews, 2020; Noble, 2020).

Figure 8. Kangaroo Island, South Australia, has been described as an Australian equivalent of the Galapagos. As this satellite image released by NASA shows, over half a million acres – almost half – of the island was burnt during the 2019–2020 Black Summer bushfires. This included 96 per cent of Flinders Chase National Park, one of Australia's oldest national parks, and a home to a vibrant diversity of species, including many endangered animals and birds. *Source:* Photo from NASA Earth Observatory images by Lauren Dauphin.

Responses to this devastation saw upwards of AU$100 million donated to animal welfare organizations (Wintle et al., 2020). For instance, a fundraiser for the New South Wales Wildlife Information and Rescue Service (WIRES) raised more than US$9 million (Simo, 2020). As of mid-January 2020, WWF Australia, in partnership with wildlife rescue and care organizations in fire-affected states, raised almost AU$15 million through their Australian Wildlife and Nature Recovery Fund, and the Port Macquarie Koala Hospital in New South Wales raised more than AU$7 million (Gibbs, 2020; McGhee, 2020). Other efforts saw Australians in fire and smoke affected areas working to ensure that wildlife that had survived the fires would not face starvation from the losses of habitat. These have ranged from state-based and consolidated efforts to individual or small local efforts to ensure surviving populations of animals have access to food and water. Thus, in the first group, the New South Wales state government organized for food drops to be delivered by air to remote areas to help stave off animal hunger. As the New South Wales Environment Minister Matt Kean has said of these food drops,

> The provision of supplementary food is one of the key strategies we are deploying to promote the survival and recovery of endangered species like the Brush-tailed Rock-wallaby. [...] The wallabies typically survive the fire itself, but are then left stranded with limited natural food as the fire takes out the vegetation around their rocky habitat. [...] At this stage, we expect to continue providing supplementary food to rock-wallaby populations until sufficient natural food resources and water become available again in the landscape, during post-fire recovery. (as cited in NSW Government, 2020)

The Victorian state government also organized for food drops and for on ground-feeding stations for surviving wildlife. In the words of the Victorian Environment Minister Lily D'Ambrosio, 'Our triage units, assessment teams and experts in the air and on the ground are working hard to get the best possible outcomes for our precious wildlife. [...] This targeted approach means food relief is being delivered to the areas that will have the greatest benefits for our macropod species' (as cited in Towell & Ilanbey, 2020).

Other efforts have been more localized, with wildlife rescue groups, volunteers and individual property owners – typically on the outskirts of fire-affected areas – also working to provide food, water, and in some cases shelter for displaced animals (see, e.g., Figure 9).

Figure 9. A volunteer in Canberra topping up water containers to support wildlife on Black Mountain, Canberra, the Australian Capital Territory (ACT), while guarding against smoke inhalation from the 2019–2020 Black Summer bush fires. *Source:* Photo courtesy of Josephine Mummery.

Thus in the Australian Capital Territory (ACT), Wildcare – a volunteer organization dedicated to the rescue, rehabilitation and release of Australian native wildlife – and the Native Animal Rescue Group, along with individual property owners, set up watering points and food stations for native animals in strategic areas. These were used to provide water and a range of foods to support native species, including grass, hay, pellets (developed for macropods and marsupials), bird seed, fruit, and native tree and shrub cuttings. Wildcare and other animal rescue and welfare organizations such as WIRES and Animals Australia also called out for people to assist in these efforts:

> When the imminent threat of the flames has subsided – the truth is, mass starvation events are now likely for these survivors. [...] Feeding of wildlife is not normally recommended, but the reality is that this is a crisis and during this emergency period, wildlife are going to need us to help them. (Animals Australia, 2020a)

In further support of people taking action, these calls are typically accompanied by further advice on how to best provide help.

> For the many people who want to help animals in areas surrounding the fires, a simple start is to put out shallow bowls of fresh water. Fruits and vegetables can also be attached to trees to help flying-foxes, birds and possums, but some foods such as bread can be harmful to wildlife and should be avoided. (Wildcare, 2020)

In the same mode, the animal welfare organization Animals Australia (2020a) published nine species-specific emergency feeding guides, whilst stressing that water is the priority and that 'wildlife should only ever be given food when there is absolutely no other option available to them'. These guides provide feeding plans for multiple sets of species.

What these various efforts make clear is that these populations and individuals of wildlife can become *visible* to us, with this visibility promoting significant efforts on their behalf. It should also be recognized, however, that this visibility and associated efforts tend to be in the aftermath of heavily reported events that make the precarious status of wildlife clear – such as the Black Summer megafires of 2019–20. Put another way, it appears that while many Australians' attitudes to wildlife might generally be in a holding pattern of abstract benevolence as the WWF report (2018f) suggests, this primarily seems to translate into actions in support of Australia's wildlife – and thereby an acknowledgement of some connectedness – when populations are shown to be visibly under threat. Thus efforts to sustain wildlife populations after fires; thus localized efforts to save particular populations or areas of wildlife habitat when they come under threat from development, logging or mining; and thus backlashes to reported proposals to cull certain populations of animals when they become inconvenient to certain human interests. As the Senate inquiry has made very clear, however, given the weakness of Australia's environmental laws, such attitudes with their peaks and troughs of activity and their dependence on problems being made visible in the first place have not been strong enough to actually protect and sustain our wildlife *out there*. Putting this another way, such efforts on behalf of wildlife tend to be reactive rather than proactive. This in turn means that instead of strong systemic promotion and support for the wellbeing and flourishing of Australia's wildlife *out there* – a focus that could foreground protection of the actual interests of wildlife to not have their habitat, food sources and broader ecosystems systematically fragmented or destroyed – business as usual is anthropocentric. Wildlife interests only gain notice and some support after they have already been damaged or further eroded – and, importantly, publicized enough to convert abstract benevolence into responsibility and concrete action.

The next sections of this chapter will consider these issues in more detail, exploring these mixed relations – abstracted and activated – that most Australians have with the animals *out there*. In particular, we examine some of the narratives and modes of human-animal relation associated with three main groups of these animals: native, feral, and dangerous. As noted, each of these groups of animals tend to remain beyond the notice of many Australians until they are presented to us as either being under threat or as a problem themselves. Also as noted, these are all animals that we generally imagine fend for themselves, that mostly remain outside the frames of our action and consideration. This is the reminder that in being heavily urbanized and typically 'separated from our wildlife' (WildArk, n.d.), Australians generally have to make some effort to have encounters with wildlife (barring those species, mostly birds, commonly visible within the urban environment). Even if a majority of Australians do care abstractly about the wellbeing of Australia's wildlife, being distanced from the natural environment and having limited contact with native species means that they have little to no immediate connection or relationship with these animal populations – unless they have been made visible to us in a way that activates our sense of responsibility.

Native

As has been noted already, Australia's native animals are considered to live *out there*, embedded within and adapted to natural ecosystems, and considered able to fend for themselves – at least under normal conditions. *Belonging* to ecosystems rather than to humans, these animals are neither *property* in the way of livestock animals nor typically recognized as friends or kin outside of First Nations perspectives, and are mostly considered to exist beyond human oversight and consideration. At the same time, the uniqueness and diversity of Australia's native wildlife provides one of the focuses of Australia's tourism industry, is a key driver of Australia's conservation movement (Australian Wildlife Conservancy, n.d.), and has contributed to ideas of Australian national identity. This is the point that 87 per cent of our mammal species, 93 per cent of reptiles, 94 per cent of frogs and 45 per cent of our bird species are found only in Australia. Indeed, better known native animals and plants – such as the kangaroo, the emu, and the golden wattle – have become symbols of Australian nationhood and part of a cohesion principle for many Australians (Aslin & Bennett, 2000; Morton, 1991). It is important to note, however, that while these factors – uniqueness, diversity, and symbolic status – are all markers of value, they all stand for abstracted values. This means that while such values certainly support the abstract benevolence for wildlife *out there* mentioned earlier, they also do not, by themselves, turn into action towards protecting the welfare or wellbeing of either individual animals or populations. Abstracted values and relations are, by definition, ill-equipped to drive actions supporting the concrete needs and interests of individuals or individual populations – there always needs to be a further trigger for action.

This is the reminder that in being considered to primarily live *out there* these animals generally live on the outskirts of most Australians' lives and experience and typically only come to *matter* in non-abstract ways when they impinge on our lives and experiences – either through personal experience or by becoming visible through public reportage. As has been noted, one significant instance that encompassed both of these methods was the Black Summer megafires. It is fair to say that no one within Australia, regardless of whether they were directly impacted or not, would not have heard about the catastrophic impact of these fires on Australia's ecosystems and wildlife. The scale and impact of these fires was broadcast across multiple media channels and is still being analysed. They have been estimated as affecting nearly 80 per cent of Australians either directly or indirectly (Biddle et al., 2020; Climate Council of Australia, 2020), and were *apocalyptic* in their effects on wildlife (Serna & Rust, 2020). Their aftermath continues to impact upon multiple communities, human and animal. Although the fires reportedly devastated all species within the fire zones and have caused a catastrophic loss of habitat, their impact on one native species was given particular prominence in reporting: koalas.

Koalas

Koalas have long been a key species in post-settlement narratives about Australia. They are an important 'face of Australian tourism', and a photograph with a koala is often a 'must-have souvenir' for many international tourists (Markwell, 2020). Koalas were also, horrifyingly, the *face* of the Black Summer fires (Mathews, 2020), and came to symbolize all the precious and largely unique Australian native species being destroyed or pushed to the edge of survival (McGhee, 2020). Slow moving, living in and dependent on highly flammable eucalyptus trees, koalas were burnt where they lived, unable to escape. In the words of one commentator,

> Along with the rest of the world, I gazed transfixed with horror at images of blackened koalas, their still-living faces mutilated and scorched. I listened to them cry out in terrified little voices as they scrambled over glowing red ground on bare feet, their fur on fire. My heart cracked as I watched desperately parched koalas, in video after video, approach a human stranger for help, reaching out to hold their benefactor's hand as they drank from the proffered water bottle. (Mathews, 2020)

These are devastating stories, and these and the images and stories shared through broadcast and social media of people rescuing burnt and injured koalas after the fires led to a sustained national and international outpouring of concern and care and various forms of support – with koala sanctuaries and wildlife hospitals receiving substantial donations, for example. Through our seeing, hearing, as well as sharing these terrible images and stories of burnt bodies and pain and death, the fate of koalas in these fires made two things clear to us all. The first is that while these animals might live *out there* and out of sight for many of us, they too are sentient and deserve our consideration on that basis. These stories and images thus drove home to us all the essential narrative of sentience – that 'animal terror and torment' is 'experienced in exactly the same way we would experience such agonies ourselves' (Mathews, 2020). The second point concerns a different narrative of identification, connection, and consideration. As Freya Mathews (2020) continues in her commentary, what these fires have also driven home to her is that living in this brown and burnt land, of being under threat from fire and flood and facing an uncertain future, is something that we share with our native wildlife. Living in this land connects us all, humans and animals.

> While the rest of the world wept for the helplessness of the adorable koala, we wept because Koala – along with our entire extended wildlife family – is not only helpless and adorable but, beneath the skin, *our deepest kin*. (Mathews, 2020, original emphasis)

Koala stories from the fires, in other words, are stories that have foregrounded kinship connections between humans and animals. They are stories in which we can learn to recognize not only our connections with life *out there* – that we are kin – but the imperative to help these kin with whom we share this land.[9]

Estimates suggest that the fires have caused koala populations in both Queensland and New South Wales to collapse by at least 70 per cent (Lane et al., 2020; Rubbo & Wellauer, 2020; Wallis et al., 2020), leading to an increase in calls by conservation and animal welfare groups for the koala to be listed as endangered in Queensland, New South Wales (NSW) and the Australian Capital Territory (ACT) (WWF, 2020e). These are important issues and should mark a strong call for reactive consideration and care of remaining populations, but it is also important to note that koalas have been recognized as being under threat of population collapse since early in the twentieth century (Markwell, 2020). That is, the majority of their populations has already been identified as being under stress due to habitat destruction and fragmentation. Particular stresses include, for example, our anthropocentric prioritizing of agriculture, urban development, mining and forestry over wildlife and wildlife habitat, the impacts of drought and disease, and the perils for koalas of living in and near human developments: namely, being killed on the roads and by dog attacks (McGee, 2020).

> In western Queensland, koala numbers declined by 80% between 1995 and 2009, mainly because of habitat loss, drought and heatwaves.
>
> In southeast Queensland, where threats from urban development are a key factor, some populations fell by 55–80% per cent between 1996 and 2015.
>
> The picture is similar in NSW, where there is evidence for a substantial decline in koala numbers over the two centuries from European settlement in 1788 to the first major statewide survey in 1986–87. Subsequent studies have identified steep declines and local extinctions of koala populations in the past 30 years. (Rhodes et al., 2016)

Remaining habitat is also considered under severe threat by climate change, given its anticipated generation of more extreme droughts, heatwaves, and bushfires. All of which means additional stress again on surviving populations and on habitat (Wallis et al., 2020).

It is in this context that the koala fire stories of connection, identification, kinship, and care must also be considered. To be clear, koala populations had been identified as being in significant decline since well before the fires, but the reasons for such declines either were not addressed in effective ways (the weak law issue) or were ignored (Foley, 2020). This longstanding lack of effective action can be attributed to the ongoing disinclination to place constraints on anthropocentric interests – a disinclination exemplified by the weakness of the Australia's EPBC Act. For instance, despite multiple inquiries and reports examining the health and status of koalas and calls for the preservation of koala habitat (see, e.g., Hosking, 2020; Rhodes et al., 2015), land clearing – a fundamental contributor to habitat loss and declines in koala numbers – continues across both Queensland and New South Wales. In fact, the rate of loss of koala habitat has actually shown to have increased in the last decades due to an ongoing favouring of land clearing and urban development over halting declines in koala populations (Rhodes et al., 2017). A further problem is our ongoing tendency to not connect habitat loss with actual wildlife deaths. This is a point that conservation and wildlife photographer Doug Gimesy stresses. As he puts this, we need to 'stop using passive language to mask how humans are responsible for destroying wildlife. Don't call it "habitat loss", we didn't misplace any habitats, it was no accident. Call it "habitat destruction". Don't call it "extinction", call it "extermination"' (Gimesy as cited in art thy neighbour, n.d.).

There is an important point here, which is that with regards to koala lives and deaths it seems that stories of connection, identification, responsibility, care, and action come most easily to us when there is no strong demand for us to shift out of our default position of anthropocentrism. That is, while responding to calls for help to support wildlife after the Black Summer fires has clearly helped in the provision of wildlife care, providing monetary support or even providing food and water to surviving wildlife are easy responses. What is much more difficult is taking a stance and actively participating in the kinds of political, economic and social reimaginings needed to really convert our care and horror regarding animal deaths and population losses from the Black Summer fires into actually halting declines in koala and other native animal populations. Such a step requires koala conservation – and koala habitat conservation – to be given precedence over any and all human demands to use land otherwise (whether for urbanization, agriculture or mining). And whilst this might seem a simple decision given the stakes for koalas, it is also clear that our abstract concern for koala lives *out there* has not deemed them to matter more than human interests in land clearing and development. The question will be whether the koala stories of fire and suffering can activate that care into actions of thinking and living differently with animals.

Kangaroos

This kind of conflict between native animal interests and those of human development is exemplified also by stories of kangaroos in this country. Unlike koalas which can only flourish within a habitat including very select kinds of eucalypt trees, kangaroos flourish in a range of habitats, including semi-arid plains, grasslands, woodlands, and open forests. They are also very mobile, able to adapt their range to climatic and vegetation conditions. Such factors have led to a commonplace view that there are plenty of kangaroos *out there*. On the other hand, because anthropocentric interests in agriculture have taken over from many natural ecosystems and previous kangaroo habitat, many of Australia's estimated 30 million kangaroos now range into what has become agricultural land – thus bringing about a conflict in interests. In the terms of this conflict, kangaroos ranging onto agricultural land have been and still are conventionally narrated as a cost to farmers insofar as the presence of kangaroos is perceived to limit the numbers of livestock that might otherwise be carried as well as damage crop production. Kangaroo presence has thus been said to result in reductions of livestock carcass weights and wool production. Kangaroo presence is also perceived as disrupting the spelling of paddocks from being grazed, damaging existing fences, and increasing the costs of fence construction due to trying to exclude them from using scarce resources (Viggers & Hearn, 2005; Wilson & Edwards, 2019). Kangaroos, in other words, are strongly framed as a *pest* species by farmers.[10] They have also been named as pest when – typically driven by drought and habitat fragmentation – they move into and through suburban and urban settings and cause hazards for drivers (Green-Barber & Old, 2019). *Pest* animals, then, are those perceived to have negative impacts on things that people value (Fisher et al., 2008), and preventing these negative impacts is thought to mean that these animals need to be *managed*, most commonly through lethal means. The term *pest* is, however, itself an anthropocentric concept used to describe the relative value (for us) of a particular species in a particular place (Marks, 1999).

Seeing kangaroos as pests – and thus as needing (lethal) management – has been and still is a widespread story, and one that is still used to justify kangaroo culls across Australia (Berlin, 2019; Boom et al., 2012). It has resulted in kangaroos being killed both commercially and privately, the argument for both being that this is necessary to preserve agricultural interests as well as to keep kangaroo numbers from increasing to 'plague proportions' (Berlin, 2019; Gall, 2020). This latter concern is further associated with the idea that high kangaroo populations could also cause the degradation of sensitive ecosystems and stymie some conservation efforts (ACT Government, 2020). This is a view often cited in defense of what are often called *conservation culls*, as by the ACT Director of Parks and Conservation, Daniel Inglesias:

> Ensuring the grasslands and woodlands are not overgrazed will help to protect our grasslands and woodlands, which provide habitat for creatures such as lizards and ground-feeding birds, and will avoid excessive soil loss whilst still maintaining sustainable numbers of kangaroos. (as cited in Pryor, 2018)

The killing of kangaroos is thus also narrated as a conservation and land management tool. This has seen the setting of annual kill quotas that are determined with reference to an extrapolated population (quotas are typically set at less than 20 per cent of this extrapolated total population). National Codes of Practice have also been developed to guide the humane shooting of kangaroos and wallabies for both commercial and non-commercial purposes, and some licensing requirements have been set for commercial shooters (but not for private, non-commercial ones) (Wahlquist, 2019a). According to this story, then, human interests in agriculture and human ideas regarding the *proper* management of native ecosystems are considered able to override those of kangaroos for inhabitation of the land they are native to. In this story human interests receive high levels of consideration, while kangaroos are attributed very little value and consideration at either the individual or population level. Framed as *pest*, kangaroos in a lot of instances are not even recipients of Australians' abstracted benevolence towards native wildlife.

The kangaroo as pest narrative is, however, not the only narrative of kangaroos being told in Australia's public sphere. And these other stories position kangaroos quite differently, with three such counter-narratives presenting kangaroos as victim, resource, and national icon. Unsurprisingly the story of kangaroo as victim is most often told by animal advocates in response to the kangaroo as pest narrative, and typically increases in prominence during periods of culling. According to this story – and there are versions of this story directed at both the kangaroo as agricultural pest narrative and kangaroo as conservation problem narrative – both pest narratives unjustly target kangaroos as a problem. For instance, regarding the story of needing to kill kangaroos for conservation reasons, it has been argued that it is inappropriate that kangaroo kill quotas are set without evidence being required of an ecosystem actually being under stress (RSPCA, 2019c). The lack of robust evidence regarding the past effectiveness of conservation culls for ecosystem preservation has also been emphasized, with research by the Commonwealth Scientific and Industrial Research Organisation (CSIRO) finding in fact that 'relatively large changes in kangaroo densities' are not consistently correlated with changes to the relevant ecosystem (Vivian & Godfree, 2014). Such evidence is a reminder that kangaroos are only ever one (native) component in a native ecosystem, meaning that the narrative of possible ecosystem degradation that drives conservation culls tends to blame kangaroos without considering the impacts of other factors such as prevailing climatic conditions, previous grazing history (including by rabbits and livestock), previous fire and drought conditions, soil conditions and nutrients (Taylor, 2017; Vivian & Godfree, 2014). There are also questions concerning the perception that kangaroos must be killed to ensure the success of agricultural interests, whether these are in the growing of crops or the raising of livestock. Research by CSIRO has shown, for instance, that few kangaroos wander more than 400 metres from native vegetation, and that kangaroos prefer native vegetation to crops (Arnold, 1990). Of course, if there is little to no native vegetation left – due to land clearing for agricultural purposes – then kangaroos will clearly try to access crops rather than starve. CSIRO research has also shown that kangaroos prefer to avoid areas used by sheep (McLeod, 1996), and that sheep and kangaroos are only in competition for grass during drought or due to a loss of native vegetation (Dawson, 2012). Like koalas, then, and all other native animals in Australia, kangaroos are being placed under pressure from land clearing and the consequent fragmentation of native habitat, but are then blamed – and killed – for accessing land that we have turned to our agricultural interests. Whereas koalas die due to their habitat losses, kangaroos are actively sought out and killed for trying to survive despite habitat losses.

Other important components of the kangaroo as victim story concern the brutality and cruelty used in killing them. Like us, kangaroos are feeling and thinking creatures; they are sentient and, as advocates argue, should thus have their interests considered and supported. (The narrative of sentience and its importance for the treatment of animals in ways that promote not just their welfare but their wellbeing has been outlined in detail in Chapter 2.) In addition, although commercial shooters are supposed to meet certain requirements regarding *humane* killing – single shot to the brain, and targeting males so as to minimize the chance of orphaning dependent joeys and young-at-foot (which would then need to be killed as well) – this is not always possible given shooting conditions (Ben-Ami, 2009), and non-commercial shooters are not regulated at all with regards to their competence (RSPCA, 2019c). In addition, the dispersed nature of kangaroo killing, and the fact that it often takes place at night, means that it goes largely unchecked (Boom et al., 2013).

> There is no enforcement of even the few, flimsy protections provided by the code: no police, no vets, no welfare agencies. The killing is completely unmonitored, except by protestors. This year there has not been even a single government ranger in sight while shooting has been underway. (Seymour, 2020)

Furthermore, although it is illegal to shoot kangaroos without a permit, farmers can gain permits for non-commercial kills where the number of kangaroos killed can be determined by the extrapolated kangaroo population and property size rather than by evidence of harm caused by kangaroos. The actual number of animals killed is also self-reported – a system clearly open to abuse. Landholders can hold what are in effect private shooting parties to kill kangaroos – with no requirements of shooting competence or humaneness – and then just mass bury or leave the bodies to rot (see, e.g., Crane, 2015; Taylor, 2018). Sadly these appear to still be common practices. Not only are the numbers of non-commercial culls increasing (Wilson & Edwards, 2019), but there is a strong likelihood of illegal killings of kangaroos being carried out throughout Australia (Kangaroo Industry Association of Australia, 2019). That such activities are allowed to continue, mostly unregulated, demonstrates the ongoing refusal by the Australian agricultural sector and politicians to seriously consider the non-anthropocentric interests of kangaroos and other Australian wildlife. (This refusal is unpacked and made visible in *Kangaroo: A Love-Hate Story* (2017), directed by Michael McIntyre and Kate McIntyre Clere.) It is also worth considering the broader impacts of a killing system that, unlike natural selection, deliberately targets the largest and healthiest males. Such a regime disrupts kangaroos' 'natural social organization' and 'population dynamics' and does not align with any of the actual aims of long-term wildlife conservation (Ben-Ami, 2009; Croft, 1999; Taylor, 2018).

Running alongside these stories of kangaroos as victims are those that suggest reframing kangaroos as a resource. These stories are typically based on the healthiness of kangaroo meat as a product and that kangaroos – unlike introduced species of cattle and sheep, etc. – are native species and thus cause substantially less land degradation than do standard livestock species (Cushing, 2019). According to this story, if farmers saw that kangaroos could be a resource with a monetary value as opposed to a pest that supposedly limits potential profits, this might improve both their treatment of kangaroos and place a stronger focus on their conservation (Wilson & Edwards, 2019). This is the anthropocentric idea that human consideration and treatment of animals is always dependent on their value – increase that value, and treatment could be expected to improve. (Of course, as we have already noted in the context of the status of livestock and discussed in Chapters 2 and 3, this value does not necessarily mean consideration of the interests of individual animals.) The final story that intersects with all of these others is that of the kangaroo as national icon.

> To outsiders, the big-footed, fat-tailed, perky-eared creatures are a stand-in for the country itself: Australia means roos, and roos mean Australia. There may be no animal and nation in the world more closely identified. (Berlin, 2019)

This is the story of the kangaroo as both symbol and tourist attraction. As symbol, kangaroos also populate our social imaginary of being Australian. Kangaroos thus 'star in movies and TV shows, poems, and children's books. Their images adorn the country's currency, coat of arms, commercial airlines, naval vessels, Olympic insignia, and athletic uniforms' (Berlin, 2019). They (and other Australian native animals) are often the inspiration behind public art installations. Australians have also nicknamed many of national sporting teams after them – the Wallabies, Socceroos, Hockeyroos, etc. Kangaroo as national icon clearly does not damage the economy, rather – along with koalas – it is a story that plays a role in bringing tourists to the country. If tourists visiting Australia view cuddling a koala as a desirable activity, seeing kangaroos in the wild is surely another.

Whilst these three kangaroo counter-narratives – as victim, resource, and icon – do not all neatly align with each other in their challenging of the kangaroo as pest story, what they do each draw attention to is the point that in being native to this country, kangaroo wellbeing *out there* should be much better protected. They were, after all, living *out there* before we gave priority to our modern anthropocentric agricultural interests, and these and other anthropocentric interests have already significantly cut into their capacities to flourish and move throughout rangelands as conditions dictate. That is, whilst the kangaroo as pest narrative contends that Australia's turn to agriculture has greatly extended kangaroo habitat – creating the kangaroo *plague* – no one could say that kangaroos are permitted to flourish in these environs. Being fenced off from water and grazing and left to die or herded up against exclusion fencing and then shot makes that very clear (Crane, 2015; Taylor, 2018). As with koalas, then, truly considering the interests of these animals means that we have to stop giving anthropocentric interests top and/or only billing. After all, as Nikki Sutterby, president of the Australian Society for Kangaroos, has noted, 'The kangaroos were here first, so why can't you give up ten percent of your land for your native animals? [...] The kangaroos are struggling too, but no one has any sympathy for them, everyone just wants them out of their property' (as cited in Taylor, 2018).

Feral

If Australians tell conflicting narratives about native wildlife that show how abstract and thin their benevolence is towards both individual animals and their species and how easily wildlife interests are overridden by anthropocentric ones, then this is compounded in Australian narratives about feral animals. These are typically understood to be non-native animals that have established themselves *out there* in the wild. Most feral animals are either domestic animals that have gone wild (from cats and dogs through to horses and camels) or animals that came into Australia accidentally (rats and mice) or were introduced into the wild for recreational use or pest control (from rabbits, foxes and deer through to cane toads). Regardless of the basis for their entry into the wild – often although not always through human means – these animals have become referred to as pest animals or invasive pest species. This is because, with the scarcity of Australia's large native predators (itself a result of human activities), many of these animals have flourished in Australian conditions. The conventional narrative that is thus told and retold by government and conservation science is that the presence of feral animals reduces biodiversity and harms the resilience of native species through predation, increased competition for food and shelter, and by destroying habitat, causing soil erosion, and spreading diseases (Australian Government, n.d.-a). They are also positioned as damaging human interests in agriculture, built environments, public health, and productivity (Commonwealth of Australia, 2017). As such, Australia's feral animals – as with its purported plagues of kangaroos – are considered pests that need active *management*, most often again through lethal means.

> The objective for managing the majority of established feral animals is to reduce the damage caused by pest species in the most cost-effective manner. This may involve localised eradication, periodic reduction of feral numbers, sustained reduction of feral numbers, removal of the most destructive individuals or exclusion of feral animals from an area. (Australian Government, n.d.-a)

Feral animals, then, comprise individuals and species that are unwanted. They are not native animals being managed *for their own good* – usually so that they do not exceed what is considered to be the carrying capacity (meaning the population of animals a specific habitat can support) of remnant habitat and then impinge on human interests, as is a standard narrative for kangaroos. Feral animals are rather species that are considered either extraneous or dangerous to native ecosystems and species. They are also typically framed as impinging on human interests (as a non-native animal introduced to Australia, rabbits have for instance been framed as damaging Australian agriculture). This narrative of causing harm dominates the Australian public sphere to the point that many conservation actions prioritize the exclusion and killing of these animals (La Canna, 2017). As with kangaroo culls, these are actions that largely ignore issues of animal sentience, and that align to a narrative held by many conservationists that animals are nothing more than 'functional units of ecological processes', and that neither the interests nor lives of individual animals typically matter (Lynn, 2015). According to this narrative, only people and/or ecosystems should be attributed value, with the result being that lethal measures, such as hunting, trapping, and poisoning, can be accepted as unproblematic tools for achieving land management and conservation goals, no matter the levels of cruelty involved (Lynn, 2015). This is a dominant story within Australia – and one that has been and still is used to justify what would be inhumane practices if used with reference to other (valued) species – but, as it will be seen, there are other stories too.

Brumbies

If kangaroos are caught in the crosshairs of conflicting narratives of value and consideration, then Australia's wild horses – colloquially known as *brumbies* – are in a similar position. Another Australian cultural icon, brumbies are evocative and powerful within the settler-colonial Australian imaginary – think Banjo Paterson's poem 'The Man from Snowy River' (first published in the *Bulletin* in April 1890) and the 1980s film of the same name, and Elyne Mitchell's *Silver Brumby* children's book series (from 1958). Horses were introduced to Australia in 1788, brought with the first British colonial fleet. As colonial settlements spread, horses both escaped and were released into the wild, becoming the brumbies that now range across Australia. Australia in fact has the largest wild horse herd in the world, maybe 400,000 or more horses, spread across nearly every region from the tropical north to the arid centre to alpine areas – most (in)famously in the Kosciuszko National Park (e.g. Figure 10; Adams, 2017). There – due in part to Paterson's mythologizing – brumbies have come to 'symbolise freedom and are a part of Australia's cultural identity', as put by one High Country cattleman and ardent brumby advocate (as cited in Burdon, 2016).

Figure 10. Brumbies in the Snowy Mountains, Kosciuszko National Park, Australia.
Source: Photo by Christine Mendoza on Unsplash.

The Kosciuszko brumbies have further been positioned in these narratives as an important 'part of the cultural fabric and folklore of the high country', and one that deserves protection, to use the words of a former Deputy NSW Premier John Barilaro (as cited in Lowrey, 2018). As described in a 2015 report prepared for the NSW National Parks and Wildlife Service:

> the Snowy River mythologies associated with the works of Paterson and Mitchell, and a plethora of other Australian poets and bush writers, inserts these wild horses into the wilderness as their 'rightful' place. In popular culture, the wild, majestic beauty of the horses and the wild, majestic beauty of the landscape were drawn together in a kind of natural association. (Context Pty Ltd, 2015, p. 29)[11]

At the same time, as an introduced species, narratives are also told about brumbies that position them as pest animals that are destructive of fragile ecosystems – particularly those of sub-alpine and alpine regions (Albeck-Ripka, 2020). According to this narrative, brumbies are described as (a) increasing soil erosion by killing vegetation, disturbing the soil and creating paths along frequently used routes; (b) destroying native plants through grazing and trampling; (c) fouling waterholes; and (d) causing the collapse of wildlife burrows. They also (e) compete with native animals for food and shelter; and (f) compete with livestock for pastures – particularly during periods of drought (NSW Government, 2019). Under this story, brumby populations should be significantly reduced or removed from these ecosystems, whether by culling (via aerial or ground shooting) or trapping and relocation. Such proposals are vehemently rejected by those who see brumbies as an important part of Australian imaginary, with such proponents describing them as 'an attack on Australia's heritage' (Albeck-Ripka, 2020). Appeals to animal welfare issues abound in both narratives, stressing respectively the inhumanity of their shooting and the welfare problems that the horses can experience *out there* during drought or from the after-effects of fire when, if not culled, they might starve or die of dehydration.

Although the debate around brumbies follows some similar lines to that raging about the kangaroo – with both species positioned simultaneously as pests and national icons – it is worth noting that in some respects brumbies are thought of more positively. More specifically, it seems that Australians find it easier to admire, relate to, and advocate for horses as opposed to kangaroos. Both species might live *out there*, but they are not similarly *visible* in our cultural imaginary. Kangaroos might be native, *proper* inhabitants of Australian ecosystems – even when *managed* as a conservation pest – but apart from Skippy they are not manifest in our stories of human-animal relations in the same way as horses are. While brumbies are described, individually and collectively, as majestic, awe-inspiring, fitting representations of wild Australia (and, indeed, of settler culture), 'noble emblems of the toughness and fighting spirit that characterises Australians', as well as outstanding and versatile riding horses when *tamed* (Burdon, 2016), kangaroos receive little such approbation beyond their being native. If public favour in the pest/icon debate regarding the status and consequent treatment of brumbies tends to be given to icon, this is still not the case for kangaroos – at least within Australia where the pest narrative continues to receive support despite its dubious footing. Two reasons for this are perhaps that horses, whether *tamed* or feral, are already considered important under an anthropocentric imaginary, and have been so considered for millennia, and horses – unlike kangaroos – are not consistently presented in the public sphere as a problem of plague proportions threatening Australia's economic future.

Cats

Feral cats, however, are consistent recipients of bad press in Australia. Another species introduced early in the period of colonial settlement, cats quickly formed a feral population, acclimating into Australia's ecosystems. They are thought to have come to inhabit most of continental Australia by 1890, if not earlier (Abbott, 2002; Flannery, 2006). The actual number of this population is unknown, with feral cat density and abundance considered to vary greatly across Australia due to the way their population fluctuates with prey availability and climate. In describing this population, however, media reports have previously used figures of 15 million, 15–23 million, and 20 million, with a figure of 18 million in use in some scientific and official sources, although it has also been shown that there is no identifiable basis for the number (ABC, 2014). More recent figures are significantly more conservative, with estimates of 2.1 million (up to 5.6 million after good rain) feral cats living *out there* in the bush, and another 0.7 million estimated to be living as strays in urban settings (Australian Government, 2019a). They are, however, consistently described as a pest species causing severe to catastrophic effects on Australia's native wildlife, and presented as a primary factor in the decline and extinction of a number of native mammal species in Australia (Woinarski et al., 2014). For instance, in one emotive statement by a past Environment Minister, Greg Hunt, feral cats were described as 'tsunamis of violence and death for Australia's native species' (as cited in Hepburn, 2015). Gregory Andrews, one of Australia's Threatened Species Commissioners (2014–17), has similarly described feral cats as the single biggest threat to Australia's native animals (as cited in Lauder, 2015).

This narrative has received a lot of public attention and support and has led to the launch of multiple efforts – federal, state-based, and local – to address the *feral cat problem* and thus save native species from their predation. Measures include the large-scale use of poison baits (usually laced with a poison that Australia's native wildlife have evolved some tolerance to), shooting, trapping and euthanizing, as well as the construction of exclusion fencing around wildlife reserves once feral cat populations have been eradicated (see, e.g., O'Brien, 2018). With some of Australia's First Nations People having a history of hunting cats for food, proposals have also been made that expert trackers and hunters could be deployed to other areas looking to control feral cat numbers (see, e.g., Hose & Deacon, 2018; Taylor, 2015). Vigilante *cat buster* groups have also arisen in both rural and urban areas in Australia, and these are variously shooting cats directly on sight or trapping cats to be taken to the pound for euthanasia. The vigilante argument is that because cat owners are legally required to keep their cats within their property boundaries at all times, any cats found off property or *out there* should be caught and killed (Eddie, 2016). Under this argument, the best cat is a dead cat. As has been noted about other pest animals whose populations and actions are considered *out of control out there*, the narrative here is that the interests of (some) native animal species and ecosystems should be consistently prioritized over the individual interests of cats. If cat welfare is considered at all in this conservation narrative, it is via calls for their *humane* destruction. There is, however, some amendment being made to this narrative of needing to go to 'war on feral cats' (Lynn, 2015). This concerns the recognition that while cats are indeed expert hunters, their presence and actions *out there* are not in fact the only or even necessarily the major drivers for Australia's high rates of faunal extinction (Flannery, 2006). Also stressed is that – given the near impossibility of removing feral cats completely from Australia's ecosystems – it is important to consider other measures of supporting native animal populations and ecosystems.

One key counter narrative to the 'war on feral cats' story is that this latter is once again a distraction from the more important issue of habitat loss and fragmentation (Doherty et al., 2019; Wintle & Bekessy, 2017).

For instance, while the Australian government's (2015) *Threatened Species Strategy* report mentions feral cats more than seventy times, the issue of habitat loss is only mentioned twice, and land clearing not at all. While the new *Threatened Species Strategy 2021–2031* (Commonwealth of Australia, 2021) does broaden its focus from individual species to also include 'priority places' and acknowledges the importance of responding to 'habitat loss, changed fire regime[s] and climate change', its aims will not be achievable without increased funding and significantly strengthened national environmental laws (Readfern, 2021). This is because, as we have already noted in the contexts of koalas and kangaroos, halting habitat loss is a politically sensitive topic given that land clearing is what makes human interests and activities in agriculture, urban development, and mining possible. Positioning feral cats as being almost single-handedly responsible for the loss of Australia's wildlife *out there* whilst ignoring the impacts of human development is thus another clear example of the strength and persistence of our anthropocentrism. This is ironic because it has also been found that if their habitat is not degraded – through human actions directly (land clearing) or indirectly through the activity of feral livestock (including brumbies) – then Australia's native animals do show the capacity to practice anti-predator behaviour and may even learn to live alongside feral cats (Ritchie et al., 2020; Stobo-Wilson et al., 2020; Wintle & Bekessy, 2017), as they once lived alongside dingoes and other native predators. It is also worth noting that it appears that dingoes and Tasmanian Devils can help keep feral cat populations in check through both direct predation and competition, and also because the presence of these animals can scare cats away from habitats in which they would usually hunt (Letnic & Feit, 2019; Lynn, 2015). Despite such points, our persistent anthropocentrism makes it easier to maintain narratives that reactively blame – and consequently *manage* – another species for Australia's faunal extinction rates than to proactively take action that would limit or alter our own interests and activities in changing Australian ecosystems. Once again, the lives and well-being of animals *out there* – whether they are native or feral – take second place to our own interests.

Dangerous

The final group of animals *out there* to be considered in this chapter are those we might typically class as dangerous. These would include all animals with the capacity to kill us (some Australian snakes, for instance), but our key focus here is with animals that we consider have the capacity to predate upon us: sharks and saltwater crocodiles, or *salties* as they are colloquially known. These are animals that may live *out there* but like brumbies they already have space in our cultural imaginary, although this time usually as animals to be feared rather than admired. This is arguably due primarily to their presentation in mainstream media and movies. The effects of such presentations – in movies like *Jaws* (1975) and its various sequels and *Deep Blue Sea* (1999) for sharks, and *Rogue* (2007) and *Black Water* (2007) for crocodiles – can linger, sparking anxiety even when we are not in any danger from either species. Such anxiety is further fed by media reports – often front page or breaking news – of attacks on humans by either species.

Sharks

Australia has the world's highest diversity of sharks and rays, with around 180 of the 509 known shark species inhabiting Australian waters (Australian Government, n.d.-b). Given, then, the longstanding Australian love of coastal recreation (for instance, beach swimming, underwater diving and surfing) and living – with some 85 per cent of Australians living within 50 kilometres of the coast (Chapman, 2017) – sharks and humans do interact, and sharks are manifest in several narratives circulating through the Australian public sphere and cultural imaginary. Historically, sharks have been perceived as a threat that should be exterminated, a view that has led to a lack of concern for their welfare (Winton, 2018). Although this view is changing, the threat of shark attack – and the associated fear of sharks – is a common narrative in Australia, and one that is given additional circulation through feature films, documentaries, YouTube videos and the often sensationalist media reporting of shark sightings and attacks (Heathcote, 2019; Smee, 2019). This is despite the fact that shark attacks on humans are comparatively very rare, with the odds of being killed by a shark in Australia described as 'one in 8 million […] the same likelihood of being killed by a kangaroo' (The Nature Conservancy Australia, n.d.). Nevertheless, this narrative of fear regarding shark attack continues to drive – and is driven by – the range of actions typically taken by Australian authorities to try to protect beachgoers against such attack. These take two forms – either (a) directly targeting sharks to remove or reduce the threat of shark attack or (b) providing warnings to beachgoers of the presence of sharks. Actions targeting sharks have included large-scale culling programs in the past, but now typically include the use of exclusion or shark capture netting around some beaches, the killing of sharks following some shark bite incidents, and the use of both lethal and non-lethal drum lines to capture sharks for either killing or relocation. Actions to warn beachgoers of shark presence take the form of beach and aerial surveillance connected with various models of warning system. Beachgoers can also download mobile apps that allow them to track shark activity off Australia's coastlines (see, e.g., SharkSmart). While warning systems are passive with regards to their effects on sharks, actions targeting sharks directly are often lethal to them, and also kill other species including sea turtles, dolphins, and seals (The Nature Conservancy Australia, n.d.).

Despite an ongoing assumption of the need for Australian governments to continue to *manage* sharks in support of human recreational interests – an assumption given new life by every story of shark attack and associated call for the government *to do something* (Smee, 2019) – a narrative oriented to shark welfare is also starting to arise. Research shows, for instance, that even in the immediate wake of shark bite incidents, there is now little public support in Australia for lethal policy responses (Pepin-Neff & Wynter, 2018a, 2018b). Shark bite incidents are further being described as 'tragic accidents' and 'as natural, ungovernable events' (Pepin-Neff & Wynter, 2018b, p. 6; see also Parkes-Hupton, 2022), rather than being framed as the actions of monster or rogue sharks (à la *Jaws*) that need to be killed for the protection of beachgoers. This kind of counter narrative regarding shark attack is thought to be informed by a better understanding of shark behaviour (Pepin-Neff & Wynter, 2018a). This would include the recognition that not only do the vast majority of shark species – and the vast majority of human-shark interactions – present little or no threat to humans, but that there are lots of strategies that can minimize the risk of beachgoers having a negative encounter with a shark (Chapman, 2017). There are other issues that may also be informing this counter narrative. These may include a growing understanding of the precarious status of many shark species in our oceans, with approximately 25 per cent of shark species being classified as at risk (meaning vulnerable, endangered or critically endangered), and increasing awareness concerning the importance of sharks in healthy marine ecosystems. The unsustainability – and sheer inhumanity and cruelty – of several of the commercial fishing practices targeting sharks globally (e.g. for the shark fin industry) is also becoming more broadly known. Such points, for instance, are included in several public awareness campaigns circulating within the Australian public sphere. These have been instigated and sponsored by Sea Shepherd, the Australian Marine Conservation Society (AMCS), Shark Conservation Australia, the Humane Society International, as well as by local activist groups. These challenges to the perception that sharks must continue to be *managed* in the interests of human safety stress that despite their potential danger as an apex predator – and despite our continued anxiety with respect to the prospect of shark attack – sharks nonetheless deserve respect and protection at both the individual and species levels. They suggest that there is another narrative possible – one of humans and sharks cohabiting – and that this is possible given a stronger focus on public (and tourist) education.

Salties

This mixture of anxiety and respect, and the idea that humans can come to successfully cohabit with an apex predator, is also set out in some of the narratives starring Australia's saltwater crocodiles, or *salties*. Like sharks, salties are apex predators able to predate upon humans; unlike sharks, however, which can be found in all of Australia's coastal waters, salties are restricted to the northern regions of Australia, Australia's *Top End*. This means that while salties may be perceived as living *out there* by the inhabitants of the regions of southern Australia, and as thereby mostly out of both sight and mind, Top End Australians have developed and live very different relationships with them. There salties have been positioned as a valuable commercial resource, generating wealth and employment which in turn promotes their conservation. Indeed, salties have come to define the Top End, being seen as 'integral to the cultural identity and [...] as much a source of pride as a source of anxiety' (Baker, 2016). Put another way, the Top End can be described as a test case for a model of human-animal cohabitation insofar as the (human) inhabitants of the Top End have 'done the unthinkable in modern times – they've ceded their coastal wilderness to the wild. Wild that can eat you' (Baker, 2016). (Of course, it must be noted that this description of Australia's Top End human population *being prepared to live alongside salties* is strongly flavoured by post-European occupation and settlement assumptions regarding the *proper* management of nonhuman animals, particularly those identified as dangerous – whether to human life or human interests. Australia's First Nations Peoples have lived alongside Australia's nonhuman animals and seen them as kin in Country for millennia, and continue to do so.) Under this Top End commercialized conservation narrative and associated way of life, salties have rebounded from their near extinction (due to overhunting and culling), and now arguably outnumber humans in their population. Such a way of life has also enabled some unconventional ideas and practices, with salties also considered possible, albeit dangerous, pets in this region (Thompson, 2019).

Although this could be read as an altruistic story focused on supporting another species' interests and intrinsic value, this would be to understate the economic incentives that have driven this deal. The end result might be salties flourishing in Top End Australia, but this has been through the development of a multifaceted and far-reaching industry that requires their conservation. Under the aegis of this industry, (a) people collect and sell crocodile eggs to local crocodile farms; (b) landowners receive a royalty from egg collection on their property (compensating them for livestock lost to crocodiles and motivating them to retain habitat); (c) farms raise the hatchlings, sell their skins and meat, and play a role in the Top End's tourist industry; (d) rangers manage crocodiles in public areas; and (e) independent permit holders make a living collecting animals from the wild, either for a farm's breeding program or for trade (Baker, 2016). Because the mortality of crocodile eggs in the wild can be very high (up to 75 to 80 per cent), and hatchling survival is also density dependent, egg collection is not considered to have a significant impact on the wild population (Taplin, 2017). Further, it has been argued that the economic value of the crocodile egg harvest also results in broader environmental gains due to improved management practices for weeds, feral animals and fire by landowners in their attempts to preserve crocodile nesting habitat (Saalfeld et al., 2014). Human and saltie populations can thus both thrive according to this economic narrative, even as some of their forms of interaction as individuals continue to be harmful or even lethal to the other. A focus on economic interest, in other words, has been able to convert the status of salties from being simply considered a threat (albeit with some economic value) – and thus as needing to be eradicated – to being highly valued and thus conserved, at least at a species level. This kind of anthropocentric narrative of value with regards to animal populations is not unique to relations with salties – it has also been proposed, as noted earlier, as one way of enabling better protections and welfare for kangaroos. Such a view sums up the strongly anthropocentric point that 'it's not realistic to expect a community to conserve an animal out of appreciation for its intrinsic value alone' (Baker, 2016). It is also suggestive that assigning an economic value – to salties in this instance – has the capacity to convert not only the standard abstract benevolence of many Australians for the animals living *out there* but also a concrete anxiety regarding these same animals into a vested, if cautious, interest in their welfare, and a clearer understanding about the requirements of living alongside them.

Conclusion

What this chapter has shown is that while many Australians arguably might care about the continent's wildlife, this does not easily convert into action that supports animal interests. Indeed, our relations with the animals living *out there* – whether native, feral, or dangerous – are highly complex, showing multiple points of disjunction and conflict in our narratives of them and our relations with them. For instance, native animals might matter to us in the abstract and sometimes also in concrete ways as with the fate of koalas and other native animals from the Black Summer fires, but this concern rarely converts into ongoing support of their populations and wellbeing – particularly if such action could be considered to result in a constraining of human interests. Thus the presentation of kangaroos as a plague for agriculture and needing lethal management. Thus the longstanding disinclination to protect koala habitat – or any native animal habitat – in the face of human interests in development despite increasing species declines and extinctions. Thus ongoing calls to kill those animals might harm human interests. And thus the ongoing blaming of feral cats for the population collapses of native animals whilst refusing to consider the very real consequences of habitat loss and fragmentation on these same animal populations due to our own land-clearing practices and preferences. And thus instances of animal and population protection when such practices do align strongly with human interests – the protection of brumbies in the light of a particular understanding of Australian identity and culture, and the protection of salties in the light of their economic value.

These issues suggest two main things. The first is that while Australians may express a general concern for the welfare of native animals, this is rarely articulated into systemic action supportive of native animal populations and ecosystems. Rather, these animals, their populations, and their native ecosystems are all continually ranked and re-ranked in accordance with their perceived value under anthropocentric narratives. Such a focus is exemplified by the weaknesses of the *Environmental Protection and Biodiversity Conservation ACT 1999* (EPBC Act) and the lack of other legal protections for native animals in the face of not just human interests in economic development but climate change. Our cultural imagination regarding the perceived value of some animals can further work to trump or protect animal interests, regardless of whether they are native or feral and of how they fit into existing ecosystems. The second point is that there is clear evidence that human-animal narratives and relations change as human interests change. This foregrounds the possibility of developing ways of life and living and models of human and animal cohabitation that might not have been conceivable under previously dominant narratives. Such changes can be more or less supportive of the interests and lives of animals that live in close relationships with us as well as *out there*. The question concerns how a majority of Australians might learn to care and to act so as to live in ways that support human-animal relations and cohabitation, even with these animals that do not fit into easy food or friend categorizations and that are not part of our everyday lives.

Notes

1   As we have noted, from the point of European settler occupation nearly all animals in Australia have been seen through a strongly anthropocentric and instrumental lens. This has led to the normalizing of attitudes that an animal's value is determined by its economic value as a resource. We have considered this issue already in the context of livestock animals (see Chapters 2 and 3), but it also applies to so-called working animals (such as the dogs represented in *Send in the Dogs Australia*, or the camels in *The Camel Boy*). For these animals, if an animal's labour is useful and provides an economic benefit, then this value will protect its welfare and well-being to at least some degree. Once an animal can no longer labour its value decreases, however, and its labour value may be overridden by its value at slaughter and its conversion into animal products. Such attitudes align with the broad point that 'a body that cannot produce is a body that is not valued' (Wrenn, 2017, p. 220), and in the Australian context an animal's loss of economic value has sadly still been shown – often through activist covert surveillance and whistleblowing – to leave it open to a complete lack of respect, even for its sentience. For example, activist surveillance of abattoirs suggests that Racing Australia's claims about less than 1 per cent of retiring or unwanted horses being sent to slaughter are a gross

Neither Food nor Friend 193

    underestimation (see, e.g., Dole, 2019; McGowan, 2019). Inquiries into greyhound racing in Australia have also shown ongoing industry acceptance of high rates of euthanasia for *unwanted* animals (see, e.g., Knaus, 2017; McEwan, 2020). While this narrative of framing animals in terms of their instrumental value still holds traction within Australia – particularly within industries that rely on animal bodies and labour – it is starting to be challenged, and public outcry over reports of so-called animal *wastage* is increasing.

2   Australia is regularly cited as one of the most highly urbanized countries in the world, even if its cities are characteristically less dense than those in other countries (Spencer et al., 2015). What this means is that, as of 2020, over 86 per cent of Australia's human population are recorded as living in urban and suburban settings (Trading Economics, 2021). In addition, most people who migrate to Australia are recorded as moving to major cities, with suburbs in Melbourne and Sydney being particularly popular for persons with non-English speaking backgrounds (IbisWorld, 2021). Conversely, 62 per cent of Australia's First Nations Peoples – who represent just over 3 per cent of Australia's total human population – live outside Australia's major cities, including 12 per cent in areas classified as very remote (Gately & Kendall, 2020). As we have noted, these demographics impact the levels of Australians' direct interactions with different groups of nonhuman animals, whether companion, livestock, or free living (including both native and feral animals).

3   It is also important to note here that in some Australian states and territories wildlife carers need to be licensed and registered. Standard advice given in all states is that any injured wildlife found should be turned over to licensed carers or vets or that these bodies should be alerted as to the animal's status and location.

4   One example of this kind of direct catastrophic impact on a species concerns green sea turtles. Green sea turtles are in grave danger because the animals hatching in the northern areas of the Great Barrier Reef – and in other hatchery areas around the world – are increasingly female due to the warming of the ocean. (In the northern Great Barrier Reef, research is already revealing that 99 per cent of hatchlings are female.) Rising land and sea temperatures are also predicted to mean higher egg and hatchling mortality. The complete 'feminization' of the population may occur in the very near future with potentially disastrous consequences for the species (Jensen et al., 2018).

5   Mountain pygmy possums are already responding to climate change by waking earlier from hibernation, meaning that they are facing food shortages due to their emergence before that of bogong moths, a key post-winter food source for them, and a species also in decline from climate change (Lee et al., 2015).

6   The Gay'Wu Group of Women is the *dilly bag women's group*, a deep collaboration between five Yolngu women and three non-Aboriginal women over a decade. The group members are Merrkiyawuy Ganambarr-Stubbs, Banapuy Ganambarr,

Djawundil Maymuru, Laklak Burarrwanga, Ritjilili Ganambarr, Sarah Wright, Kate Lloyd and Sandie Suchet-Pearson.

7   Our point here relates to what has been called humanity's *moral circle*, meaning the imaginary boundary we humans draw around those we consider worthy of moral consideration (see, e.g., Singer, 1981/2011). Introduced by historian William Lecky in the 1860s and popularized by philosopher Peter Singer in the 1980s, the idea of the moral circle is a fundamental concept among philosophers, psychologists, activists, and others who think seriously about what motivates people to do good. The basic idea of the moral circle is that moral consideration – and an acceptance of responsibility – is easy for those we have immediate contact with and care for, for instance one's family, friends, followed by in-group members. As contact and care levels are diffused, however, moral consideration and the acceptance of responsibility typically diminishes. Importantly, the breadth of those included in one's moral circle can change – a fundamental idea for theories of moral progress. As Singer reminds us, 'The circle of altruism has broadened from the family and tribe to the nation and race, and we are beginning to recognize that our obligations extend to all human beings. The process should not stop there. [...] The only justifiable stopping place for the expansion of altruism is the point at which all whose welfare can be affected by our actions are included within the circle of altruism' (1981/2011, p. 120). Our point is that one's moral circle is expanded through coming to care for others, and this is most typically triggered by the expansion of one's knowledge and understanding and the development of a sense of connectedness to others. Such expansion in turn typically rests on whether others possess characteristics we recognize and value, on cognitive bias and other motivational influences, as well as on some individual characteristics (see Crimston et al., 2018).

8   One controversial proposal regarding this problem suggests that Australians should be allowed to keep native Australian animals as companion animals (something only allowed, as of 2021, in South Australia and Victoria). As the author of this proposal, a paleontologist at the University of New South Wales, Mike Archer, explains with reference to this proposal, 'Any animal we put our arm around and look after is guaranteed a future' (as cited in Wynne, 2021). As Archer also explains, 'Unless we set up breeding colonies and enable people to have a pet quoll instead of just a cat, we're not going to get that that bonding that we need in the next generation, to care about the conservation of our Australian mammals' (as cited in Wynne, 2021). 'These are animals that are ready to cross the line and become your best friend. We've never given them this chance. We've said the only animals that are suitable for being companions with us are alien invading animals that we brought in from Europe. And this is crazy' (as cited in Wynne, 2021). Such a proposal aligns with some of the stories discussed in Chapter 4.

9   Such stories align with some of the understandings of kinship and responsibility that inform the relations to Country seen as foundational by Australia's First Nations Peoples.
10 Here from the perspective of frame analysis (Goffman, 1974), framing kangaroos as pests by settler-culture agriculturalists presents a certain framework as an organizing theme to make sense of the massive culling of kangaroos for urban dwellers. Thus framing here is a cognitive short cut for framing of content in mainstream and social media to lead readers/audiences to the perception and preferred conclusion that kangaroos – at other times framed as Australian national icons – are really pests.
11 It is interesting that both in the poem and the film *The Man from Snowy River* (1984), the brumby, as an historical icon, has alignment with a valued model of settler masculinity. For instance, the Australian settler male's desire to be a rugged individual who freely charts his own destiny is evident in the analogy made between the Australian brumbies' wild nature and the Australian settler male's passion for autonomy.

CHAPTER 6

Stewardship, Sustainability, and Protecting the Great Barrier Reef

Although Australians generally express a care for animals and many align themselves with calls for their better protection, it is very clear at this stage in the book that there are a range of inconsistencies in the articulation of this care. One important issue is clearly illustrated in Chapter 5 and refers to the way different animals have been attributed different levels of value, usually on the basis of their apparent usefulness to us. Certainly, it is clear from previous chapters that such views – typical of anthropocentric beliefs – are being challenged via recognition of animal sentience and our own capacities to see (some) animals as companions or kin. However, as these chapters also show, converting this recognition into change with regards to the actual treatment of animals continues to be difficult. Key problems here are the continued foregrounding of human interests in development and profit over animal interests and wellbeing to the point that Australia has been described by the World Wide Fund for Nature (WWF) as 'marching [our native] mammal species towards extinction faster than any other nation' (WWF, 2021). Such analyses clearly mark our failure to accept responsibilities towards animals that we are not obviously involved with. This is the point that while there is an increased awareness as to our care and responsibilities towards those animals that contribute to our lives – companion as well as livestock animals, for example – it still appears difficult to extend such consideration to the animals that live *out there*, outside of the remit of our everyday lives.

This difficulty is compounded by two factors. First, as made visible in Chapter 5, while a stress on sentience is effective for promoting responses to events of animal suffering, such as from the Black Summer fires, it is much less effective in prompting the protection of animal wellbeing via the protection of animal habitat and ecosystems. This is particularly the case for habitat and ecosystems that come into conflict with standard economic visions of profit via industry or development. The second point is also to do with perceived limits to our ideas of responsibility but in this case ties into longstanding assumptions that animals living in wild ecosystems – such as Australia's native animals – should not have their lives and communities interfered with unless it concerns the restoration of the natural balance of the ecosystem, or unless it is to redress the impact of direct human activity (Petty, 2019; Wilkinson, 2016). This is a common view that wild animals and places should be allowed to live and develop on their own with as little interference from humans as possible (Gamborg et al., 2012). Although this view certainly stresses the need to respect the integrity of wild animals and their ecosystems, it can also overlook the stresses that anthropocentric activity is placing even on seemingly wild and natural ecosystems. Put together these various issues illustrate a discrepancy between (a) clear increases in levels of care regarding animal sentience in Australia and (b) the ongoing systemic degradation of biodiversity and loss of animals via anthropocentric development practices. As is becoming apparent, the achievements of the first – despite their significance – do not seem able to alter the entrenchment of the second.

These issues come to the forefront in this chapter. In particular, the aim here is to consider what drivers and models of responsibility might bring us not simply to respond to explicit animal suffering but also to consider and actively work to enable animal wellbeing through the protection of habitat and broader ecosystems. As such this chapter marks a change in focus from mapping the course of the narrative of sentience towards reshaping human-animal relations to exploring ideas and courses of action towards an extension of responsibility with regards to animal lives and wellbeing. Such an extension of responsibility, after all, would enable (a) the protection of not just animals (or animal sentience) but also their habitat and broader ecosystems, and (b) the achievement of a more equal balance of human with animal interests. Such a focus in turn stands in contrast to both overriding animal interests (typically the case under anthropocentrism) and bringing about more attention to (some) animal interests through increasing our recognition of animal sentience. This latter issue – explored through the previous chapters – is certainly fundamental for challenging the assumptions and standard practices of anthropocentrism. However, drawing out some of the methods that would enable Australians to value and take a wider ranging consideration of animal interests – including with regards to the protection of animal habitat – will be key for fully taking up the challenge of the Anthropocene. Given this focus, this chapter traces several ideas which could help us extend our ideas of responsibility and care beyond the scope of sentience so as to include broader issues of habitat and ecosystem. These are the ideas of environmental stewardship and sustainability. These are both relational ideas with the capacity to remind us that our interests and lives are always interlinked with those of animals; and they can help us develop ideas for living alongside animals in ways that exceed anthropocentric conventions. Also considered in this chapter will be Australian First Nations' ideas of land and sea management, also known as *caring for Country* (also see Chapter 4).

As with other chapters, we also consider how these ideas might work in context: in this instance through examining the status and protection of the Great Barrier Reef (GBR), one of Australia's cultural and natural heritage icons and an exemplary ecological community (or, more accurately, multiple interlinked communities). However, before beginning our discussion of post-settlement Australians' various relations with and attitudes to the GBR and of how these illustrate some understandings of how ideas of environmental stewardship and sustainability can function to extend our care practices, it is important to first outline several issues. These include what we mean by these ideas, why they are important, and how they might inspire broader responses from us than do our considerations of the lived experiences of individual animals or even of specific species or cohorts of animals. This discussion comprises the next two sections of this chapter. It will be followed by a detailed examination of our relations with and attitudes to the GBR with reference to these ideas. The chapter concludes with an exploration of how considering the needs of not just animals but their habitats via reference to ideas of environmental stewardship and sustainability can help us identify a new balance between human and animal interests – a balance potentially showing some alignment with Australian First Nations' ideas of *caring for Country*. As we will show, this has the capacity to open some different trajectories for the thinking and living of human-animal relations. These possibilities will be considered in more detail in Chapter 7 with reference to the development of practices supportive of multispecies cohabitation and community.

Care Beyond Sentience: Protecting Animals and Habitat

Animals in the wild live within the ecosystems that comprise their habitats. Without appropriate habitat, animals in the wild are not able to live. These seem obvious points to make, but it is clear from the previous chapter that while there is growing recognition in Australia of the need to better protect native animal species like koalas, for example, there is substantially less commitment to protecting the habitat these and other species need to live and sustain their populations. This lack of commitment can be traced back to two main factors: ongoing and aggressive commitments to land clearing and development (Kilvert, 2020a), and the assumption that wild living animals will be able to relocate themselves away from degraded or destroyed habitat. First, as was set out in Chapter 5, a key issue underpinning both animal and habitat loss in Australia is *land clearing*, typically to enable agriculture, plantations, mining or urban development. A distinctly Australian term, *land clearing* most often involves attaching a heavy chain between two bulldozers and dragging it across the land; it is commonly followed by burning (Barham et al., 2018). Although typically justified with reference to development needs, land clearing has in fact been identified as causing long-term costs to farmers, governments and society insofar as it hastens erosion and reduces soil fertility, increases the risk of soils becoming saline, exacerbates drought and also reduces the numbers of native pollinators and many wildlife species (such as woodland birds and insectivorous bats) that control agricultural pests (McAlpine, 2009; Nulsen, 2012; Watson et al., 2018). Land clearing has also been identified as having contributed to a similar number of extinctions in Australia (sixty-two species) as introduced animals such as feral cats (sixty-four) (Ritchie et al., 2021). These issues have led to over 300 Australian scientists calling upon Australian governments at all levels 'to return to, or pass new, effective legislation, supported by sound regulations, to protect native vegetation from broad-scale land-clearing' (Ecological Society of Australia, 2020).

Regarding the extent of land clearing, it has been shown that just 50 per cent of Australia's forests and bushlands remain intact compared with pre-European settlement; the other 50 per cent were either permanently destroyed and replaced with another land use, or classed as degraded or fragmented (Wilderness Society, 2021). In the case of one Australian state, for example, assessments have shown that while 61 per cent of New South Wales still has native vegetation cover, only 9 per cent of that is considered to be in close to natural condition. The remaining 52 per cent of native vegetation is already degraded by land use, principally grazing, while the last 39 per cent of the state has been cleared or converted to intensive land use (Barham et al., 2018). Indeed, Australia's rates of land clearing have led to Australia being the only nation in the developed world included in WWF's global list of deforestation hotspots (Cox, 2021). Land clearing has furthermore been identified since 2001 as a Key Threatening Process under the Federal Government's *Environmental Protection and Biodiversity Conservation ACT 1999* (EPBC Act) (as cited in Lindenmayer & Burgman, 2005). Unfortunately, as Chapter 5 indicated, the EPBC Act has proven ineffective at protecting either threatened species or ecological communities in the face of land clearing, and indeed the vegetation communities that underpin habitat generally receive no federal protection until they have already been extensively cleared (Tulloch et al., 2016). Thus, while there is increasing awareness of the biodiversity losses being sustained from land clearing due to extensive reportage and campaigning by both scientists and activist organizations (including the Wilderness Society and the World Wide Fund for Nature) as well as by local groups (such as the West Australian Forest Alliance) striving to protect local forests, there continues to be a gap between this awareness and effective action. This gap arises due to the ongoing anthropocentric determination of land as resource, meaning that native vegetation communities and habitat continue to be easily overridden by interests in development and economic profit.

Another factor contributing to Australia's ongoing commitment to land-clearing practices is the priority given in many of Australia's conservation policies to what have been called *biodiversity offsetting schemes*. These are schemes in which developers are required to compensate for anticipated habitat loss by improving or establishing it elsewhere (e.g. such requirements are embedded in the *NSW Biodiversity Conservation Act 2016*). Unfortunately, however, when a vital part of habitat is cleared, it can clearly no longer support the local wildlife, and protecting or replacing habitat somewhere else will not save those specific populations (Hosking, 2020). As described by Deborah Tabart, the chief executive of the Australian Koala Foundation, despite their prevalence, offsetting schemes do not deliver effective conservation measures:

> Think about it from the point of view of a koala [...] Your tree has been chopped down. What are offsets going to do for you? You have a pretty good chance of getting hit by a car or mauled by a dog as you are pushed out of your home. If you can manage to survive the initial eviction, you still have to figure out where you are supposed to go to find this new offset area, this promised land. And then, once you have managed to get to this reservation, it's still going to be years before the new trees are of any use to you, and even longer before the ecosystem is completely restored, if ever. I mean, who thought this up anyway? It is ridiculous. (as cited in Milman, 2014)

Given their total inability to protect specific animal populations, offset schemes can at best protect generic ideas of animal populations and their habitat. The point here is that offsets do not actually require a *like-for-like* replacement of suitable habitat, and in some cases development projects that have not identified suitable offsets have still been approved on condition that the company would 'look for an offset down the track' (Kilvert, 2020b).

A further problem lies in the *hands-off* view that animals living *out there* in wild ecosystems should generally be left to their own devices and not interfered with (Gamborg et al., 2012). While such a view is certainly important for maintaining the integrity of wild ecosystems, it also makes it easy to assume that wild living animals will be able to relocate themselves away from degraded or destroyed habitat. This, however, is a highly problematic assumption for several reasons. Firstly, land clearing itself kills animals. They may be killed or injured when native vegetation is cut and removed or soil and debris are shifted (see, e.g., Finn, 2017; Readfearn, 2020). Those that survive the clearing process will be left in an environment that is typically unfamiliar, hostile, or unsuitable. Because many of our native animals are territorial or have small home ranges, they can have strong associations with small areas of habitat and can end up starving when their old ranges now lack appropriate resources (Finn, 2017). Secondly, the fragmentation and reduction of habitat and the presence of barriers such as roads and fences make it extremely difficult for animals to successfully relocate when their current habitat is encroached upon (through development) or becomes unable to support them (due to overcrowding or changed conditions due to climate change or other disasters like the Black Summer fires). For instance, in the case of koalas, the destruction and fragmentation of their habitat means koalas must spend more time on the ground moving from tree to tree. This makes them, as noted above, much more vulnerable to being hit by cars and attacked by dogs, while elevated levels of stress make them prone to sickness and disease (WWF, 2018e).

Calls for animals to be forcibly relocated are also problematic, however, even given such calls can be attempts to ensure that wildlife is not simply killed to make way for development. Thus, in several instances in the outer suburbs of Perth, Western Australia, developers have been pressured to relocate rather than kill kangaroo populations deemed to be in the way of development plans (see, for instance, Barry, 2019; Predovnik, 2015; Swan Valley Wildlife & Environment Advocacy Group, 2021). The difficulty is knowing where animals should be relocated to, given that remnant habitat is typically already inhabited, as well as the fact that relocation processes are highly stressful to wild animals – particularly for large animals like kangaroos (Germano et al., 2015; Stuparyk et al., 2018). Indeed, studies of one large mob of western grey kangaroos (initially comprising 154 individuals, fifty-two with one pouch young) who were forcibly relocated in 2019 to make way for urban development in the outskirts of Perth show that over 80 per cent of these animals died in the first month following their relocation – primarily due to the stresses associated with their capture and relocation (Cowan et al., 2020). It is also worth noting that of this mob, only 122 kangaroos actually survived their initial capture (Cowan et al., 2020). In addition, the kangaroos that survived capture were relocated to a habitat of Jarrah and Wandoo forest that was very different from the open grassy understorey with scattered tuart trees they were used to, meaning that they needed to navigate different experiences from those to which they were accustomed (Cowan et al., 2020). Even if sufficient resources are present in the new habitat to support an increase in population, animals new to that habitat may not be able to access them safely (Stamps & Swaisgood, 2007). It thus appears evident that the stresses of trying to learn a different environment with different resources and relationships also contributed to the high death rates of the relocated animals.

What these points make clear is that the relations between animal populations and their habitat can be neither simply re-made nor replaced. Animals displaced from their habitat through development projects and who are not relocated extremely carefully will most likely die, one way or the other. The additional point here is that habitats are always shared, with different species interacting with and depending on each other – for food or shelter, for instance – as well as being in complex interactions with water sources and other natural features. That is, habitats stand for multispecies *ecological communities*, meaning the group of species that co-exist in a specific type of habitat and habitually interact with each other. More than that, ecological communities are understood to have a natural composition and function which, if upset, can place that community as a whole at risk of extinction (Jackson et al., 2016). Envisaging ecological communities, then, is a cogent reminder that animals *out there* do not live in isolation or in a vacuum, and that acting with reference to any particular group of animals should also consider the impacts of such action on habitat and relevant ecological communities too. Put another way, species conservation efforts will ultimately fail if they are not accompanied by substantial efforts to protect habitat and ecological communities (Collard et al., 2019). Unfortunately, even when the importance of considering habitat and ecological communities along with specific species of animals is understood, this knowledge may still be ignored in the face of our desires to favour and prioritize development interests (see, e.g., Figure 11).

Stewardship, Sustainability, and Protecting the Reef

Figure 11. Land clearing in north Queensland. Queensland remains a deforestation front in Australia with most clearing in the state linked to the cattle industry. *Source:* Photo courtesy of Bill Laurance.

A key point here is of course the lack of effective legislation that can challenge economic development imperatives. As has been noted in the previous chapter, existing federal and state laws that are supposed to prevent native animal species and populations from being injured, or their overall presence endangered through loss of habitat, are extremely weak, as well as poorly implemented and regulated (Keck, 2020). As described by the Australian Koala Foundation (2021),

> The Australian Koala Foundation is frequently contacted by members of the community concerned about developments taking place in their local area. And, like those that contact us, we want to see the trees and the Koala protected. But the sad

truth is that the laws of this country do not protect Koalas. The AKF cannot help stop a development that will have a significant impact on Koalas because there is currently no law that will let us help.

This is the point that while there is recognition that land clearing can cause such harms as the loss, fragmentation, or degradation of habitat, there are no regulations in Australia that evaluate how land clearing harms animals. That is, no Australian state or territory has developed a clear framework to evaluate this harm, let alone minimize it in future development proposals (Finn, 2017). This lack of legislation means that attempts to protect animals, habitat, and ecological communities – even those already identified to be under threat – typically face an uphill battle against policies and practices that single-mindedly promote development (see, e.g., Rendall, 2020). This situation has been worsened by the financial fallout from the coronavirus pandemic which has increased angst about economic growth. On the one hand, then, public awareness is increasing as to the devastating impacts of the Black Summer (and other) bush fires on Australia's wild living animals and ecological communities, as well as the extent of species and biodiversity decline in Australia. This, in turn, is driving recognition of the need for substantially stronger conservation practices (Commonwealth of Australia, 2019; WWF, 2018a, 2018b). On the other hand, despite extensive reports about plummeting koala populations and the need to rein in land development, authorities in New South Wales are continuing to fast-track new land development projects so as to boost the economy (Callari, 2020; Howell & Witt, 2020).

What stands out across these various points is that calls to protect animals, habitat and their ecological communities can fall into a conflictual pattern in which (a) development and desires for economic growth and profit can only stand in contrast to (b) the conservation and protection of our native wildlife and habitat. Such binarized thinking is clearly illustrated in comments appended to a public petition to save wildlife started by the Swan Valley Wildlife & Environment Advocacy Group (2021):

> There are too many greedy land developers who have no consideration for the wildlife affected by their actions. We already have enough urban sprawl & it needs to

> stop. Councils & Govt should learn how to say NO to these developers & think of the environment before thinking of $$$$$$.
>
> More and more habitat removal, just where are Australia's icons supposed to go. Your tourist industry is built on the backs of these iconic animals yet the disregard shown for their existence by Government is astounding. Profit over the welfare of native Flora and Fauna. Yet more laws being introduced to make it easier for habitat to be used for profit by big businesses that don't care about locals [sic] opinions or how it negatively impacts lives of animals and humans.
>
> It breaks my heart when I visit Henley Brook to visit family to see dead kangaroos struck by vehicles and kangaroos with reduced habitat. It is a disgrace how areas are completely razed and wildlife corridors and parklands are not provided and part of the overall design. Shame on the developers and council for allowing blatant habitat loss just to create a sea of houses on ridiculously small blocks. There has to be a better way for all.

Establishing this 'better way' is, however, the challenge. As we have noted across the preceding chapters, this way starts from giving greater consideration to the interests of animals. Nevertheless, while such consideration can be broadly agreed upon in the face of the immediate and visible suffering of animals, it appears to be more difficult to extend consideration of animals to include the protection of habitat and ecological communities. This is perhaps because such an extension marks the site of an uncomfortable insight – that we are all, each and every one of us, the beneficiaries of anthropocentric development. Our homes and suburbs are all native animal graveyards, as are all development sites. To change this and to manage our suburbs so that they might allow native animals to thrive again will mean that we need to implement a range of changes with regards to not just how development is carried out but how we live in our homes. As was noted in Chapter 3 with regards to the problem of the *meat paradox* (Bastion & Loughnan, 2017), however, simply learning that there are reasons for a change in behaviour does not necessarily make that change happen. That is, while the assumption that changes in knowledge automatically produce changes in behaviour is a basis for much activist and advocacy work, it is never failsafe. Indeed, it has been found that in situations with polarized viewpoints – such as is often framed between development and animal advocacy – awareness and knowledge campaigns can sometimes lead to denial and the further entrenchment of previously held views. This is particularly the case if these views relate to one's sense of identity and custom. More specifically, people can become motivated to ignore evidence that goes against a view that they want to hold or find comfortable. This has been dubbed the 'backfire effect' (Zhou, 2016, p. 802).

What this means is that maintaining strict differences between views and positions, seeing them as polarized and thus requiring people to shift from one view to the other, can in fact stymie mobilization for social change. Indeed, such a divide can lead to suggestions that in their attempts to campaign on behalf of the interests of animals, animal advocates are unreasonably and unjustifiably anthropomorphizing animals and trying to overturn *natural* differences between humans and animals (Flint, 2013; Munro, 2004). Here the accusation is that animal advocates are trying to overturn the *natural* order and push others to believe that animal interests could matter as much as or more than human interests. Such views are clearly present in the following comment which was posted anonymously in response to one animal advocate's essay (see Bekoff, 2011) critiquing the fur industry for its record on animal cruelty:

> I'm tired of the endless anthropomorphizing of animals in our society. I honestly don't care if people wear fur or if animals suffered so those garments can be made.
>
> The product is useful and valuable to humans, so humans will continue to want/make these products.
>
> Animals suffer all the time at the hands of people, that's because they're lower on the food chain. Almost everything you own or do has harmed an animal in some way either directly or indirectly. I'm not just talking about pork chops. Habitat encroachment via urbanization, deforestation, collateral damage from natural resource utilization, pollution, all of it.
>
> Animals are food not friends. (Anonymous, 2011)

While such responses to animal advocacy and activist campaigns are also visible in the Australian public sphere (see, e.g., Rodan & Mummery, 2019), our interest here is to explore modes of consideration of animal interests that do not fall into this kind of binary thinking that opposes human interests to animal protection. That is, rather than assuming that better protecting animal interests automatically requires overriding human interests, we consider here what it might look like to consider both human and animal interests, albeit giving a higher weighting to animal interests than has been typical of standard anthropocentric thinking. To explore this possibility of protecting not just human interests but also embodied animal interests and animal habitat, we suggest that the introduction of ideas of environmental stewardship and sustainability can be productive.

Environmental Stewardship and Sustainability

With its roots in cultural traditions and religions worldwide, and most broadly meaning our acts of caring for what we value (Beavis, 1994; Berry, 2006; Palmer, 2006), the idea of being a *steward* of nature has been understood to mean making good use of, but also valuing and taking care of nature. *Stewardship*, in this context, thus broadly refers to an approach to the natural world that emphasizes conservation of and respect for our natural environment (Callicott, 2013; Leopold, 1949). More specifically, environmental stewardship means the responsible use and protection of the natural environment through conservation and sustainable practices to enhance ecosystem resilience and human well-being (Chapin et al., 2010). In other words, it describes the ethical responsibility and associated action that grow out of recognizing our reliance on and interconnection with the natural world. In effect, such an idea of stewardship stands for an ethic that stresses 'the responsible use (including conservation) of natural resources in a way that takes full and balanced account of the interests of society, future generations, and other species, as well as of private needs, and accepts significant answerability to society' (Worrell & Appleby, 2000, p. 263). Such an ethic is thus understood to support a variety of activities, with environmental stewardship being used to refer to such diverse actions as creating protected areas, replanting trees, limiting harvests, reducing harmful activities or pollution, creating community gardens, restoring degraded areas, or purchasing more sustainable products. It is applied to describe strict environmental conservation actions, active restoration activities and/or the sustainable use and management of resources. Stewardship actions can also be taken at diverse scales, from local to global efforts, and in both rural and urban contexts. (Bennett et al., 2018, pp. 597–8)

At the same time, the broad tradition of environmental stewardship should be understood as still inflected toward anthropocentrism. This is because it coalesces around the idea that we can and should actively shape ecological systems in order to (a) enhance (our ideas of) ecological resilience and (b) ensure that the natural environment continues to function to our benefit. This refers to the capacity of ecological systems to provide those services that are integral to all life. These include the production of atmospheric oxygen, soil formation and retention, and nutrient and water cycling (Chapin et al., 2009). In other words, while environmental stewardship is certainly a call for the conservation and protection of the natural world, it is also simultaneously a call oriented to the meeting of human interests. However, unlike tendencies to prioritize short-term economic interests in development, environmental stewardship also considers such long-term interests as conserving and protecting the natural world for future generations, for example. Importantly, despite its anthropocentrism, environmental stewardship also has the capacity to recognize more than just human interests. That is, it describes a model of care able to identify and consider responsibilities not just towards other individuals – human and animal – but also towards ecological and environmental systems, processes, and communities.

Put another way, the practices of environmental stewardship are driven by an ethic of environmental sustainability which stresses the importance of responsibly interacting with the planet to maintain natural resources and not jeopardize the ability for future generations to meet their needs. The essential point here is that all human – and animal – generations are dependent on the various goods and services of nature for survival. This in turn is the reminder that the natural world is not limitless in either its own resilience or its capacities to provide us with services and resources, a fact that climate change is driving home to us all (Davidson, 2020; Harris et al., 2018; Rosane, 2020). Further, we note that human economic interests are always playing out within environmental constraints. In the words of the former United Nations Secretary-General Ban Ki-moon, we can have 'no Plan B' with which to ensure the lives and wellbeing of future generations – whether human or animal – 'because there is no planet B' for us to draw from (Sekhri, 2013). The combined ethic of environmental stewardship and sustainability thus stresses the point that to care for our own future is also to care for the future of the ecological systems and communities we rely upon. This means also that we can come to care for the future of the animals that inhabit and contribute to those ecological systems and communities. Such insights have thus led to environmental stewardship being advocated in the United Nations Millennium Declaration (2000), with stewardship being broadly presented as 'a role every morally decent person ought to adopt towards nature, without specific appointment' or any expectation of 'remuneration' (as cited in Welchman, 2012, p. 5).

While these ideas of environmental stewardship and sustainability suggest the possibility of rebalancing human and animal interests whilst still accepting of a broadly anthropocentric perspective, they can also be considered in light of Australian First Nations' ideas of *caring for Country*. As we have noted previously, caring for Country has its origins in the holistic relationships between traditional Australian First Nations societies and their customary land and sea homelands – *Country* – that have evolved over millennia. When unpacked in terms of land and sea management, *caring for Country* thus includes a wide range of environmental, natural resource and cultural heritage management activities that are undertaken in accordance with custom by First Nations individuals, families, groups and organizations across Australia, sometimes in partnership with other stakeholders (including governments, scientists, producer groups, conservationists and others) (Hill et al., 2013; Jarvis et al., 2018). These activities include formal government-supported natural resource management projects (Roughley & Williams, 2007), First Nations and co-managed protected areas (Muller, 2003; Nursey-Bray & Rist, 2009; Ross et al., 2009), initiatives in biodiversity conservation and restoration (Nursey-Bray, 2009), water planning processes (Jackson & Altman, 2009) and the pursuit of cultural objectives (La Fontaine, 2006). They might further involve customary or cultural resource management (hunting, gathering, burning, ceremony, knowledge sharing), commercial economic activities (bush harvest for sale, pastoral, management, art) as well as threat abatement (weed and feral animal control, fire management, threatened species management, revegetation) (Davies et al., 2010; Hill et al., 2012; Muller, 2008). The main point is that while these activities clearly resonate with the focuses of both environmental stewardship and sustainability, they are also an expression of the forms of identity, kinship, customary law rights, responsibilities and obligations towards *Country* that comprise *caring for Country* (see, e.g., Aliento, 2020; Woodward et al., 2020).

Turning now to examine how these relationships and commitments might work in practice, we move to consider the increasing calls to protect and conserve the Great Barrier Reef (henceforth GBR). As an ecological community (or, more precisely, a multitude of interlinked ecological communities) that is recognized as being not simply highly valuable and extremely fragile, but under threat, the GBR provides a productive standpoint from which to assess the usefulness of ideas of stewardship and sustainability. More specifically, considering the GBR makes it clear how aims of protecting all Australians' interests are fundamentally entangled with those of conserving not only (a) habitat and ecological communities but (b) individual species or cohorts of animals. Recognizing such entanglement is fundamental for our development and living of a social imaginary where human-animal relations and modes of living are no longer primarily driven or dictated by anthropocentrism.

The Great Barrier Reef: Value and Threats

Recognized as a World Heritage Area and as one of the Seven Wonders of the World, the GBR contains the world's largest collection of coral reefs. Covering an area of 348,000 square kilometres off the northeast coast of Australia and stretching over fourteen degrees of latitude, the GBR is recognized as the 'world's largest living organism' (Deloitte Access Economics, 2017, p. 53). Within the GBR there are 3000 individual reef systems, 760 fringe reefs, 600 tropical islands and about 300 coral cays (WWF, 2018d). Indeed, while its coral reefs were what initially made the GBR famous, they only comprise about 7 per cent of the GBR Marine Park and associated World Heritage Area. The rest of the Marine Park is an extraordinary variety of marine habitats, ranging from shallow inshore areas – such as seagrass, mangroves, sand, algal and sponge gardens, and inter-reefal ecological communities – to deep oceanic areas more than 250 kilometres offshore. These comprise the habitats for a huge number of animals and plants, including a variety of rare and endangered species. Species include over 500 species of coral, 1,500 species of fish, 4,000 types of mollusc, over 130 varieties of sharks and rays, more than thirty species of whales and dolphins, as well as half of the world's diversity of mangroves and many seagrass species. The GBR is further home to a range of other natural phenomena, including annual coral spawning, whale migration, turtle nesting, and significant spawning aggregations of many fish species (UNESCO, n.d.). It is this diversity of species and habitats, along with their interconnectivity, that makes the GBR one of the richest and most complex natural ecosystems on earth. It is, in other words, an extremely complex ecological community – or, better, an interlinked series of ecological communities – in which multiple species interact and live together interdependently. These species include ourselves, given the high levels of our interactions with, our economic investments in, and our impacts on the GBR.

This richness, along with the fact that its ecosystems shelter many threatened species, has seen the GBR recognized as possessing 'enormous scientific and intrinsic importance' (UNESCO, 2016, p. 35). UNESCO indeed rates the GBR as having 'outstanding universal value' – stating that 'no other World Heritage property contains such biodiversity' (UNESCO, n.d.). More specifically, its inclusion on the World Heritage List was based on the criteria that the GBR exemplifies 'a major stage of the earth's evolutionary history', further demonstrating 'significant ongoing geological processes, biological evolution' and the long history of human engagement with the natural environment (Lucas et al., 1997, p. 16). This latter point is the reminder that the connection of some of Australia's First Nations Peoples to the GBR spans over 60,000 years, and that its natural features are deeply embedded in Indigenous culture, spirituality and wisdom (Deloitte Access Economics, 2017).[1] In addition, the GBR 'contains unique, rare and superlative natural phenomena, formations and features and areas of exceptional natural beauty' (Lucas et al., 1997, p. 16). In the words of Sir David Attenborough, the GBR is 'one of the greatest and most splendid natural treasures that the world possesses' (as cited in Deloitte Access Economics, 2017, p. 3). It goes without saying that as such a treasure the GBR should be carefully protected – indeed, that it requires stewardship.

The GBR further possesses immense cultural importance through being seen, both nationally and internationally, as one of the natural assets contributing most powerfully to Australia's global brand. The extent of the value of the GBR to Australia is captured by Deloitte Access Economics (2017) in their report *At what price: The economic, social and iconic value of the GBR*. In economic terms, as this report describes, in 2015–16 the GBR brought AUD 6.4 billion to the Australian economy through supporting 64,000 full-time jobs (direct and indirect) (Deloitte Access Economics, 2017; see also Climate Council, 2018; GBRMPA, 2018). Examining the GBR's relevance for (a) tourism; (b) commercial fishing and aquaculture (fish and seafood farming); (c) recreational activities, and (d) scientific activities, the report valued the GBR more broadly as an Australian asset at $56 billion. As such, it is the reef's economic value to Australia that is most often used to make a clear case for its ongoing protection and careful management – for its need for stewardship. It was this value, then, that led to the establishment of the GBR Marine Park, the *Great Barrier Reef Marine Park Act 1975* and the appointment the same year of an independent government agency – the Great Barrier Reef Marine Park Authority (GBRMPA) – to explicitly take on stewardship of the reef. This framework has led to the main objective of the *GBR Marine Park Act* being to 'provide for the long term protection and conservation of the environment, biodiversity and heritage values of the Region' (Dyer et al., 2020, p. 9). Human uses of the GBR, whether to do with recreation, education, or science, are in turn framed as needing to be subject to this main objective; namely, such activities should only be pursued so far as they are consistent with the long-term conservation of the reef. Finally the GBRMPA is tasked every five years with reporting on any challenges the reef might face.

Unfortunately, as is the case with many Australian ecosystems, the GBR is under threat and highly vulnerable. The *2019 GBR outlook report* thus acknowledges that while the GBR is 'retaining its outstanding universal value as a World Heritage Area, its integrity is being increasingly challenged' (GBRMPA, 2019, p. iii).

> Cumulative pressures, predominantly from climate change, combined with the time required for the recovery of key habitats, species and ecosystem processes, have caused the continued deterioration of the overall health of the Great Barrier Reef. The accumulation of impacts, through time and over an increasing area, is reducing its ability to recover from disturbances, with implications for Reef-dependent communities and industries.

These pressures include multiple mass coral bleaching events (e.g. in 2016, 2017, 2020, see GBRMPA, 2018, 2020), along with increasingly frequent extreme weather events. Also noted are impacts from coastal development (particularly industrial development), and declining water quality. Together these pressures have substantially degraded the health of the GBR to the point that it has been assessed as having lost more than half of its corals since 1995 (GBRMPA, 2018, 2020; see also BBC News, 2020a, 2020b; Moore, 2020; Readfearn, 2019). In the 2019 report to UNESCO's world heritage committee, the Queensland and federal governments admit the reef is 'an icon under pressure with a deteriorating long-term outlook' (as cited in Australian Government, 2019b). Climate change – specifically sea temperature rise, ocean acidification and the altering of weather patterns – has been identified as the most serious and pervasive threat to the GBR and all coral reefs globally (Climate Council, 2018; GBRMPA, 2015; Union of Concerned Scientists, n.d.). That is, sea temperature rise is a key cause of coral bleaching events (see, e.g., Figure 12) and subsequent declines in coral spawning. Referring to the change in seawater chemistry due to increased levels of carbon dioxide in the oceans, ocean acidification in its turn reduces the capacity of corals and other calcifying organisms to build skeletons and shells, which reduces their capacity to create habitat and support other populations.

Figure 12. Coral bleaching severity survey on Orpheus Island 2017, Great Barrier Reef, Queensland. *Source:* Photo courtesy of ARC Centre of Excellence for Coral Reef Studies Flickr account (CC BY-ND 2.0).

Climate change is also contributing to the altering of weather patterns, including the increased occurrence of more frequent and/or intense cyclones, floods and heatwaves, all of which impact on the Earth's oceans and reefs (GBRMPA, 2018, p. 1). While natural, such events, when severe, can extend recovery times of coral ecosystems by up to twenty years, and their increase in occurrence will further weaken the GBR's resilience (Day & Heron, 2019). Other key threats have been identified as the modification or destruction of coastal habitats from urban and industrial development and the continued flow of sediment and land-based pollutants from agriculture and mining activities into GBR waters (Association of Marine Park Tourism Operators, n.d; Australian Government, 2019b; GBRMPA, 2015, 2020). These include pesticides, industrial fertilizers, concentrations of coal pollutants, farm chemical pollution, and animal faeces. Each of these runoffs reduces water quality, damages sea grass and corals, and fuels mass outbreaks of the coral-eating crown-of-thorns starfish. As the report to UNESCO's World Heritage Committee admits, without addressing these threats, the maintenance of a healthy reef for future generations is extremely unlikely (as cited in Australian Government, 2019b).

Protecting the Reef: Environmental Stewardship and Sustainability

The visible nature of both these threats and the degraded health of the GBR, and the associated potential loss of value of the reef to Australia – made public through a multitude of reporting mechanisms – has led to urgent calls for the GBR to be better protected. In the words of the Great Barrier Reef Marine Park Authority (GBRMPA):

> Only the strongest and fastest possible actions to decrease global greenhouse gas emissions will reduce the risks and limit the impacts of climate change on the reef. Further impacts can be minimised by limiting global temperature increase to the maximum extent possible and fast-tracking actions to build reef resilience. (as cited in Moore, 2019)

As well as demands for the Australian and the Queensland governments to commit to effectively addressing climate change, there have been multiple calls to stop or prevent other kinds of actions that are seen as also having the capacity to further damage or erode the GBR. These calls take the form of campaigns against the development of proposed coal mines and associated infrastructure (such as those backed by the Adani Group, now named Bravus Mining and Resources, and Queensland businessman Clive Palmer); campaigns against other proposed developments as well as dredging and shipping within or near the reef environs; campaigns to reduce chemical and sediment pollution; and campaigns for wetland rehabilitation and generally improving water quality; and for the protection of endangered species. Such campaigns are proposed and promoted by a variety of organizations including the Australian Marine Conservation Society (AMCS, n.d.), Association of Marine Park Tourism Operators (n.d.), Greenpeace (n.d.), Mackay Conservation Group (n.d.), North Queensland Conservation Council (n.d.), the Wilderness Society (n.d.), Wildlife Preservation Society of Queensland (n.d.), the World Wide Fund for Nature (WWF, n.d.), and others. Some of these campaign aims are further carried into the development and promotion of citizen science projects such as the one developed by Citizens of the Great Barrier Reef (CGBR) to assess the condition of the GBR, and recovery projects such as green turtle nesting areas by Great Barrier Reef Foundation (GBRF). Together campaigns and associated initiatives all stress our shared responsibility to protect the GBR. They thus call on all Australians to 'fight for our reef' (AMCS, 2021b), 'fight for the reef' (WWF, 2018c), 'unite for our reef' and 'commit to a brighter future for the reef' (CGBR, n.d.-a), as well as to 'help save' endangered species who are threatened by development activities that would impact on the GBR (see, e.g., GBRF, 2021).

As should be clear, these various campaigns and initiatives tend to be organized around and promote the principles of environmental stewardship and sustainability, with such aims set out as being supportive of both human and other interests and futures, including those of the GBR itself and its inhabitants. More specifically, these campaigns and projects stress the importance of raising public awareness with regards to the major threats faced by the GBR. They also strive to use mainstream mobilization – through rallies, petitions, lobbying, raising funds, recruiting citizens into science projects and other activities – to not just demand that strong federal and state government action be taken in support of the GBR, but to achieve a range of specific actions. Five forms of public action are typical. First is crowdfunding (through, for instance, the Citizen's Reef Fighting Fund supported by GetUp!), most often to fund legal challenges to damaging development activities (Rodan & Mummery, 2018). The second is lobbying major banks to persuade them against funding damaging development activities – in particular the proposed Adani Carmichael coal mine (see, e.g., Wangan & Jagalingou Family Council, n.d.). Concerned with the need to increase public awareness of the poor environmental records of organizations such as Adani involved in proposed developments, the third action has taken the form of the making and screening of short films and advertising campaigns by GBR advocates (Rodan & Mummery,

2018). The fourth action is the lobbying of both the federal and state governments to act to better protect the reef through switching, for instance, to renewable energy (see AMCS, n.d.; North Queensland Conservation Council, n.d.). Several of these aims are visible in this statement from the Australian Marine Conservation Society (AMCS, 2021b):

> Together we can stop the Adani Carmichael coal mine and other mega mines, and insist that the federal government urgently switches to clean, renewable energy. Hundreds of thousands of people are already taking action to #StopAdani and new coal mines. Our leaders have a legal and moral responsibility to protect our Reef and together we will hold them to it.

A fifth form of action has been the development and promotion of citizen science projects such as the Great Reef Census. Undertaken in October 2020 after a successful pilot program in 2019, the Great Reef Census was a reef-wide, collaborative project of data collection and analysis aiming to not only 'deliver vital insights for the conservation of this iconic ecosystem', but 'engage the global community in the future of the Reef through education, storytelling and action' (CGBR, n.d.-a; CGBR, n.d.-b). This involved first putting together a makeshift research fleet of all available boats (including tourism boats, island ferries, fishing charters, private vessels, luxury superyachts, and research ships) which then supported teams of recreational divers and snorkelers who committed to surveying 182 ecologically significant sites by 'swimming along a reef and taking a photograph every 10 metres' (Kim, 2020). The second step was for these photographs – 12,762 were taken in total – to be uploaded for citizen scientists around the world to help analyse (CGBR, n.d.-b). Specific aims were to 'trial innovative ways of capturing reconnaissance data from across the Reef' while helping scientists and GBR managers 'locate some of the most important sources of coral recovery' (CGBR, n.d.-b). In the words of University of Queensland marine ecology professor Peter Mumby, who helped devise the survey plan, this work is necessary because there is 'no way that scientists are able to keep tabs on [the] very large number of reefs' comprising the GBR (as cited in Kim, 2020). Such an exercise is also considered important for its mobilizing of a global community in the ongoing protection of the GBR:

> Not on the Reef? No problem! You can get involved in the Great Reef Census whether you're in an office in Sydney or a classroom in Amsterdam. In November, we'll be recruiting citizen scientists from around the world to help analyse the images captured from across the Reef, so stay tuned! (CGBR, n.d.-b)

Led by Citizens of the Great Barrier Reef, the Great Reef Census brought a wide range of participants to work together such as members of government, industry, research, and the broader Australian public. More specifically, the project was funded through a partnership between the Australian Government's Reef Trust and the Great Barrier Reef Foundation, the Prior Family Foundation and the Reef & Rainforest Research Centre. Delivered in collaboration with the University of Queensland, the Great Barrier Reef Marine Park Authority, and the Australian Institute of Marine Science, with support from James Cook University, the University of Technology Sydney and the University of Tasmania, and a variety of tech companies, the project involved thousands of other individuals in the work of collecting and/or analysing data. The organization, Citizens of the Great Barrier Reef, hopes that the mobilization around and results of the 2020 census will enable further scaling up of the census in 2021 to reach 300 to 500 reefs. There is also hope that such work will form part of 'the vanguard' of a twenty-first century scalable approach to conservation challenges whereby science, industry, tourism and grassroots participants collaborate to solve problems, and 'foster a sense of stewardship for the Reef' (Dickson-Smith, 2020; Kim, 2020).

These are all actions that stress stewardship of the GBR over time, acknowledging variously its outstanding universal value, its importance for biodiversity and in providing habitat, its work – common to all reefs – of recycling carbon dioxide from the atmosphere, as well as its immense economic and cultural value for Australia and Australians. The promotion of broad levels of mobilization and collaboration is also pertinent here. This is the point that environmental stewardship and sustainability projects are rarely *purist* in their aims or in the actions they promote. They rather need to be understood as encompassing diverse actions taken in partnership 'by individuals, groups or networks of actors, with various motivations and levels of capacity' (Bennett et al., 2018, p. 599). While such actions must align with protecting, caring for and/or responsibly using the environment, they are also always influenced by the social-ecological contexts of actors and will thus produce both environmental and social outcomes (Bennett et al., 2018). This broad church focus is evident in the set-up of the Great Reef Census as well as in the framing of conservation work by the Great Barrier Reef Marine Park Authority (GBRMPA, 2021a). For instance, the GBRMPA stresses that it works in partnership with 'Traditional Owners, other Australian and Queensland government agencies, industry, community organisations, and individuals to help achieve protect and manage the Reef' (GBRMPA, 2021a). This is particularly exemplified in its Reef Guardian stewardship program which, initiated in 2003, explicitly aims to promote community stewardship of natural resources (Evans, 2011) by recognizing the 'good environmental work undertaken by communities and industries to protect the Great Barrier Reef' (GBRMPA, 2021b). As the GBRMPA (2021b) outlines,

> The program demonstrates [that] a hands-on, community-based approach can make a real difference to the health and resilience of the Reef. Reef Guardians are taking on voluntary actions beyond what is required by law and sharing information. These actions will help to improve the economic sustainability of industries operating in the Great Barrier Reef Region and ensure the environmental sustainability of the Marine Park.

Environmental stewardship and sustainability thus promote the work and stories of strategic collaboration and knowledge sharing. Such work is further considered effective with regards to enhancing community understanding and public awareness, as well as supporting shared learning – all work that can further grow commitment to environmental stewardship and sustainability.

What is also important about the work of environmental stewardship and sustainability is that it allows for consideration and protection not just of the interests and needs of particular species over time, but of the broader resilience of ecological communities and, indeed, of habitat. That is, the work of environmental stewardship and sustainability explicitly recognizes habitat as being of value as well as the members of multiple species – a focus that has been missing in many of the other calls for action to protect animals and animal populations considered in previous chapters. This, of course, is an easy consideration with the GBR in that much of the *habitat* in need of protection – the hard corals comprising the GBR itself, for example – is itself a vulnerable animal population, and one also recognized as being of immense importance. What also stands out with regards to the formulation of calls for stewardship and sustainability regarding the GBR is that while there is acknowledgement of the importance of protecting endangered species, for example, these also cannot be considered independently of the broader ecological communities – even, indeed, *Country* (or *Countries*) – they interact with and exist within.[2] Indeed, using the framework of environmental stewardship and sustainability as the basis for an ethic of care and responsibility for other species enables a broad consideration of needs and potential harms, and makes subsequent ideas regarding responsibility and action immediately visible. Such a focus enables a broader consideration than is typically made apparent by an ethic of care and responsibility derived from a focus on sentience, insofar as this can tend to narrow its focus to individual animals. One of the reasons for this broader consideration is, of course, that environmental stewardship and sustainability, by definition, consider a wide range of interests, human and other, over time. Whilst they can certainly be anthropocentric in their framing of ecological communities (including habitat) as resources of significant economic value, this framing in turn identifies these same communities and their habitat as in need of protection – as resources, but also as holding some intrinsic value – in a way that is able to engage with broader ideas of timescales and relationality than is typical under anthropocentrism. Of course, when considered in the terms of *caring for Country*, even this adapted anthropocentrism is too narrow a description for the relationalities described through care.

Rebalancing Human and Animal Interests through Stewardship

One of the questions that has thus been driving this book and which has been explicitly foregrounded in this chapter is: how can human-animal relations be imagined so that they do not end up protecting human interests to the detriment of the interests and lives of other species? This question is important for two reasons. First, of course, we face the very real risk of the irrevocable loss of Australia's unique biodiversity to the toxic combination of our anthropocentrism, resultant development practices, and climate change – if these are left unchecked. The second reason concerns the discrepancy between Australians' willingness as individuals to express our care for animals and to act – in at least some ways – on behalf of that care, and the difficulty of bringing this care to bear on issues of habitat loss and the disruption and collapse of ecological communities. Part of this issue is of course the ways these issues so often end up framed as binaries between development and a care for animals and, as with all binaries, overcoming them is difficult. Positions tend to be easily reified into identity categories and value conflicts, and challenges can in turn promote unhelpful reactions of defensiveness and resistance.

One possibility explored through this chapter for bypassing this binarism between development and care for animals has been the engagement of ideas of stewardship and sustainability. As we have demonstrated, ideas and practices of stewardship and sustainability are effective in focusing not just on individual species of animals, but also on their habitat and the intrinsic relationality of the different species that comprise ecological communities. Such ideas also provide a mechanism for overcoming some of the short-term and often binary thinking typical of anthropocentrism – specifically that which automatically promotes human interests in economic growth over the interests of other species and the resilience of our ecological communities. That is, both environmental stewardship and sustainability encourage longer term thinking that recognizes (a) that we as humans are reliant on the ecosystem services and resources provided by resilient ecological communities, and (b) that nonhuman species and ecological communities add value to our modes of living. Such views remind us that while models of environmental stewardship and sustainability do promote the protection of other species and ecological communities, they do not frame this work as being in opposition to the need to also provide for our own (human) interests. Collapsing the binarism between both human and animal interests is important because any dreams we might have for imagining, developing and sustaining multispecies communities will always need to include meeting our own interests. Here, of course, Australian First Nations Peoples' practices of *caring for Country* stand out in particular as driven not by any presumed opposition between human and animal interests – or even by the need to overcome such opposition – but rather by deep recognitions of kinship and interconnection wherein all of Country must flourish for any part of it to flourish. That is, *caring for Country* stands for a holistic approach wherein 'landscape, water, fire, soil, animals, people, plants and cultural teachings are interrelated' (Aliento, 2020). This suggests that one highly productive possibility for learning to reimagine human-animal relations and living might be to build stronger engagements and collaborations between First Nations knowledge holders and the rest of Australia (see, e.g., Aliento, 2020; Frew, 2019; Gay'Wu Group of Women, 2019). This and other possibilities will be the focus of the next chapter.

Notes

1   There are some seventy First Nations Traditional Owner groups with authority for Sea Country management in the Great Barrier Reef Marine Park. Torres Strait Island Traditional Owner groups are also connected to the Reef and hold cultural knowledge of their traditional use of the Great Barrier Reef region more broadly. Three groups are directly connected to Raine Island. For details on the specific groups comprising the Traditional Owners of the Great Barrier Reef, see the Great Barrier Reef Marine Park Authority (GBRMPA, 2021c).
2   Note that such focuses would appear to be refined further in stewardship programs tied into the *caring for Country* work carried out by Australia's First Nations Peoples. This is the point that *caring for Country* is very clearly more than the sum of the protection of the interests and needs of particular species, the strengthening of the

resilience of ecological communities, and the building of practices of strategic collaboration and knowledge sharing. Rather, the work of *caring for Country* has also been recognized as fundamental for both individual wellbeing and cultural and communal identity. The holistic focus inherent in *caring for Country* can be seen in the definition of 'Aboriginal health' asserted by the Australian National Aboriginal and Community Controlled Health Organisation (NACCHO), the peak body for Aboriginal community controlled health services in Australia: 'Aboriginal health means not just the physical wellbeing of an individual but refers to the social, emotional and cultural wellbeing of the whole Community in which each individual is able to achieve their full potential as a human being thereby bringing about the total wellbeing of their Community' (NACCHO, n.d.). As we have noted, understandings of *Country* and kinship mean that this Community is always more-than-human. For Australia's First Nations Peoples, then, stewardship is consequently always much more than just managing the physical environment (Larson et al., 2020).

CHAPTER 7

Different/Together: Building More-Than-Human Communities

As has been discussed in preceding chapters, post-settlement Australia's commitment to anthropocentric modes of thinking, living, and storytelling has played out in many ways as a *war on animals*. This has seen commitments to inhumane models of animal agriculture as well as the actual declaration of *war* on a variety of animal species that have been perceived as pests, including kangaroos and dingoes, as well as feral cats, rabbits, foxes, cane toads, starlings, sparrows, and even wedge-tailed eagles.[1] Such actions have most often been driven by the perceived need to protect agricultural and other anthropocentric interests, a need that has also been a main driver for the eradication of native ecosystems. Such practices have led to Australia recording increasing losses in biodiversity and logging the highest rate of mammal extinction in the world over the last 200 years (Spring & Earl, 2019). And, yet, as we have shown in previous chapters, Australians do care about animals and there are stirrings of change with regards to many conventional engagements with animals. Indeed, new ideas and practices with regards to animals are coming into use and gaining traction within the Australian public sphere, and they may further form the basis for alternative modes of living and an alternative social imaginary regarding human-animal relations. For instance, responses to the narrative of sentience in animals have led to calls for stronger considerations of the welfare and wellbeing of livestock animals, and the rise of new norms for eating and consumption. Further ideas have involved increasing recognition of what it might mean to build friendships and kinship with animals, as well as growing insights into the need to better protect and cohabit with not just the members of different species but ecological communities. These, then provide a basis for modes of thinking and living in which animals and animal populations could be considered and valued in ways other than through an anthropocentric lens. We will consider such possibilities in more detail later in this chapter.

This chapter continues the trajectory that was started in Chapter 6. Specifically, this is to engage some of these previous insights and points of change with regards to outlining the possibility of a multispecies or more-than-human community where human interests would not automatically be considered at the expense of the interests of other animal species. Bearing some alignment with Australia's First Nations Peoples' ideas of *caring for Country*, such an idea has the potential to enable less anthropocentric forms of inter-relation and futures, but these are also not inevitable. That is to say, some ideas and practices of a multispecies community can – and do – also play out within an anthropocentric imaginary, although these do also always remain open to alternative shaping. It is this kind of potential regarding the shaping of a multispecies community that comes into focus in this chapter. However, before beginning our discussion of these possibilities, it is important to provide more detail to what we mean by this idea of a multispecies or more-than-human community, and to explore how such ideas might inspire different, more-than-human focuses. This comprises the first two sections of this chapter. Adding to discussions carried out in previous chapters regarding the emergence of alternative narratives and practices concerning the various species we engage with in their different contexts, the third section identifies some of the practical ideas for supporting more-than-human life that are gaining traction both with reference to broader ecological communities as well as to how we live in our own homes and backyards, and broader communities. We conclude the chapter with an outline of some of the principles for thinking and living that would seem fundamental for building and supporting multispecies communities and that have the potential to gain some traction in the Australian context.

Multispecies Thinking and Community

There is no form of human life or experience or community that has not arisen in dialogue with a wider world: 'All living beings emerge from and make their lives within multispecies communities' (van Dooren et al., 2016, p. 2). That is, regardless of our explicit ethical, political, and social commitments, we have always shared our lives and our communities with all sorts of other species: animal, plant, and microbial. While this shared life certainly demonstrates our living in – our being entangled in – a multispecies community, it is our modes for living alongside other animal species that is our specific concern throughout this book. (We recognize, of course, that living alongside other animal species automatically means we also always live in community with plant and microbial life). Of the other animal species that we live with and alongside, these include: those we rely on (e.g. livestock animals); those we cohabit with willingly (our companion animals); those we attempt to manage with an eye to our own comfort and convenience (wildlife and members of feral animal species and what we often designate as pest species); and those which move in and live alongside us regardless of our wishes (some wildlife and members of pest species again). This is the case no matter where in the world we might live. Having said this, however, what also stands out is that only a limited number of other species are actively welcomed into Australian communities. With reference just to animal species, for example, while many Australians do strive to ensure their companion animals are comfortable within our – and their – homes (Petcare Information Advisory Service, 2010), other species are made to fit around and within the constraints of our interests, and others are actively denied any hospitality. This does not necessarily mean that their interests are explicitly overridden (although it can if we think about how we tend to treat unwanted species, for instance), rather that their interests are just not considered in our community planning. Our communities may not be anthropocentric and speciesist in their actual practice, but they certainly, systematically and routinely, give preference to human interests and needs in both their planning and their practice. They are consequently human focused in their intention.

This is to the point that, at least in our contemporary urban environment, we have become used to only thinking and practising community in human terms. Indeed, conventional definitions of *community* stress that it involves a group of *people* who share something in common – a location, common interests, and attitudes. This commonality is then typically thought to further imply some level of fellow feeling, some sense of inter-relatedness and shared identity, which is considered able to dictate inclusion and exclusion. Indeed, community is understood in terms of that which is held – and valued – in common (Devadas & Mummery, 2007). This might refer to a belief and value system, or to a sense of place or locale, or some other factors held in common. Admittedly, given the historical reality that communities have been subject to disruption and dislocation, communities are rarely, perhaps never, completely homogenized, and united. They further remain porous to the influx of new experiences and memories. What this suggests is that *community* can be most usefully understood as an imaginative category that seeks to gather and connect diversity together into a broader identity (Anderson, 2006; Jacobs, 1996). Put another way, communities stand for 'cultural constructions' that provide 'important symbolic as well as practical frameworks to life' (Olwig, 2002, p. 125). Both constructions and frameworks have, however, tended to be anthropocentric. That is, as with our traditions of justice and ideals of ethics, community too has been typically understood as only pertaining to human beings and human interests. This is down again to the anthropocentric division of humans from other species that underpins our current modern social imaginary. In this case the assumption is that only humans can share in the interests and recognize the commonalities that underpin a community, meaning that only humans (and their property, including thereby some animals) should be protected in that community. Certainly, there are other ideas of community in use in various fields. For instance, there are the ideas of ecological community prevalent in the biological sciences and discussed in the previous chapter and, as we have noted, the practical reality of community is always that of multispecies entanglement (van Dooren, 2019). In addition, as we have noted, Australian First Nations

Peoples' ideas regarding *caring for Country* also foreground the importance of multispecies entanglements. Commonplace Australian ideas and practising of community, however, do continue to place human interests front and centre.

Anthropocentric thinking and practice have been highlighted throughout the preceding chapters, with the interests and lives of livestock animals, native and feral animals, all shown to typically be subordinated to human interests in accordance with the anthropocentricism of our social imaginary. It is also clear, as was discussed in Chapter 4, that while our models of friendship, companionship and kinship do require recognition of interests other than our own, these are still dependent on their alignment with our interests. That is to say, any other who cannot be recognized within the context of our interests – as at least sharing some of the interests we give value to – is not going to be identified as kin, let alone as a friend. And it is also the case that many of those living with companion animals expect those animals to abide by human models of engagement – even when such behaviour runs counter the animal's own species-natural behaviour (Starling & McGreevy, 2018, 2020). Certainly, our earlier chapters have also striven to point out instances and possibilities for overcoming or exceeding such anthropocentric assumptions in our relationships with animals, and have traced some of the alternative frameworks proposed for human-animal relationships: such as sentience, care, friendship, responsibility, stewardship, as well as the Australian First Nations' idea of *caring for Country*. However, as should also be evident, while such narratives and frameworks are gaining traction to some extent in Australia, to bring about the changes they tentatively gesture towards requires not just individual efforts and care, but a collective revisioning of how we live. This would entail a revisioning of our ideas and practices of community.

Such possibilities are being explored at a theoretical level. There are, for instance, various outlines of past and existing hybrid or multispecies communities, reminders that our so-called human lives and communities have always included other species (see, e.g., Haraway, 2008, 2016; Tsing, 2012; van Dooren, 2019). Such outlines elaborate the point that we always already live in a 'co-emergent world based on intimate human-more-than-human relationships of responsibility and care' (Bawaka Country et al., 2016, p. 16). We are also reminded that our very human being is itself always a multispecies community – we are, fundamentally, always, and already, complex microbial communities (Hayes & Sahu, 2020; Ursell et al., 2012). There are entire fields of study being developed to examine and foreground such communities, including multispecies studies and the new forms of ethnographic and ethological inquiry these are generating (van Dooren et al., 2016). Our concern in this chapter, however, while sympathetic to such work, is rather to ask about what is happening now with regards to our thinking and practices with animals, and what these insights might suggest with regards to instituting future understandings and frameworks for a multispecies community and social imaginary. Importantly, we explore these possibilities in the context of current and changing Australian attitudes and assumptions, including the key point that Australia needs to be understood as – at best – oriented to animal welfarism rather than to animal rights-based arguments. Indeed, as we have set out in the past chapters, even existing Australian commitments to improving animal welfare become contested when human interests in development look to be under threat.

Our focus, then, is not to make the point again that we already live in multispecies communities – this is evident as soon as we consider how entangled our human lives always are with the other species we rely on and live alongside. Rather we consider what Australian communities might look like if they explicitly came to consider more-than-human interests. What they might look like if they started to consider and make room for diverse – multispecies – forms of living, expression, meaning-making, and creation. Initially this is to simply follow the questions that community always holds in itself: How should we be and live with others? How inclusive (or exclusive) should our community be? What should these decisions be based upon? Our additional questions concern what these decisions might look like if the anthropocentric investment was not so strongly adhered to, and if more-than-human interests started to be counted seriously in our community development, maintenance, and practice. As such, when we use the terms *multispecies community* or *more-than-human community*, we suggest ideas and practices of cohabitation and community which explicitly orient beyond just the human, and which actively give space and time for more than just humans. That is, we aim to suggest practices of cohabitation and community that are attentive beyond human interests, that actively recognize that while our communities already are more-than-human, they can also become much more hospitable to such others. We do stress again, however, that – at least at this stage in Australia – the practices of cohabitation that suggest at least some orienting to a more-than-human community and that are starting to gain some public traction by no means entail the offering of an equal hospitality to all species, let alone equal rights in cohabitation. We discuss these issues and the speciesism they still stand for later in this chapter.

Orienting to a More-than-Human Community

When it comes to explicitly enabling more-than-human communities, multispecies studies in its various forms stresses the importance of cultivating 'arts of noticing' and 'passionate immersion' (Tsing, 2011, p. 19) in more-than-human lives and living. It suggests that we should become better at both paying attention to nonhuman others and responding meaningfully in positive ways to their needs and interests (van Dooren et al., 2016). This is fundamental work with regards to community building insofar as the definition of community – with its focus on shared lives and interests – would suggest that those we do not meaningfully pay attention to and respond to in positive ways would be those we either overlook or deliberately exclude from community. Interestingly, this work of attentiveness is also commonplace work in the context of developing and maintaining relations such as friendship and kinship. In *Storm Boy*, for instance, Mike and Mr Percival both recognizably pay attention to and respond to the other; their friendship has them immersed in each other's worlds. In *Emu Runner*, Gem too can be characterized as being attentive to the emu she sees as kin, and she is shown as trying to immerse herself in the emu's world. *Skippy the Bush Kangaroo* not only depicts Sonny's and Skippy's respective attentiveness to each other, but critiques human-animal interactions that are marked by inattentiveness in basing many storylines around a lack of human consideration for animals and their interests. Similar points regarding the inattentiveness of many human-animal interactions are made in *Dot and the Kangaroo*.

If attentiveness and immersion help us better understand the interests of more-than-human others, then they also open questions as to how to think and actively build a more-than-human community. For instance, practices of attentiveness and immersion that show us our longstanding entanglement in multispecies communities might also make it clear that we need to question and resist the anthropocentrism that has been one of our default standards for community inclusion or exclusion. That is, they make clear to us the need to 'move away from a perspective where urban environments are for human inhabitants alone' (Clarke et al., 2019, p. 61). They are practices that have the capacity to take us 'beyond our previously known worlds' (Rose, 2007, p. 88). This, of course, is easier said than done. After all,

> how do we make the experiences of non-human others palpable? How do we hear, and how do we encourage others to hear, the non-human voices? How do we bring them into participatory processes [...]? Most important, how do we convince others, who are less familiar with such perspectives, that decentering human privilege is important [...]? (Clarke et al., 2019, p. 61)

There are multiple issues here, but a clear first step is to remember that no community – no matter how attentive it is to policing its own identity and modes of belonging – is ever completely identical to any story it might have of itself (Derrida, 1994/1997; Nancy, 1991). This is the reminder that the development of a community is always dynamic with diverse, even unexpected, agencies each exerting their own force and interacting in the shaping of outcomes (Latour, 2003). Thus, even while our historical story has been one of dismissing nonhuman animal agencies and perspectives as unimportant – beyond the minimal consideration we provide some already tolerated species – if we are attentive, we can notice that all around us nonhuman constructions of meaning and intent are already in process throughout our communities. Being attentive, then, can make other perspectives and interests more visible to us, whilst also undermining our anthropocentric assumptions regarding community enclosure and commonality. The question, of course, is how might we further develop our attentiveness?

One possibility here – prefaced in Chapter 4 – is to start from the attentive, affectionate, and embodied relationships we already build with our nonhuman animal companions. These, after all, remind us that we can and do actively build attentive relationships with nonhumans, an insight that may in turn help us build broader and further-reaching kinship relationships with other animals – relationships that may eventually enable a societally wide challenge of the animal-industrial complex that maintains use-value views of livestock animals. Although taking a stance that considers the needs and interests of individual nonhuman animals and individual species does not address the interdependencies that would need to be understood in multispecies communities (an issue also considered in previous chapters in the context of protecting ecological communities), developing sensitivity and attentiveness towards the needs and interests of individual members of other species is one way to start to move beyond an anthropocentric perspective. Of course, such sensitivity and attentiveness develop from immersion – from coming to know other animals, from spending time with them, coming to care about them. The difficulty here is that while many Australians might be said to be immersed in the lives of their companion animals and in their relations with them, this is rarely the case for other animal populations. For instance, it is true that outside of interacting with some highly visible species – birds, for example – many Australians can 'feel very separated from our wildlife' (WildArk, n.d.).

> The 'extinction of experience' is a term coined by ecologist Robert Pye over 20 years ago, and it relates well to our wildlife. Some of our mammals have become so threatened, or survive far from our urban population centres that people just don't see them. And if you don't see them you don't appreciate them. And if you don't appreciate them, well, you don't conserve them. I'd encourage everyone to stop at the next dead animal they pass on the road (if it's safe to do so!) get out and have a look at it. Appreciate it for its uniqueness. And consider the irony in the fact that most Australians only observe our wildlife when it's dead on a roadside. (WildArk, n.d.)

This sense of separation from a lack of direct experience is also common with regards to the levels of engagement many Australians would have with livestock and feral animals. Indeed, it is this sense of separation that has arguably made it easy for a range of anthropocentric practices that are harmful to animals to be assumed to be normal and necessary to the point of challenges to them being framed as unreasonable – and in some cases un-Australian (e.g. see Chapters 3 and 5). As noted above, this sense of separation is not easy to overcome while, at the same time, it marks what must be overcome if we are to be able to imagine and implement relations and engagements with animals that are not driven by anthropocentrism.

One common approach to this issue in human-animal relations of separation and associated lack of care and kinship is one that we have highlighted across multiple chapters. Typical of animal activists, this takes the form of advocacy for the recognition of sentience in animals. As we have shown, this work strives to present a basis for (a) increasing our care and attentiveness regarding other animals, (b) enabling our recognition of kinship with these others, and thus another mode of (c) stepping beyond an anthropocentric perspective. Dependent on a recognition of similarity in human and animal capacity to experience their lives, accepting animal sentience can help us move towards practices of care, attentiveness, and ethical obligation. Such care can furthermore become a basis for our preparedness to more broadly extrapolate animal needs and interests in support of multispecies cohabitation and community. It is thus this attentiveness and care that underpins the establishment of animal havens and rescue centres such as Edgar's Mission Farm Sanctuary. Scattered throughout Australia, these organizations strive to rescue, rehabilitate and either rehome or provide forever homes to animals in need. Such work builds a model of multispecies living and in addition often uses a combination of education, outreach, advocacy, community enrichment and farm tours to encourage more people 'to expand their circle of compassion to include all animals' (Edgar's Mission, 2021). Another example of how such attentiveness towards animal others can shape a more-than-human community can be seen in the inclusion of dog friendly public spaces – dog beaches, enclosed dog parks and dog agility parks – that allow these companion animals and their humans to play and exercise together (see, e.g., Figure 13), along with the promotion of dog-friendly holiday homes.

Different/Together: Building More-Than-Human Communities 237

Figure 13. Dog agility courses have been designed to allow dogs and humans to play in ways that provide dogs with mental and physical stimulation. Agility courses can be either run competitively or just for fun. Dogs would typically run and play in such spaces off-leash. *Source:* Photo courtesy of Sylvia Hamilton.

Admittedly this example is based on our cohabitation with members of a species we already value and these spaces are often fenced and limited in size, and place requirements on the sociability of users, both dog and human. Nevertheless, such possibilities are important insofar as they align with ideas of communities with regards to the importance of the coming together of individuals (dogs and humans in this instance) in attentive connection and consideration of each other. As we have noted, in already illustrating our recognition of some more-than-human interests, our stories and practices of friendship, kinship and empathy with animals can also provide insight into what coming together in multispecies connection and consideration – in cohabitation and as community – might involve.

A second possibility regarding the process of learning to identify and be attentive to the needs and interests of nonhuman others could come out of considering how other non-companion animals already interact with us and within our communities. Of particular interest would be instances of non-companion animals – for instance, members of native or feral species – actively carving out space and facilities within our communities for their own needs, and interacting through this work with us and with other species. One species stands out here with regards to both the strength of their commitment to this work and their visibility within our communities: crows, members of the genus *Corvus* (comprising crows, ravens, jackdaws, and rooks). Crows – and we use the term *crow* from now on to refer broadly to all members of the genus *Corvus* – are highly intelligent birds with exceptional long-term memory who display adaptability and highly complex social reasoning and behaviour (Black, 2013; Debus, 2012; Loretto et al., 2012). Indeed, crows generally seem to do many of the same kinds of things we do. They *talk* to each other, teach each other new skills, use tools, steal and hide things, tease members of other species, and even play (Debus, 2012). They are highly innovative birds – curious, opportunistic and prepared to experiment whilst also attentive to risk (van Dooren, 2019). For instance, as has been noted of one set of crow populations in South East Queensland, crows have not just developed the cognitive learning to use cars to crack nuts on roads but purposely conduct this technique near safe crossings to reduce their chances of being hit by oncoming traffic (Moreton Bay Regional Council, n.d.). It is arguably through their adaptability and preparedness to innovate with mixed materials and in changing conditions that crows have successfully been able to alter their behaviour so as to inhabit a wide range of habitats, including urban and suburban human communities. Indeed, research has shown that when living in close association with people, crows are actively attentive and responsive to human behaviour, to the point of 'remembering our faces and passing information about us as individuals on to one another' (van Dooren, 2019, p. 2; see also Cornell et al., 2012). They further follow the direction of our gaze, paying attention to what we are looking at – even when our focus is on something other than them (Schloegl et al., 2007). Put another way, like our companion animals, crows deliberately engage in interactive and adaptive relationships with us, regardless of our (lack of) awareness of their efforts.

With five species of native crow, crows are a ubiquitous presence in Australian communities, abundant throughout both Australia's agricultural regions and urban and suburban environments.

> [T]he Torresian Crow is the urban corvid you will usually see north of Newcastle (NSW). The Australian Raven is the most common urban corvid in Sydney, Canberra and Perth; the Little Raven in Melbourne and Adelaide; and the Forest Raven in Hobart. The Little Crow is common in outback towns. (Debus, 2012)

The ability of crows to thrive within our communities has been met with mixed reactions. While crows arguably provide a *hygiene service* to our communities by recycling rubbish and dead animals (also, historically, human corpses during times of war or plague), they have also been considered evil and dangerous scavengers (Chappell, 2006; see also Jones & Everding, 1993) and even vicious predators (Moon, 2005). They are typically deplored as aggressive, noisy and messy (van Dooren, 2019), with one respondent in a study describing them as the 'noisy, ugly black birds that poo [sic] a lot!' (Moon, 2005, p. 69). Scavenging food scraps from parks, public areas and industrial waste sites, crows are certainly highly visible community residents. It has become commonly believed that it is due to urban and suburban waste management practices that crow numbers have been able to increase so 'unnaturally' in Australian communities (Brisbane City Council, 2019; City of Melville, n.d.). Their allegedly overlarge populations have also seen them blamed for the disappearance of other native species in urban and suburban areas, variously with reference to their predation on the eggs and young of other species and/or by out-competing them for resources (Bateman, 2015; City of Melville, n.d.). Whilst such assumptions are open to dispute (see, e.g., City of Melville, n.d.; Moon, 2005), they are still commonly held.

Crows have also been negatively perceived for their ingenuity when it comes to nesting and nest building in human focused environments (van Dooren, 2019). For instance, given the removal of what would be desirable nesting trees (tall eucalypts), crows are gaining expertise in nesting in and on buildings and other structures (such as telecommunication towers). They are further gaining notice for using the rubber strips out of car windscreen wipers and shredded letters (taken from letterboxes) in their nests alongside more conventional materials (Backyard Buddies, n.d.). Although clear indicators of crows' innovation and capacities for 'cultural change' with regards to their claiming of space within Australian communities, such behaviours are not always seen as acceptable by human residents (van Dooren, 2019).

For instance, in one study carried out in Perth, Western Australia, survey respondents included a range of comments on the damage caused by crows within the urban environment (Stewart, 1997). Here complaints were that they: destroyed home gardens; stole fruit, vegetables and pet food; ripped off windscreen wipers and destroyed flywire and television antennas; pulled out window putty and window and roof insulation; pecked and scratched windows; and destroyed other pieces of property (bike seats, reticulation, furniture and unsecured backpacks, and bags). One respondent also shared how a crow caused a blackout by eating through an electrical insulation cable to a house (p. 46). Such descriptions of damage are common in other discussions of crow behaviour (Department of Biodiversity, Conservation and Attractions, 2017; van Dooren, 2019).

Similar complaints have been made in other regions throughout Australia and have led to calls to councils and other services for resident crows to be scared off or culled (see van Dooren, 2019), as well as a booming business in diverse crow deterrent devices (see, e.g., Bird Control Australia, n.d.; Crow Away, 2021). Crows are, it seems, just too visible, too loud, and too focused on meeting their own needs to be acceptable nonhuman neighbours – at least according to the anthropocentric expectation that all of our neighbours be unobtrusive or, better, pleasant company, and generally behave in accordance with our interests. Nonetheless, it is this very visibility of crows, along with their active problem solving to meet their own needs, that makes it very clear to us that there are others residing in our communities. Indeed, we do not need to be practised in attentiveness to find ourselves aware of the crows living in accordance with their own interests alongside us. Unlike many of the other species living in and around our human-oriented communities, crows make themselves at home without our help – and even sometimes in the face of our opposition to their presence. Crows make very clear to us that not all our neighbours are human and, attentive to us, they draw our attention in return.

Crows thus remind us that other species not only operate in but construct space and time in line with their own needs and possibilities. This is a broader reminder that space and time stand for social constructs with their own historicity – more specifically that they are each produced, practised and understood through the ways we all, human and animal, relate to each other and to our environments (Lefebvre, 1974/1991). This is to say that space and time are practised, experienced, and understood through the patterns and rhythms of repetitions and differences that make up the embodied practices of our everyday life as well as the time of personal and collective history (Alhadeff-Jones, 2019; Lefebvre, 1992/2004). These are vital points because they remind us that, just like us, crows and other animal species have their own practices and rhythms regarding space and time and embodied life, and that these can play out differently to ours. Orienting a community beyond human concerns thus means not only becoming attentive to the needs, interests, and practices of others but also listening to and opening community to other than human rhythms. In turn, becoming attentive to the possibilities of multispecies relational rhythms, reminds us that all embodied lives, practices and relations – whether human or not – always play out through and within their environmental contexts.

Rewilding Our Communities

There is an important point here which is that practices of attentiveness, immersion, and neighbourliness make both communities and shared worlds. We can continue to make communities and worlds that ignore or downplay other-than-human interests and sentience, or we can start to think and practise our living in more neighbourly, other-regarding ways. Crows are once again useful guides here insofar as they are not waiting for an invitation or the achievement of specific conditions to live alongside us and other species in the shared world of a multispecies community. Indeed, while crows have been considered problems for and by human residents because they live to their own rhythms, most Australian city and regional councils now try to educate (human) residents about how to live alongside crows, albeit in ways that try to discourage crows developing 'unnaturally larger than normal numbers' (City of Melville, n.d.). Thus, in the words of the City of Melville's Living with Ravens Information Sheet, while crows need to be 'accepted' as 'part of our urban environment', they – meaning individual crows, their population, and their environment – also need to be *managed* so that they do not become a *pest*:

> Reducing artificial sources of food by altering waste management practices will be the most effective means of regulating raven numbers in the metropolitan area. Bins must be fitted with self-closing lids and all litter and food scraps should be disposed of correctly.
>
> The feeding of ravens is strongly discouraged.
>
> Effective long-term control of raven numbers requires a concerted effort by the community to devise better waste management practices. It is only with good planning and application of these management practices within your suburb that raven numbers will gradually decline over time. (City of Melville, n.d.)

Whilst such information does mark a recognition of crows as *naturally* resident in our communities, and is suggestive of some level of a multispecies neighbourliness, it does still mark a very constrained hospitality. Specifically, such advice hearkens back to the deep-seated idea that ecological communities operate in some kind of *natural* balance that should be maintained and protected. Such a view is made explicit by the Brisbane City Council (2019) in the context of crow numbers: 'Brisbane City Council has adopted "working towards a natural balance" as the principle in controlling the crow population'. The problem, as noted previously, is first that this tends to foreground ideas of stability (in numbers, species) over dynamism, and second tends to ignore the impacts of our own presence. That is, this idea of balance only rarely considers the burgeoning human populace as comprising part of any such balance – another instance of our anthropocentric tendency to not factor ourselves into our thinking and actions regarding what we consider the appropriate balance for a multispecies community. Living to a balance appears, in other words, to only be applicable to the members of other species. Such points raise the following questions. If human community practices also suit crows – or can be repurposed by crows to suit their needs and interests, allowing them to thrive alongside humans – then on whose basis and for whose interests are we striving to regulate crow numbers? We call for and reward innovation and adaptability in our own practices, why are we so antagonistic towards such capacities being demonstrated by members of other species? Is our concern with regulating crow behaviour and numbers not just additional evidence of the entrenched nature of our anthropocentrism? Does it not exemplify another example of how we expect to be able to develop communities on our own terms whilst ignoring the increasing impact of our presence and our numbers on other species, native ecosystems and in multispecies communities?

Of course, this idea of balance can be used in other ways that strive to actively support the presence of species missing from a multispecies community. This view is behind the reintroduction of native species – often the smaller marsupials – back into ecological communities from which they had disappeared as well as attempts to control or eradicate the members of feral species. It is also behind attempts to control members of thriving native species – crows, for instance, or kangaroos, as noted in Chapter 5 – if their numbers are considered to hinder the capacity of other species to thrive. Thus, although these ideas of balance are highly problematic and paradoxical, they are behind a suite of initiatives gaining traction in Australia which are striving to try and *rewild* not only regional and rural Australia but also urban and suburban human-oriented communities. This work increases recognition that not just wilderness and rural Australia, but urban and suburban communities, have the capacity to support multispecies populations. Basically, *rewilding* is the idea of restoring ecosystems and species interactions to promote complexity and self-sustaining ecological communities and ecosystems (Fernández et al., 2017; Pettorelli et al., 2018).

In Australia, support for rewilding has developed from several factors. These are increased recognition of the escalating threats to biodiversity (Cresswell & Murphy, 2017), the need for new approaches to reverse species' decline, and the broad complementarity of rewilding with other conservation initiatives (Sweeney et al., 2019a). Rewilding efforts also fit well with the kinds of calls for better environmental stewardship and sustainability that we explored in the previous chapter. Rewilding efforts can be active in the sense of the deliberate reintroduction of species into areas, particularly those that allow for the limitation or eradication of non-native predators and other competitors – islands, peninsulas, other fenced enclosures, for example (see, e.g., WWF, 2019). They can also be passive in the sense of simply meaning the removal or reduction of agricultural (including fishing) practices in a particular area under the expectation that native plant and animal species can and will recolonize that area (Sweeney et al., 2019b). Rewilding also includes initiatives and programs that seek to broadly encourage biodiversity, ecosystem function and the persistence of native species in urban and suburban settings (Maller et al., 2019). In urban and suburban settings, such initiatives have also been framed in terms of *bringing nature back into cities* (Clarkson et al., 2007; Mata et al., 2020; Matthews, 2020). Such projects are typically designed to encompass both human-supported and *natural* colonization of urban environments by native species and target both public and privately owned land. This is the point that creating the capacity for native wildlife to live and move safely within urban and suburban regions is extremely difficult to retrofit into existing suburbs without the inclusion of private residences insofar as private land ownership is a primary land use throughout such regions (Perth NRM, 2021).

Outside of urban and suburban regions, rewilding initiatives in Australia are usually the province of large organizations such as Australian Wildlife Conservancy, Arid Recovery, Bush Heritage, Rewilding Australia, and WWF-Australia which may work in partnership with governments, national and regional parks, First Nations Peoples, wildlife breeding programs (often run through Australian zoos), and other large landholders (see, e.g., Arkaba Conservancy, 2017). Given such initiatives prioritize '"landscape scale" issues of species interactions' (WildArk, n.d.), projects are typically constructed so that they can meet four main aims. These are to help: (a) preserve biodiversity or threatened species as well as more broadly protect wildlife and their habitats; (b) maintain a functional natural landscape; (c) meet strategic conservation goals; and (d) support viable populations of species and ecosystems for the long term (Bush Heritage, 2021). Initiatives are also more likely to be active in nature. This is the point that, in Australia, simply removing agricultural practices from a region, for example, is unlikely to halt declines in biodiversity unless deliberate steps are also taken to control pest plants and animals and to reinstate ecological processes. Initiatives thus tend to prioritize some or all of a range of objectives. These include, (a) integrated pest management programs (ideally using native species); (b) the introduction of sustainable agriculture programs that actively strive to maintain natural biodiversity alongside their focus on agricultural productivity; (c) habitat restoration programs, including the reinstatement of ecological processes, some of which are themselves performed by native species (such as digging marsupials that play the role of soil engineers). They can also include, (d) threatened species recovery programs; and (e) economic renewal programs that aim to boost local economies through increased farm productivity and tourism (see, e.g., Menz & Sharp, 2018). A key point is that for such rewilding efforts to achieve optimal outcomes – specifically, achieving broadscale sustainable regional and rural rewilding – they require the development of longstanding partnerships and significant commitments to sustainable and rewilding practices by large landholders. Such commitments, if achieved, will change the face of Australian agriculture and other standard development practices.

It is also worth noting that smaller habitat protection and local environmental stewardship initiatives can also align well with rewilding aims. These generally arise from a specific community's aims to protect a particular habitat and thus preserve a local and vulnerable population – such as, for instance, turtles on Bribie Island which are under threat from the decision to allow increased vehicle access to Bribie's ocean-side beaches (see, e.g., Bribie Island Environmental Protection Association, n.d.; Ogden, 2021). Particular species may also be the focus of action, such as gliders which are under extensive threat from habitat loss from fire and continued land clearing (e.g. Eger, 2020) or Fairy Terns which are endangered due to increased human activity on the beaches they nest on (Birdlife Australia, n.d.). Such initiatives typically follow the path of building public awareness and support through activist campaigns and petitions (often using such sites as Change.org), and then using this support as leverage for changing the views and policies of a local council or of the appropriate government ministers. As such, campaigns and petitions tend to target specific government ministers and other decision-makers, and further encourage supporters to contact those decision-makers directly. Campaigns may also engage various models of crowdfunding to raise funds for such items as covering any legal costs or funding grassroots conservation activities (Gallo-Cajiao et al., 2018; Rodan & Mummery, 2018). Such actions align well with the increasing emphasis on including local communities and resource users in rewilding, conservation and environmental stewardship policies, programs and practices (Bennett et al., 2018). They further stress the idea that state, territory and federal governments need to be held accountable by local conservation groups with regards to the success of such measures.

This kind of local community focus also extends into urban and suburban efforts at rewilding. Efforts might be developed and supported by projects run by city or town-based organizations that engage community groups, local councils and schools as well as individual residents – see, for instance, Perth NRM (2021), NatureLink Perth (n.d.) – but may also be developed and shared through a variety of programs targeting individual residents. One such program in Australia that promotes individuals engaging in rewilding aims has been *Gardening Australia* (see e.g. Lomas, 2015–21), a long-running gardening and lifestyle television program on Australia's national broadcaster, the Australian Broadcasting Corporation (ABC) free-to-air and ABC iView with a Facebook page, social media presence, an associated magazine (569,000 readership), a plant-identifying app and other spin-off products such as books and DVDs. *Gardening Australia* airs during prime time viewing with a consistent audience reach of 800,000 plus viewers every episode making it one of the premier gardening programs on Australian television (ABC Gardening Australia Magazine, 2021). First broadcast in February 1990 and now in its 32nd season, *Gardening Australia* has consistently promoted organic and environmentally friendly gardening methods throughout its episodes. It also provides advice on growing native plants and stresses the role home gardeners can play in developing biodiversity and extending habitat and food sources for native birds, animals, reptiles, and insects. As one of the ecologists featured on episode 22 in series 30 in 2019, Mandy Bamford, states, 'Biodiversity is really important in gardens, because it helps support populations of animals within our cities' (Gardening Australia, 2019). More specifically, planting local natives – meaning the species that once grew in Australian suburbs – helps to recreate the connections that originally existed between plants and local native wildlife.

In addition to gardening tips, the show also presents information and instructions for making a variety of nesting boxes to support multiple native species. As is stressed in multiple episodes, including appropriately situated habitat or nesting boxes in home gardens is important because natural tree hollows – essential for over 300 native Australian vertebrate species for roosting, nesting, and raising young – are an increasingly scarce and valuable resource, particularly in urban and suburban areas (see, e.g., Figure 14; Gardening Australia, 2020a). This is due to the tendency for old, hollow-bearing trees to be removed in these areas either as safety risks or for aesthetic reasons.

Figure 14. Nesting boxes installed at a height of ten metres in trees at the Fremantle Arts Centre, Western Australia, during an artist's residency in the spring of 2009. The residency culminated in an exhibition entitled 'to hear the language of birds'. Nesting boxes were in use by galahs when the artist visited the site a year later. *Source:* Photo courtesy of Paul Uhlmann.

Other suggestions provided in various episodes include leaving hollow, fallen branches in place if possible as they will become valuable habitat for lizards and small mammals (Gardening Australia, 2016), including multi-level plantings so as to cater for the different local species in the area (Gardening Australia, 2012) and retaining leaf litter as it 'provides habitat for skinks and feeds fungi, which in turn feeds a huge number of insects, then many birds, marsupials and lizards eat the insects' (Gardening Australia, 2020c). Gardeners are further advised to plant local species where possible as they 'will flower and fruit at times that are predictable for native species and will help maintain local gene pools' (Gardening Australia, 2020c). They are also encouraged to create a pond and/or bog to support local frog populations (Gardening Australia, 2018, 2020c). Also stressed is the importance of providing clean water for nonhuman visitors, and of stopping or substantially limiting the use of pesticide sprays. Considered together such points are reminders that the plants we choose to grow, how we structure our garden, as well as the pest control methods in use all determine whether our gardens will be friendly to and supportive of visiting native wildlife.

It is finally worth noting that in being shared through the format of a popular television lifestyle program – and *Gardening Australia* at different times has been recognized as one of the highest rating programs after news programs (Tutty, 2018) – such advice is reaching an audience already interested in gardens and gardening. Projects such as building and mounting nesting boxes, or creating a frog pond and bog, are also small and do-able by individuals, and are presented as just another kind of productive garden work to do. Such ideas and advice are also reiterated across a multitude of other sites such as other gardening and sustainability programs and organizations, and local councils. They all share the same rewilding message that encouraging the provision of habitat in urban and suburban settings is key for the sustaining of biodiversity and the support of native species.

> The next time you are in your garden, try to imagine what your suburb might have been like prior to urbanisation. What kind of plants would have grown there and which native animals might have lived in your area? [...]

> Protecting areas of remnant native vegetation is essential to preserve and maintain biodiversity. You can help by planting local native species in your garden. (Government of South Australia, 2016, p. 4)

Residents are also advised to remove and not plant species recognized as environmental weeds given their capacity to spread into bushland and further degrade native plant communities. Local councils typically provide information on the plant species classified as environmental weeds for their location and may also provide incentives to residents to include more local plant species in their gardens.

Private gardening for wildlife initiatives remain uneven in their impacts given their high dependence on individual interest, commitment, and activity. In addition, given that properties change hands, individual initiatives can also be reversed if new landowners do not share such commitments. As recognition of the threats to Australian biodiversity spreads, however, local councils are increasing their commitments to protecting remnant vegetation, improving habitat, increasing green canopy, and creating havens for locally native species on public land. This might be through planting more local species on road verges, in parks and along waterways and coastal dunes, deliberately carving habitat hollows and/or installing habitat boxes in mature trees, implementing measures to control feral animals, and, in certain councils, requiring companion animals such as cats and dogs to be secured at night to reduce their capacities to predate native animals (see, e.g., City of Ballarat, 2019). Building safe wildlife crossing points for where main roads intersect with ecological corridors is also being trialled in some councils, including culvert underpasses, land-bridge overpasses, koala refuge and glider poles and rope bridges (see, for instance, Brisbane City Council, 2020; Hastie, 2018). Other typical initiatives include the mainstreaming of consideration of biodiversity within council and city programs, supporting community volunteers engaged in bushland restoration activities, promoting community activities to keep habitats clean and plastic free, and in some cases the use of incentive packages to encourage vegetation retention and increased use of local plants in landscaping by developers and private landholders (see, e.g., City of Canning, 2018; City of Rockingham, 2021). Local councils are increasingly also implementing a variety of measures to ensure their initiatives toward habitat protection and extension deliver improved biodiversity. Such measures typically include (a) the introduction of flora and fauna surveys, vegetation mapping and habitat assessment, (b) the identification of priority sites, supporting sites and priority fauna species, (c) the identification of potential habitat linkages to enable safe movement of wildlife, as well as (d) the identification of threats to biodiversity (see, e.g., City of Fremantle, 2020; City of Sydney, 2020). All such initiatives are thus aimed at improving habitat and building corridors for local species – work that supports the rewilding of urban and suburban communities. This is a strong message in the City of Sydney's *Urban ecology strategic action plan* (2020, p. 2):

> The City's vision for the Plan is to restore and conserve resilient urban ecosystems that support a diverse range of locally indigenous flora and fauna species, and in so doing to create a liveable City for all of its inhabitants.

Building Multispecies Communities

What these various initiatives have in common is a clear commitment to the idea that our communities can – indeed, should – be more than human. They should be open to and supportive of more than human residents. At the same time, however, these commitments are rife with inconsistencies. For instance, urban and suburban spaces should provide habitat for local wildlife as well ourselves (and our valued companion animals), but only it seems if the balance of species is considered appropriate, and only if such provision of habitat suits our preferences. This is the point that such a focus on restoring and conserving habitat and ecosystems is always partial. Such work is also typically carried out in reaction to the levels of land clearance and development standard in urban and suburban environments; ideas of conservation do not drive our urban and suburban development practices. In addition, such a focus is also led by ideas regarding the *natural* or *appropriate* presence of other species in our communities. The trouble is, despite the prevalence of such phrases as *maintaining a natural balance* and *preventing unnatural population increases*, there are no reliable metrics to use in the making of any such evaluations. Despite this, these are typical phrases in urban and suburban policies regarding crow populations and are also used with reference to other native species whose populations can be considered to get out of hand (e.g. kangaroos).

First, the belief that we humans understand the appropriate balance of species in any ecological community *out there* in the seemingly undeveloped bush is itself highly problematic. This is because of the immense direct and indirect footprint of Australia's regional and rural activities on bush habitats. Examples include, (a) the clearing of land for agriculture, mining, and other activities; (b) clear felling and other logging operations; (c) the introduction of non-native animal and plant species; and (d) the reduction of access to water sources, habitat, and requisite territory for native species. That is, direct changes in ecosystems due to human-oriented development activities – or indeed due to fire or flood events – always mean additional indirect impacts on so-called untouched habitat, whether from displaced populations of animals or the diminished resilience of remnant habitat. The result is that there are very few natural ecosystems that have not already been impacted by anthropocentrism. Thus, it is hard (and perhaps futile) to set ecological benchmarks based on some presumed unmodified state of nature (Woinarski, 2016). Second, urban Australians, including local government, are very new at the work of deliberately supporting even a minimal version of a multispecies community in urban and suburban settings. The newness of this work is further exacerbated by the effects of climate change. This is the point that many species – animals and plants – will need to be able to change their locatedness in unpredictable ways in order to thrive as climatic conditions change around them (Pecl & Hobday, 2016).

A further challenge here is to do with the pragmatics of supporting wildlife in urban and suburban settings – specifically regarding which species should be reintroduced and/or welcomed. Typically, this has been tied to the anticipated capacity of Australia's wildlife species to be able to flourish in rewilded urban and suburban spaces while not becoming seen or experienced as problems or pests. Considered in this light, the size of the species matters, along with both the size of requisite habitat and the capacity of individuals to move safely through existing wildlife corridors.

Indeed, Australian urban rewilding programs tend to focus on the reintroduction and support of pollinators (e.g. some insects, birds, and members of the flying fox and glider families), local birds, reptiles, and small mammals (Sweeney et al., 2019a). More generally, it has been suggested that non-human native species should be thoroughly assessed for their suitability for being brought back into urban and suburban environments. It is suggested that such assessments make specific reference to a species' perceived cultural value, social acceptability, conservation significance, ecological feasibility and economic viability (Mata et al., 2020). In practical terms, of course, a species' perceived cultural value and social acceptability – including both its alleged *charisma* and its potential to be disruptive of standard human rhythms of living – is going to play a major role in determining its potential welcome and support by human residents. This is, of course, a highly constrained and still anthropocentric model of rewilding, one which will be unlikely to support the reintroduction of all of the species that would have previously inhabited urban and suburban spaces – particularly larger species such as kangaroos or dingoes. The broader point here is that the success of urban and suburban rewilding initiatives – and thus of building and sustaining even partial multispecies communities – seems to rest on the important requirement of rewilding not just urban spaces but ourselves (Lezy-Bruno, 2017). And this requirement needs us to not only become open to living with other species in a multispecies community, but to start to actively live in ways that enable and support this. To help address this requirement with regards to building and supporting multispecies communities, we here foreground four lines of thought and practice drawn out of our previous discussions that will be important for any such reorientation to a multispecies community.

Sharing. Contemporary urban Australians are new at sharing with other species. Namely, we are new at recognizing that other species than just our own have the capacity to use and re-purpose our community facilities, and in accepting such practices as being as legitimate as our own. Although, we might unknowingly share our spaces and facilities with some mammals and reptiles who are nocturnal, we are new at actively sharing time, space, and environmental resources with other species, and in thinking that we should share space and such resources in this way. We are new, in other words, at being generous and compromising outside of our own interests – outside of our anthropocentrism. Our historical and ongoing lack of capacity here is of course about how it has been considered appropriate and even normal for our anthropocentric development practices to override the habitat needs of other species. This has been to the point of biodiversity and species collapse that are the markers and crises of our time. It is also the story of our heavy-handed decision-making over what the appropriate membership and balance is for a community – and our forgetfulness that we are always living in a multispecies community. As a fundamental value for a multispecies community, sharing, however, is something we can learn through recognizing that friendship and kinship can be extended beyond our own species. After all, some of us already know how to share with friends and kin. We know what it means to share our lives with loved companion animals, and the rising focus on supporting urban and suburban wildlife suggests we can certainly learn to share with at least some other animals. As we have noted, some ideas about how to do this are publicly available, are gaining traction in Australian communities, and are being implemented by individuals, councils and other community groups.

Caring. No matter how trite it sounds, sharing is based on caring, which in turn stands for a threefold focus: an emotional orientation, an ethical obligation, and a commitment to practical labour (de la Bellacasa, 2012). Recognizing our need to (more) fairly share our communities, environment and resources with other species is reliant on recognizing that we care for these others – that we care that they have a future in this country, as species and even (sometimes) as individuals. As we have shown, such care can be built from the recognition of the sentience – or, in some instances, the rights – animals share with us. In this context, and as set out in Chapters 2 and 3, care underpins attempts to change the conditions for our use of livestock animals, as well as our attempts to reconsider some of our own eating and consumption habits. In the same context, care is thus behind the gradual and future societal shifts with regards to the appropriate treatment of (livestock) animals predicted in the Futureye (2018) report. Care also drives Australians' (and others') generosity in responding to calls to fund and otherwise support wildlife hospitals in the aftermath of disasters such as the Black Summer fires. It also underpins Australian stories of responsibility and stewardship, according to which we recognize that we have an obligation to protect other than human species and ecological communities for both our own and their own benefit – an obligation that is played out in attempts to rewild our spaces, protect vulnerable native animal populations and species, and conserve existing ecological communities such as the Great Barrier Reef. Care thus underpins our capacity to make space in our understandings and practices of community for more-than-human others – namely, to be tolerant of and hospitable to other rhythms and practices of life and living even when they do not neatly fit with our own preferences. Care finally underpins our capacities to make and recognize kin and friends.

Interspecies Solidarity. Care also has the potential to promote multispecies ideas of community by underpinning ideas of an interspecies solidarity. Basically, solidarity is the political expression of care, compassion, and empathy, and involves the provision of support despite differences – in this case, support across species distinctions. The idea of interspecies solidarity is thus a reminder that others should not have to be like us for us to care about and to act on behalf of their wellbeing. It is also a reminder that care needs to become political (Cobble, 2010; Herd & Meyer, 2002; Tronto, 1993). Care needs to function to change things, to create more caring communities (e.g. Glenn, 2000; Tronto, 2013). This is the point that, with care marking not just an activity but a political value and a 'tool for critical political analysis' (Tronto, 1993, p. 172), the enactment of care as solidarity encourages 'not just personal transformation but social transformation' (Scholz, 2008, p. 61). This is not to say that such transformation must always be a collective effort. Individual acts of solidarity matter, and they can be effective in disrupting dominant perceptions and power relations (Coulter, 2016a). As we have also seen from consideration of the work of animal activists and advocates in earlier chapters, such acts can also set a domino effect in motion which can propel a broader set of processes. More specifically, acts of interspecies solidarity and a multispecies approach to the work of care and community can be read as challenges to strengthen and expand our thinking and practices to overcome entrenched anthropocentrism. They mark a call therefore to not only start to consider other, more-than-human patterns and rhythms of living, but to also strive to consider our actions and patterns from the perspectives of other animals. This is the work of recognizing that we are always in interspecies and multispecies relationships and community with others but that we also need to recognize ourselves as being called upon to be responsive and responsible in these relationships – meaning that we need to overcome our tendencies toward anthropocentrism and become properly attentive to the 'needs, interests, desires, vulnerabilities, hopes, and sensitivities' of these others (Gruen, 2014, p. 3).

Understanding. The work of solidarity is by no means simple. Every aspect of it demands not just that we come to care and to see animals as deserving of care, and that we practice that care – in the forms of recognizing kinship and more-than-human residents in a multispecies community – but that we also come to understand how to practise care and solidarity across species lines. It is only through such work, after all, that we can gain enough understanding to (a) 'foster the circumstances and trusting relationships within which animals' can display and pursue their preferences, and then (b) 'interpret the signals that animals give regarding their subjective good, preferences, or choices' (Donaldson & Kymlicka, 2012, p. 4). This is, of course, the work of attentiveness and immersion we have mentioned earlier, the work of building trust and interspecies relationality and solidarity. It requires open-mindedness, patience and persistence (Anthony, 2009, p. 196), as well as attentiveness to learning and respecting species-specific needs. And yet, this is only one level of the understanding that needs to be developed. The other concerns our own anthropocentrism and the uncaring, unsharing, and inconsistent practices it has enabled, normalized, and, as such, hidden from our critical attention. That is, we need to come to understand the full impacts of our anthropocentrism, and to recognize and challenge the various inconsistencies in the treatment of animals it has allowed. This is the point that there has been extensive and successful campaigning to increase care for livestock animals who are caught in intensive farming and live export practices that give little to no consideration to their sentience or wellbeing, and for native animals who are killed, injured or made homeless through the Black Summer fires or other disasters. Yet, at the same time, we do not register that our conventional practices in agriculture, mining, forestry, and urban and suburban development consistently spell genocide to entire ecological communities and the species populations that comprise them. Why is it still difficult to accept that it is our development practices and assumptions regarding economic growth that are feeding climate change and the biodiversity collapses this is generating? Without question, then, consideration of interspecies solidarity is a challenge to not only understand what animals are thinking and feeling, but to also change our practices in order to better respect their experiences and lives. It is the obligation to recognize that:

> Animals are inextricably and intimately interwoven with our work, lives, and futures. A just and caring society cannot condone the exploitation and oppression of others, and cannot be built atop a mass, unmarked animal graveyard. (Coulter, 2016b, p. 213)

Put another way, it is not until we develop critical awareness of the anthropocentrism entrenched in our development and living practices as well as in our assumptions regarding economic growth, that we will be able to properly build and maintain the caring, sharing and solidarity relationships and practices needed for a multispecies community.

Together these lines of thought and practice make clear that the possibilities for multispecies cohabitation and communities 'depend on designs that consider nonhuman as well as human cultures' (Roudavski, 2020, p. 732). Indeed, it is only through becoming and remaining attentive to the requirements and everyday rhythms of different species and their interactions with us and each other that we will be able to make kin and live *with* other species in interspecies solidarity and multispecies communities. More specifically, the work of building multispecies cohabitation and community necessarily involves finding ways to extend our ideas regarding participation, accessibility and inclusive design beyond human-oriented outcomes, activities and rhythms of life (see also Meijer, 2019). As we have noted, this might mean protecting and increasing habitat in urban and suburban zones, deliberately designing infrastructure – such as buildings and other structures, roads and pathways, lighting systems and waterways – to support more than our interests, and being generally more attentive to meeting the disparate needs of our wildlife, our companion and livestock animals, and ourselves. Models of inclusive design also stress the need to rethink what have been standard assumptions and practices with regards to land clearing and land use for (our) development. Along with the implementation of explicit rewilding initiatives, this might encompass several approaches: the redevelopment of agricultural and other development practices to better cater for the interests of both livestock and Australia's native wildlife; the protection of remnant habitat, giving priority to linking and extending habitat; and the development of safe passages for wildlife through artificial barriers (such as roads). Such initiatives would allow native animal populations the capacities for safe movement they will need to survive Australia's changing climatic conditions and would further provide them better resilience in facing other challenges including predation and competition by feral animal populations. Inclusive design would finally also consider the interests of the non-native animals now embedded in Australian ecological communities, and the interests of those ecological communities.

There are three further factors that inform this work. First, as we have noted, the work of building and supporting multispecies cohabitation and community is primarily driven by our willingness to be attentive and responsive to the needs and interests of nonhuman animals. We need to learn to ask who might enjoy or suffer from current and proposed community practices, and how. As we have outlined, this is based on our attentiveness – our willingness to share, care, and work towards interspecies solidarity and understanding. Second, it needs to be recognized that practices that might enable and be supportive of interspecies solidarity and multispecies community are not driven by one-size-fits-all propositions. That is, such practices and living must be based on and informed by multiple modes of interaction and engagement that remain open and responsive to the diverse needs and interests of all the stakeholders for a multispecies community, human and animal. There must also be acceptance that needs and interests can clash, and that there can be no templated solutions for resolving such clashes. Measures must rather be experimental and provisional – never taken as more than the best option for right now. This makes the third point one of recognizing that multispecies cohabitation and community requires us to both (a) actively make provision for other than human interests (given that the anthropocentric legacy is one of prioritizing human interests) and (b) accept that human interests and design expectations are not the only drivers in the development of alternative cohabitation and community spaces, practices and rhythms. This is the point that animals may themselves modify or otherwise appropriate our designs as they interact with them – and that this must be allowable. This is to remember that we are not the only ones involved in cohabiting and building multispecies communities.

There is, of course, one final point to note here regarding the various practices set out in this chapter that we have suggested might help in the development of multispecies communities and models of cohabitation in Australia. This concerns consistency. The issue here is that all of these practices mark only partial investments in multispecies cohabitation – they are still speciesist in fundamental ways even as they suggest some loosening of the hold of anthropocentric assumptions. We see two main factors in play here. First, growing awareness of Australia's biodiversity and its biodiversity crisis has strengthened the sense that there are incompatibilities between the interests of different cohorts of animals – particularly between Australia's native animals and the introduced animals that are identified as causing damage to native ecosystems and to native animal populations. For instance, extensive and highly public discussion of the *feral cat problem*, for example, has normalized a view that not all animal interests should or can be protected within the Australian context. That is, protecting the interests of feral cats can mean a failure in the protection of the interests of Australian marsupials, birds and lizards, and vice versa. Given the increased awareness of such problems and that their resolution is inevitably speciesist, inconsistencies in ideas and practices regarding the protection of animals in Australia become normalized.

Second, as we have noted throughout previous chapters, Australia – at least at this stage – is much more strongly oriented to welfarist rather than rights thinking when it comes to recognizing and protecting nonhuman animal interests. We see this as underpinned by two separate lines of thought. The first of these is that with Australia's settler culture having been very strongly oriented to agriculture and thus the protection of agricultural interests, there are longstanding narratives normalizing the use of animals. As we have stressed in previous chapters, these are still commonplace although they are undergoing some revision. Australia is still predominately a meat-eating culture; animal agriculture is still a major economic contributor. In this context, although animal rights-based and vegan arguments continue to be commonly dismissed in Australian political, economic, and public spheres as extremist (even as there is increasing interest in vegan models of eating and consumption), sentience-based welfarist arguments are gaining in traction. Thus, although welfarist arguments are enabling changes in ideas as to the acceptable treatment of animals, they also do not see that the human use of animals as unacceptable. Such a perspective thus strives to significantly strengthen animal welfare protections but does not call for the complete dismantling of the industrial animal complex, for example. The second line of thought is drawn from Australian First Nations ideas of *care for Country*. As we have noted, while *care for Country* certainly aligns to at least some extent with ideas of interspecies solidarity, it also does not reject practices of animal use. Rather, both animals and animal products are extremely important in the continuation of Australia's First Nations cultures. As resources but also as part of *Country* and First Nations culture, animals must, however, be treated with respect. Species must however be managed for their preservation in *Country*, as their loss would irreparably damage the relationships comprising *Country* and culture.

> The impact of the loss of our totem animals is enormous. When our totem dies, our connection with the spirit is compromised. The spirit and totem are as one and once our totem dies, a bit of our spirit dies as well. We feel the whole of Country in ourselves and its loss is felt in our whole spirit, not just the body that carries the spirit. (Victorian Aboriginal Heritage Council, 2020, p. 33)

Although levels of understanding of Australia's First Nations cultures are still uneven among non-Indigenous Australians, there is increasing recognition and appreciation of Australia's First Nations' Cultural Heritage (Victorian Aboriginal Heritage Council, 2020). Indeed, as we have noted earlier (see Chapter 5 in particular), Australian First Nations' perspectives are starting to gain traction in the work of land and sea management in Australia. Although it is clearly incorrect to describe the work of *care for Country* as welfarist given that it is driven by a very different set of foundational beliefs to that of sentience-based animal welfarism, it is apparent that the work of *care for Country* aligns better with welfarist rather than rights-based views.

While these points together make clear that even when beginning to orient to considering animal interests and strengthening ideas and practices towards multispecies cohabitation and community, Australian patterns of thinking and living remain speciesist and inconsistent. Of course, this is not a purely Australian problem. Indeed, a review of various multispecies arguments has also indicated that common and promoted conceptions of the *multiple* still tend to quietly exclude 'highly commodified and instrumentalized animals' such as livestock (Arcari et al., 2021, p. 951).

> What does it say about a city that aims to become 're-enchanted', 're-natured' or 're-wilded', in reflection of an idealized conception of entanglement and conviviality with 'nature', if within that same city other 'natures' are still slaughtered, bred, traded, confined, raced, tested on, put to work, abused and killed? (2021, pp. 954–5)

These are important points, and certainly any ambitions of fully realizing a multispecies community would require humans to consider their practices of animal use extremely carefully (whilst also recognizing that such a community will inevitably still encompass some patterns of predation). What needs to be stressed, however, is that such ambitions are a very long way from being realized in the Australian context. Hence, whilst we certainly recognize the inconsistencies in the various ideas and practices that have been explored in these chapters with regards to their protection of animal interests, we remind readers that these ideas and practices still need to be understood as marking steps towards change. They illustrate, in other words, points where counter-narratives of sentience, care and kinship would appear to be gaining in public traction with regards to their challenging of anthropocentric norms. They illustrate at least some steps towards the revisioning of Australia's anthropocentric social imaginary, thus contributing to the development of some beginning patterns of multispecies cohabitation.

Note

1   As we have explained in Chapter 5, there have been multiple claims within the Australian public sphere for the need for a *war* on feral cats. Described as 'public enemy number one' (Hollingsworth, 2019), cats are commonly presented as needing to be eradicated or otherwise controlled in order to protect Australia's native wildlife. In this line, as we have outlined previously, there is broad acceptance of the need to both cull feral cats and increasing awareness of the need to better regulate the lives of domestic cats through cat registration, curfews and spaying and neutering so that they do not predate on native species (see Commonwealth of Australia, 2020; Ham, 2021). While cats are the current target of this rhetoric of a need for *war*, such descriptions of going to war against the members of particular species have been and remain common in Australia's settler culture, particularly in connection with the perceived need to protect agricultural interests. While much of the killing of members of so-called pest species (in this instance, thylacine, kangaroos, dingoes, wedge-tailed eagles, as well as rabbits and foxes) has historically been carried out by farmers and those they contract, it is worth noting that many of these *wars* have been government sponsored, with some even involving the Australian armed forces. 1932, for example, saw the Seventh Heavy Battery of the Royal Australian Artillery embarking on actions using machine guns to attempt to eradicate emus in the Western Australian wheatbelt after complaints from farmers – actions now known ironically as the *Emu Wars* (see Gore, 2016). Western Australia was also the base for a war on sparrows in the 1950s, with government *Sparrow Rangers* armed with shotguns to eliminate sparrows on sight, and another government sponsored war on starlings in 1971 (Australian Associated Press, 2006). Similar patterns of violence against designated *pest* species have been common in all of Australia's other states and territories.

POSTSCRIPT

Making Change

Having outlined some of the moves towards practices of interspecies solidarity and multispecies cohabitation that are gaining in visibility in Australia – including some of those embodied in Australia's First Nations' ideals of *caring for Country* – what stands out is how far Australians generally are from realizing ideals of interspecies solidarity and multispecies community. And yet, attentiveness and concern for better protecting animal interests and enabling their wellbeing are taking hold in Australia. This is shown not just by the multitude of calls to action on behalf of animals that are being shared throughout the Australian public sphere but by the diverse forms of support such calls are receiving. As we have seen, this support can be manifested in a variety of ways. These include (a) responding to activist and fundraising calls, lobbying politicians, and participating in direct actions calling for improvements in animal welfare; as well as (b) making changes to everyday shopping and eating habits. Many Australians are also (c) striving to support wildlife and taking part in community activities, stewardship and citizen science projects aimed at supporting ecological communities (like the Great Barrier Reef) and the rewilding of urban and suburban spaces. Concern for animals is also shown through (d) an increase in challenges to the inconsistencies and cruelties anthropocentrism has permitted in the treatments of animals in Australia. Such efforts illustrate a broadscale stepping up in support of calls for the better protection of animals (and their habitats) from systemic cruelties and disregard. They further illustrate growing interest in practices supporting interspecies solidarity, and multispecies cohabitation and community, including those drawn out of First Nations understandings of *caring for Country* (see, e.g., Weir et al., 2011; Woodward et al., 2020). Such interest might also support a shift away from a thin, abstracted benevolence toward animals to recognizing a thicker, more concrete and care-full interconnectedness with animals. Also illustrated is the development of new ways of thinking and even the development of new norms indicative of a potential shift in social practice. Together these points suggest that assumptions and practices regarding human-animal relations are slowly starting to change. Such manifestation of public support for animals to live better protected lives is, of course, in alignment with the predictions set out in the 2018 Futureye Report regarding progress towards mainstream recognition of sentience in animals. It also marks an extension of those predictions given that the focus of this report was primarily with Australians' engagements and relations with livestock animals.

Our question is whether these various actions and shifts in attitude indicate the beginning of a new social imaginary regarding human-animal relations in Australia, one that is starting to divest from anthropocentric assumptions and practices. This is a difficult question to answer. As we have outlined at the beginning of this book, a social imaginary stands for a collective understanding about how the world should be and how we should live. Standing for the common sense accepted by members of a society, a social imaginary reflects the modes of valuing, doing and social relation that are normative for the society in question. As we have shown, a dominant feature of post-settlement Australia's social imaginary has been its anthropocentrism. Indeed, the view of nonhuman life as resource has underpinned many of the 'most basic cultural narratives' (Plumwood, 2010, p. 32) that Australians have lived by. However, although this anthropocentrism is still clearly visible in many standard individual and collective practices, these chapters have also shown that Australians are starting to imagine that human-animal relations can take a different form and that we can come to care, act, and live in different ways. These are not ways that as yet illustrate a collective shift in understanding – as would be expected in the development of a new social imaginary – but they are examples of changes in attitude and of how a range of counter-narratives and associated practices with regard to animals and human-animal relations are gaining traction across Australia.

What is clear is that coming to live in less anthropocentric ways demands, for example, that we consider the wellbeing and lives of the cohorts of animals we engage with. For instance, whilst we might strive already to meet some of the interests of our companion animals in our models and practices of community, this consideration clearly needs to be extended to the animals we class and use as livestock – whose lives go into so many of the products consumed daily by Australians. As we have noted in Chapter 2 this means, as a minimum, implementing not simply the requirement of minimizing livestock pain and suffering but, further, of maximizing their wellbeing. This means supporting not just minimum requirements regarding nutrition and health but also considering how health and wellbeing are enabled more broadly through environment and through relationality, through being able to carry out species-specific behaviours, and through concern for mental state. As we have shown, calls for such consideration are gaining in momentum as increasing numbers of Australians demand 'real animal welfare improvements' for livestock (Bradshaw, 2019). In some ways, it is easy to recognize the necessity for improvements in welfare and wellbeing for livestock animals insofar as, even under anthropocentrism, they do start with some value to us and the welfare inadequacies in many of their standardized treatments are obvious. The challenge lies with the animals that live *out there*. As we have noted in the context of native animals, for example, these are animals and populations that many Australians only rarely have contact with and typically do not see as a responsibility or as needing intervention and care.[1] This is despite the fact that our lives and expectations constantly impinge on the capacities of wildlife to live well, particularly with regards to the retention of the habitat they need to live. While this general lack of care is being challenged, maintaining a high or consistent level of care for these animals appears difficult. Coming to accept animal habitat as worthy of protection – particularly in the face of development interests – is also an ongoing problem.

As we explored in Chapter 5, further challenges lie with the animals living *out there* that have been identified as invasive or as pests – including not just introduced animals but those considered to cause harm to human interests. These are cats, foxes, rabbits, rats, cane toads, to name a few, as well as brumbies and feral livestock animals, but also some native animal populations considered to be in *plague* proportions. These animals are typically blamed for native animal population collapses, for agricultural losses, as well as for the degradation or loss of ecological communities. As we have outlined in Chapter 5, these animals have typically been treated in violent and lethal ways. And yet, it needs to be remembered that these animals are also sentient and thus in need of consideration under the basis at least of a broad kinship in interests (to not suffer, to live a good life in accordance with species-specific needs). It should also be remembered that many of these animals are already part of Australia's ecological communities to the point that they cannot simply be 'scraped off the top of "real" native ecologies' (van Dooren, 2011, p. 290). The problem is that little effort has so far been made to figure out what it would mean and might look like if these animals and their interests were also considered a part of our communities.

Managing Conflict

Certainly, as we have outlined, multiple animals' lives can be significantly improved through paying attention to their sentience and their interests in their own wellbeing. This, indeed, is a key tenet underpinning much of Australian welfare-based animal advocacy and activism and has meant that a significant portion of this work is focused on enhancing mainstream Australians' willingness to share, care, and actively work towards interspecies solidarity and understanding (see, e.g., Animals Australia, n.d.-d). Although organizational aims regarding this work are not all neatly aligned, they do all tend to foreground issues of sentience, wellbeing, and care. Thus, earlier chapters outline some of the initiatives that are being undertaken using these ideals to better protect the animals Australians rely on and which – invited and uninvited – live alongside us. As we have noted in the previous chapter, however, such initiatives will require ongoing revision as conditions change. More problematic is that the path towards interspecies solidarity and multispecies cohabitation remains filled with conflicts between divergent interests and values. These are apparent, for instance, between human and animal interests, as well as between the interests of native and introduced or feral animals, and native and livestock animals.

As we already know with regard to the conflict between human and animal interests, such conflicts are not easily resolved. That is, there are no hard and fast rules for deciding among the competing interests of equally sentient individuals. Such conflicts do, however, draw attention to questions regarding hierarchies and the *appropriate* balance of interests. They remind us that all choices, actions, and clashes are marked by 'a tangle of diverse attachments, values, commitments, needs, available options and temporalities' (Celermajer, 2021, p. 161). In all of our thinking regarding the possibilities of interspecies solidarity and multispecies cohabitation, we thus need to become attentive as to where potential conflicts in interests might arise. Complicating this situation further, it is also important to remember that some conflicts between different animal needs are a *normal* part of an ecological community. Indeed, ecological communities always operate in accordance with intricate models of relationship, including those of 'living and dying, killing and being killed' (p. 156). There are, nonetheless, two key points that need to be remembered here. The first is to remember that the interactions of animals in their ecological communities do not 'play by our rules of fairness'. This is where it is fundamental that humans develop attentiveness towards other species' rhythms and lives. The second is to remember that 'our alterations' of the environment we all share bring 'a great deal of death and suffering to animals', and that we should all be doing much more to address these harms (Petty, 2019).

This second point marks an imperative that is squarely within our remit and essential to stepping away from our tendencies to automatically prioritize anthropocentric decision-making. This is our need to adjust our thinking from the assumption that our priority should be (near) continuous growth of (our human) population and of (our) economic profit. This is the challenge to not let anthropocentrism – or ignorance or indifference – determine our choices and actions. The focus rather needs to be on exploring what human lives and living might look like in multispecies communities (and economies) where nonhuman lives and wellbeing also hold value and are given consideration. This is difficult work. It requires the recognition that the anthropocentric mindset prioritizing growth and human interests is fundamentally unsustainable of biodiversity and has, further, been catastrophic for all nonhuman species. It is this mindset, after all, that has driven population collapses or extinctions in the case of Australia's wildlife, as well as the normalizing of cruelty towards both livestock and those animal populations considered to be invasive or pests or otherwise unwanted. We need, in other words, to examine our politics of care, to think through what it means to care across species lines and live together with nonhuman others in ways that enable multispecies cohabitation and community.

Building Collectivity

What is clear is that in order to contribute to the development of a new social imaginary, new ways of living need to inspire interspecies solidarity and collectivity. As animal advocates and activists have realised, both solidarity and collectivity can be built from moral shock, outrage, and care, and achieving such responses is consequently integral to much of their campaigning. As we have also pointed out, solidarity and collectivity can arise from the processes of developing respect, friendship, and kinship across species lines – and remembering that we can build such relationships with far more animals than just designated *companion* animals. What is also important to keep in mind is that solidarity and collectivity do not need to be built on deeply shared values and identities but can also arise from what we have elsewhere called 'weak ties' that do not require participants in social change actions to explicitly commit to or recognize themselves as part of a collective identity (Rodan & Mummery, 2018, p. 170). This is important because while some values and identities might be easily shared – such as those based around *care for Country*, respect, care and environmental stewardship, for example – others may be narrower and the cause of dissension (veganism is one such example, see, e.g., Rodan & Mummery, 2019). The key point, however, is that collective calls for change and even instances of solidarity and community do not need to be based in strong collective identities and values to be effective but can rather arise from strategic alliances or precarious coalitions that change with needs and conditions (Mummery, 2017).

Such a view supports not just the strategic alliance of individuals but of diverse local groups who might possess varying aims but can still come together with each other, and larger scale bodies, to demand, leverage, and enable change with regards to a shared area of concern. As we noted with reference to environmental stewardship activities, multi-stakeholder partnerships between local communities (including First Nations communities), civil society organizations, non-governmental organizations (NGOs), public agencies, and other funding bodies can be extremely effective (Connolly et al., 2014; Romolini et al., 2016). This is particularly the case when alliances come together and are shaped in response to local concerns and care-based attachments (Chapin et al., 2012). It is also important to remember that for the building of modes of kinship, cohabitation and community that interest us in this book, alliances and coalitions must also include and involve nonhuman participants in ways that listen to and support their diverse needs and interests. This might be through building modes of care, kinship, and cohabitation via an ethic that strives to be responsive to nonhuman interests, but may also come to the point of envisaging and implementing new models of citizenship and sovereignty able to support some nonhuman rights (see, e.g., Donaldson & Kymlicka, 2011). It goes without saying that these models would benefit from attentiveness to and respectful engagement with some of the understandings of multispecies kinship and care inherent in First Nations' ideals of *caring for Country*. Ideas of weak ties, precarious coalitions, and strategic alliances matter in each of these instances as they can help us step away from the divisions and hierarchies set by anthropocentrism. They can also support us in making further change as interests and conditions change and our levels of understanding grow regarding nonhuman needs.

Bearing Witness

The final dimension in challenging our anthropocentrism is to remember that we are called to bear witness. Media reportage, activist campaigning and the work of diverse cultural intermediaries have brought all Australians to bear witness not only to Australia's collapsing biodiversity – from unchecked land clearing and development, climate change, wildfires, marine heatwaves and other such events – but also the inhumane treatment so many animals receive. We also bear witness when we see and call out instances of cruelty to animals for being outside of the Australian social license. We finally bear witness to events we might have striven unsuccessfully to change. For instance, despite petitions and lobbying, kangaroo culls continue to be carried out under the name of conservation. In the Australian Capital Territory (ACT), for example, despite ongoing activism, 1,505 grey kangaroos were killed in 2021 under the umbrella of reducing 'grazing pressure and help[ing] our nature reserves maintain resilience against the effects of climate change now and into the future' (Lindell, 2021). This is despite the thinness of evidence as to the necessity and actual effectiveness of such culls. Similarly, despite public pressure mounting against the inherent cruelty of duck shooting and concerns regarding declines in waterbird populations (McLennan, 2021), hunters from the age of 12 are still permitted in several Australian states to shoot frightened birds in the name of *fun* over a period of several weeks. These killings and others will doubtless go ahead for some time to come, but such actions are also being witnessed and reported with the aim of increasing and leveraging public pressure against them to the point of making change. This kind of 'strategic witnessing' in the form of 'radical record-keeping' is sometimes all that can be done in the face of what we consider cruel and unsustainable treatments of animals and their ecological communities, but it is also a powerful tool towards making change (Jarvie et al., 2021, pp. 363, 368). Supported by our understanding that sentience and need for care crosses species barriers, that cruelty is always abhorrent, and that stewardship is integral to avoiding whole system collapses, bearing witness is an important step in the extension of our care and responsibility agenda.

Conclusion

Anthropocentrism is not easy to see past. Given its hold on us all, no matter where we live, it is not easy to imagine that changes can be made in the relationship forms long taken for granted with regard to animals. And yet, as we have traced throughout this book, there are multiple initiatives occurring throughout Australia that promote other ways of living with nonhuman animals – ways that are more attentive to their interests and needs alongside our own. These are ways that do not make the mistake of thinking that only humans matter, and that remember that animals are never themselves one homogenized group. These are ways that might remain inconsistent, strategic, and precarious, but it is also their precarity that keeps them responsive and responsable, open to new formulations. Told through stories that Australians (and others) can connect with – of sentience, friendship, kinship, care, the importance of stewardship, the possibility of multispecies cohabitation – such ways have become not just imaginable but achievable. It is as such that they are possible harbingers for a kinder, more caring set of human-animal relations and a social imaginary able to support more-than-human interests and new forms of multispecies cohabitation and community. Given that the cost of turning away from such possibilities is too high for all of us, animals and humans, it is time we remake our rhythms of living and community – differently but indisputably together.

Note

1   The exception is of course when these animals are made visible to us as being in need, most often from the impact of disasters like the Black Summer fires or the 2022 floods.

Bibliography

AAP. (2016, December 2). *Reality cooking shows could be a powerful weapon against obesity*. Health Times. <https://healthtimes.com.au/hub/public-health/50/news/aap/reality-cooking-shows-could-be-a-powerful-weapon-against-obesity/2158/>.

Abbott, I. (2002). Origin and spread of the cat, *Felis catus*, on mainland Australia, with a discussion of the magnitude of its early impact on native fauna. *Wildlife Research*, *29*(1), 51–74.

ABC. (2014, November 20). *Fact check: Are feral cats killing over 20 billion native animals a year?* ABC News. <https://www.abc.net.au/news/2014-11-13/greg-hunt-feral-cat-native-animals-fact-check/5858282?nw=0>.

ABC Four Corners. (2011, May 30). *A bloody business*. [Transcript]. <https://www.abc.net.au/4corners/4c-full-program-bloody-business/8961434>.

ABC Gardening Australia Magazine. (2021). *2021 Media kit*. Nextmedia. <https://www.nextmedia.com.au/media-kits/gardening-australia-media-kit.pdf>.

ABC Landline. (2013, June 16). *Animals Australia under the microscope*. [Transcript]. <https://www.abc.net.au/landline/content/2013/s3782456.htm>.

ABC News. (2013, July 18). *Animals rights activists 'akin to terrorists', says NSW minister Katrina Hodgkinson*. <https://www.abc.net.au/news/2013-07-18/animal-rights-activists-27terrorists272c-says-nsw-minister/4828556>.

ABC News. (2019, April 8). *Vegan protesters charged after Melbourne's CBD brought to a standstill during peak hour*. <https://www.abc.net.au/news/2019-04-08/melbourne-vegan-protest-blocks-trams-traffic-causes-chaos/10980056>.

ABC Rural. (2013, October 4). *Jamie Oliver partners Woolies to stop selling caged-hen eggs*. ABC News. <https://www.abc.net.au/news/rural/2013-10-04/nrn-woolies-caged-eggs/4998380>.

Ackland, L. (2019, June 24). *Would you eat meat grown from cells in a laboratory? Here's how it works*. The Conversation. <https://theconversation.com/would-you-eat-meat-grown-from-cells-in-a-laboratory-heres-how-it-works-117420>.

ACT Government. (2020). *2020 kangaroo cull*. Environment, planning and sustainable development directorate – environment. <https://www.environment.act.gov.au/home/home-news-listing/2020-kangaroo-conservation-cull>.

Adams, M. (2017, October 13). *Friday essay: The cultural meanings of wild horses*. The Conversation. <https://theconversation.com/friday-essay-the-cultural-meanings-of-wild-horses-84198>.

Adams, M. (2020). *Anthropocene psychology: Being human in a more-than-human world*. Routledge.
Adams, S., Blokker, P., & Doyle, N. J. (2015). Social imaginaries in debate. *Social Imaginaries*, *1*(1), 15–52
Adams, S., & Smith, J. C. A. (Eds.). (2019). *Social imaginaries: Critical interventions*. Rowman and Littlefield International.
Admassu, S., Fox, T., Heath, R., & McRobert, K. (2020). *The changing landscape of protein production: Opportunities and challenges for Australian agriculture*. The Australian Farm Institute. <https://www.agrifutures.com.au/wp-content/uploads/2020/02/20-001.pdf>.
Agribusiness Australia. (2020). *2020 State of the industry: Implications for the Australian agriculture sector*. <https://www.agribusiness.asn.au/documents/item/575>.
Ahern, P., & Edgar's Mission. (2017). *Cooking with kindness*. Affirm Press.
Aiello, G., & Parry, K. (2020). *Visual communication: Understanding images in media culture*. Sage Publications.
Aitkin, D. (1985). 'Countrymindedness' – the spread of an idea. *Australian Cultural History*, *4*, 34–41.
Albeck-Ripka, L. (2020, June 28). Majestic icon or invasive pest? A war over Australia's wild horses. *The New York Times*. <https://www.nytimes.com/2020/06/28/world/australia/brumbies-horses-culling.html>.
Alhadeff-Jones, M. (2019). Beyond space and time – conceiving the rhythmic configurations of adult education through Lefebvre's rhythmanalysis. *Zeitschrift für Weiterbildungsforschung*, *42*, 165–81.
Aliento, W. (2020, January 31). *Listen to the original land managers to learn about fire, country and community*. The Fifth Estate. <https://thefifthestate.com.au/articles/listen-to-the-original-land-managers-to-learn-about-fire-country-and-community/>.
AMCS (Australian Marine Conservation Society). (n.d.). *Campaigns*. Retrieved February 25, 2021, from <https://www.marineconservation.org.au/>.
AMCS (Australian Marine Conservation Society). (2019a). *Aquaculture assessment criteria*. <https://goodfish.org.au/resource/aquaculture-assessment-criteria/>.
AMCS (Australian Marine Conservation Society). (2019b, November 13). *Australia's leading sustainable seafood guide now includes restaurants, new species and a new look*. <https://www.marineconservation.org.au/australias-leading-sustainable-seafood-guide-now-includes-restaurants-new-species-and-a-new-look/>.
AMCS (Australian Marine Conservation Society). (2019c). *GoodFish: Australia's sustainable seafood guide*. <https://goodfish.org.au/>.

AMCS (Australian Marine Conservation Society). (2019d). *Wild capture assessment criteria*. <https://goodfish.org.au/resource/wild-capture-assessment-criteria/>.
AMCS (Australian Marine Conservation Society). (2021a). *The Australian Marine Conservation Society (AMCS) is the voice for Australia's oceans*. <https://www.marineconservation.org.au/about/>.
AMCS (Australian Marine Conservation Society). (2021b). *Fight for our reef*. <https://www.marineconservation.org.au/fight-for-our-reef/>.
AMCS (Australian Marine Conservation Society). (2021c). *Fisheries*. <https://www.marineconservation.org.au/fisheries/>.
Anderson, B. (2006). *Imagined communities: Reflections on the origin and spread of nationalism*. (Rev. ed.). Verso Books.
Anderson, K. (2007). *Race and the crisis of humanism*. Routledge.
Andrews, M., Sclater, S. D., Rustin, M., Squire, C., & Treacher, A. (2003). Introduction. In M. Andrews, C. Squire, & A. Treacher (Eds.), *The uses of narrative: Explorations in sociology, psychology, and cultural studies* (pp. 1–10). Transaction Publishers.
Animal Charity Evaluators. (2017). *Animals Australia comprehensive review*. <https://animalcharityevaluators.org/research/charity-review/animals-australia/#comprehensive-review>.
Animal Medicines Australia. (2016). *Pets in Australia: A national survey of pets and people*. Newgate Research. <https://animalmedicinesaustralia.org.au/report/pet-ownership-in-australia-2016/>.
Animal Medicines Australia. (2019). *Pets in Australia: A national survey of pets and people*. Newgate Research. <https://animalmedicinesaustralia.org.au/report/pets-in-australia-a-national-survey-of-pets-and-people/>.
Animals Australia. (n.d.-a). *Animals don't belong here*. Retrieved January 15, 2021, from <https://secure.animalsaustralia.org/take_action/live-export-shipboard-cruelty/?ua_s=BLE.com>.
Animals Australia. (n.d.-b). *Investigations*. Retrieved September 15, 2021, from <https://www.animalsaustralia.org/investigations/>.
Animals Australia. (n.d.-c). *There is only one way to expose hidden animal abuse …* Retrieved September 15, 2021, from <https://www.animalsaustralia.org/investigation-appeal/>.
Animals Australia. (n.d.-d). *Who is Animals Australia?* Retrieved May 19, 2021, from <https://www.animalsaustralia.org/about/>.
Animals Australia. (2011). *Petition exceeds 200,000 signatures*. <https://www.animalsaustralia.org/features/live_export_petition_tally_exceeds_200000.php>.
Animals Australia. (2012). *Make it possible campaign*. [Video and transcript]. <https://www.youtube.com/watch?v=fM6V6lq_p00>.

Animals Australia. (2016a, June 30). *Animals Australia chosen as a 'standout charity' on world stage.* <https://www.animalsaustralia.org/features/standout-charity-listing.php>.

Animals Australia. (2016b). *Save Babe campaign.* <https://secure.animalsaustralia.org/appeal/save_babe/>.

Animals Australia. (2017). *Taste for life: Eat kindly, tread lightly, live well.* ABC Books.

Animals Australia. (2020a, February 4). *A guide to feeding hungry wildlife survivors of bushfires.* <https://animalsaustralia.org/features/feeding-hungry-animals-in-bushfire-zones.php>.

Animals Australia. (2020b, December 8). *Somewhere.* [Video]. <https://www.animalsaustralia.org/features/animals-australia-launches-new-campaign-to-help-farmed-animals.php>.

Animals Australia. (2020c, May 12). *Why we're excited to see plant-based meats in the meat section.* <https://www.animalsaustralia.org/features/why-were-excited-to-see-plant-based-meats-in-the-meat-section.php>.

Animals Australia. (2021a, March 19). *Salmon farms are emptying our oceans to feed their fish.* <https://animalsaustralia.org/features/fish-farms-eating-up-our-oceans.php>.

Animals Australia. (2021b, March 23). *10 things you need to know before you buy farmed fish.* <https://www.animalsaustralia.org/features/fish-farms.php>

Animals International. (n.d.). *When did it become okay to treat animals like this?* Retrieved January 15, 2021, from <https://www.animalsinternational.org/take_action/live-export-global/>.

Ankeny. R. A., Phillipov, M., & Bray, H. (2019). Celebrity chefs and new meat consumption norms: Seeking questions, not answers. *M/C Journal of Media and Culture*, 22(2). <https://doi.org/10.5204/mcj.1514>.

Anonymous. (2011, March 10). What does this have to do. [Comment on the article: "Victims of vanity: Wearing animals is donning pain and suffering"]. *Psychology Today.* <https://www.psychologytoday.com/us/blog/animal-emotions/201103/victims-vanity-wearing-animals-is-donning-pain-and-suffering>.

Anthony, L. (2009). *The elephant whisperer: My life with the herd in the African wild.* Pan Macmillan.

Arbon, S., & Duncalfe, Z. (2014). Food, animals, and the law: Do we have a moral obligation to protect them from the suffering that the law does not? *Griffith Journal of Law and Human Dignity*, 2(1), 199–220.

Arcari, P. (2019). The ethical masquerade: (Un)masking mechanisms of power behind 'ethical' meat. In M. Phillipov & K. Kirkwood (Eds.), *Alternative food politics: From the margins to the mainstream* (pp. 169–89). Routledge.

Arcari, P., Probyn-Rapsey, F., & Singer, H. (2021). Where species don't meet: Invisibilized animals, urban nature and city limits. *Environment and Planning E: Nature and Space, 4*(3), 940–65.

Arkaba Conservancy. (2017). *Arkaba: Saving wildlife through immersive outback safaris.* <https://www.arkabaconservancy.com/>.

Armstrong, P. (2017). Dinesh Wadiwel. The war against animals. [Review]. *Animal Studies Journal, 6*(2), 237–47.

Arnold, G. (1990). Can kangaroos survive in the wheatbelt? *Journal of the Department of Agriculture, Western Australia, Series 4, 31*(1), 14–17.

Arnold, S. (2019, April 16). Exclusive: Senator Rice details Australia's faunal extinction emergency. *Independent Australia.* <https://independentaustralia.net/life/life-display/exclusive-senator-rice-details-australias-faunal-extinction-emergency,12575>.

art thy neighbour. (n.d.). *Doug Gimesy puts faces to the animal victims of human activity, and his photos might just change minds.* Retrieved May 14, 2021, from <https://www.artthyneighbour.com/artists/wildlife-photographer-doug-gimesy-puts-faces-to-the-animal-victims-of-human-activity>.

Arvanitakis, J., Cage, C., Groutsis, D., Kaabel, A., Han, C., Hine, A., Hopkins, N., Lattouf, A., Liu, I. J., Lo, I., Lumby, C., Rodrigues, U., Soutphomassane, T., & Weerakkody, S.U. (2020). *Who gets to tell Australian stories?* Media Diversity Australia. <https://www.mediadiversityaustralia.org/wp-content/uploads/2020/08/Who-Gets-To-Tell-Australian-Stories_LAUNCH-VERSION.pdf>.

Aslin, H. J., & Bennett, D. H. (2000). Wildlife and world views: Australian attitudes toward wildlife. *Human Dimensions of Wildlife: An International Journal, 5*(2), 15–35.

Association of Marine Park Tourism Operators. (n.d.). *Home.* Retrieved February 25, 2021, from <https://www.ampto.org/home.html>.

Australian Associated Press. (2006, June 2). War on cane toads. *The Age.* <https://www.theage.com.au/national/war-on-cane-toads-20060602-ge2fvw.html>.

Australian Bureau of Statistics. (2017). *Discussion paper: From nature to the table: Environmental-Economic accounting for agriculture, 2015–16.* <https://www.abs.gov.au/ausstats/abs@.nsf/7d12b0f6763c78caca25706100cc588/755e5e7876f483deca2581e6000fb200!OpenDocument>.

Australian Bureau of Statistics. (2021). *Value of agricultural commodities produced, Australia.* <https://www.abs.gov.au/statistics/industry/agriculture/value-agricultural-commodities-produced-australia/latest-release>.

Australian Conservation Foundation. (2018, March 6). *New research reveals Australia's critical habitat laws are broken.* <https://www.acf.org.au/new_research_reveals_australia_s_critical_habitat_laws_are_broken>.

Australian Conservation Foundation. (2020). *The extinction crisis in Australia's cities and towns: How weak environment laws have let urban sprawl destroy the habitat of Australia's threatened species.* <https://d3n8a8pro7vhmx.cloudfront.net/auscon/pages/17703/attachments/original/1596500683/Extinction_crisis_in_cities_and_towns.pdf?1596500683>.

Australian Government. (n.d.-a). *Feral animals in Australia.* Department of Agriculture, Water and the Environment. Retrieved April 6, 2021, from <https://www.environment.gov.au/biodiversity/invasive-species/feral-animals-australia>.

Australian Government. (n.d.-b). *Sharks in Australian waters.* Department of Agriculture, Water and the Environment. Retrieved April 6, 2021, from <https://www.environment.gov.au/marine/marine-species/sharks>.

Australian Government. (2011). *Australian animal welfare strategy (AAWS) and national implementation plan 2010–14.* Department of Agriculture, Water and the Environment. <https://www.agriculture.gov.au/animal/welfare/aaws/australian-animal-welfare-strategy-aaws-and-national-implementation-plan-2010-14>.

Australian Government. (2015). *Threatened species strategy.* <https://www.environment.gov.au/system/files/resources/51b0e2d4-50ae-49b5-8317-081c6afb3117/files/ts-strategy.pdf>.

Australian Government. (2019a). *The impact of cats in Australia.* [Fact sheet]. <https://www.environment.gov.au/biodiversity/invasive-species/feral-animals-australia/feral-cats>.

Australian Government. (2019b). *State party report on the state of conservation of the Great Barrier Reef World Heritage Area (Australia).* Commonwealth of Australia. <https://www.environment.gov.au/system/files/resources/bfcd4506-2d94-4dc4-9eab-2cc97b931fac/files/gbr-state-party-report-2019.pdf>.

Australian Government. (2021). *Australian animal welfare standards and guidelines.* Department of Agriculture, Water and the Environment. <https://www.agriculture.gov.au/animal/welfare/standards-guidelines>.

Australian Koala Foundation. (2021). *Koalas and development.* <https://www.savethekoala.com/about-koalas/koalas-and-development>.

Australian Wildlife Conservancy. (n.d.). *Wildlife.* Retrieved January 11, 2021, from <https://www.australianwildlife.org/wildlife/>.

Backyard Buddies. (n.d.). *Crows and ravens.* Retrieved February 24, 2021, from <https://www.backyardbuddies.org.au/backyard-buddies/crows-and-ravens>.

Bagaric, M., & Akers, K. (2012). *Humanising animals: Civilising people.* CCH Australia Limited.

Baier, A. (1986). Trust and antitrust. *Ethics, 96*(2), 231–60.

Bainbridge, A., & Branley, A. (2021, June 23). *Battery hens to be outlawed by 2036 under new national proposal.* ABC News. <https://www.abc.net.au/news/202106-23/caged-eggs-phased-out-by-2036-under-national-proposal/100236246>.

Bakan, S. (2021, September, 20). *Aussies are spending $21.9 billion on pets. These are the hidden costs.* The New Daily. <https://thenewdaily.com.au/finance/dollars-and-sense/2021/09/20/pets-hidden-costs/#:~:text=These%20are%20the%20hidden%20costs,-Aussies%20are%20spending&text=Australian%20pet%20owners%20are%20spending,pets%20than%20themselves%20in%20lockdown>.

Baker, S. (2016, April 28). Crocodiles rising. *Hakai Magazine*. <https://www.hakaimagazine.com/features/crocodiles-rising/>.

Barham, D., Gray, L., Hall, S., Loane, C., Panegyres, J., Walker, G., Blanch, S., Taylor, M., & Sweeney, O. (2018). *Towards zero deforestation: A plan to end deforestation and excessive land clearing in NSW.* Nature conservation Council of NSW, the Wilderness Society, WWF-Australia and the National Parks Association of NSW. <https://www.nature.org.au/media/355843/181109-tzd-report-final.pdf>.

Barnes, A., & White, R. (2020). Mapping emotions: Exploring the impact of the Aussie Farms Map. *Journal of Contemporary Criminal Justice, 36*(3), 303–26.

Bar-On, Y. M., Phillips, R., & Milo, R. (2018). The biomass distribution on Earth. *Proceedings of the National Academy of Sciences, 115*(25), 6506–11.

Barrowclough, N. (2015, June 13). How activist Lyn White rewrote the rules on animal welfare. *The Sydney Morning Herald.* <https://www.smh.com.au/good-weekend/how-activist-lyn-white-rewrote-the-rules-on-animal-welfare-20150610-ghkmar.html>.

Barry, H. (2019, January 30). *What to do with the roos? Relocation plan deemed not viable for Baldivis development.* WA Today. <https://www.watoday.com.au/national/western-australia/what-to-do-with-the-roos-relocation-plan-deemed-not-viable-for-baldivis-development-20190129-p50uev.html>.

Barua, M. (2019). Animating capital: Work, commodities, circulation. *Progress in Human Geography, 43*(4), 650–69.

Baskin, J. (2015). Paradigm dressed as epoch: The ideology of the Anthropocene. *Environmental Values, 24,* 9–29.

Bastian, B., & Loughnan, S. (2017). Resolving the meat-paradox: A motivational account of morally troublesome behavior and its maintenance. *Personality and Social Psychology Review, 21*(3), 278–99.

Bateman, B. (2015, November 17). Are urban ravens really 'thugs' and 'murderers'? *Western Web.* <https://westernweb.net/2015/11/17/are-urban-ravens-really-thugs-and-murderers/>.

Bauer, J., & English, A. (2011). *Conservation through hunting: An environmental paradigm change in NSW*. Game Council NSW.

Bawaka Country., Wright, S., Suchet-Pearson, S., Lloyd, K., Burarrwanga, L., Ganambarr, R., Ganambarr-Stubbs, M., Ganambarr, B., Maymuru, D., & Sweeney, J. (2016). Co-becoming Bawaka: Towards a relational understanding of place/space. *Progress in Human Geography, 40*(4), 455–75.

BBC News. (2020a, October 14). *Great Barrier Reef has lost half of its corals since 1995*. <https://www.bbc.com/news/world-australia-54533971>.

BBC News. (2020b, March 26). *Great Barrier Reef suffers third mass bleaching in five years*. <https://www.bbc.com/news/world-australia-52043554>.

Beavis, M. A. (1994, March 11–12). *Environmental stewardship: History, theory and practice workshop proceedings*. [Occasional paper 32]. Institute of Urban Studies, University of Winnipeg. <https://winnspace.uwinnipeg.ca/handle/10680/1043>.

Bekoff, M. (2011, March 10). Victims of vanity: Wearing animals is donning pain and suffering. *Psychology Today*. <https://www.psychologytoday.com/us/blog/animal-emotions/201103/victims-vanity-wearing-animals-is-donning-pain-and-suffering>.

Bekoff, M. (2013, June 20). A universal declaration on animal sentience: No pretending. *Psychology Today*. <https://www.psychologytoday.com/au/blog/animal-emotions/201306/universal-declaration-animal-sentience-no-pretending>.

Bell, J. (2020, August 13). *Warranggal Warruwi towards Maal circle*. IndigenousX. <https://indigenousx.com.au/warranggal-warruwi-towards-maal-circle/>.

Bembridge, C. (2016, January 20). *Australia Day lamb ad, starring Lee Lin Chin, attracts dozens of complaints from vegans*. ABC News. <https://www.abc.net.au/news/2016-01-11/vegans-lodge-complaints-over-lamb-ad/7081706>.

Ben-Ami, D. (2009). *A shot in the dark: A report on kangaroo harvesting*. Animal Liberation NSW. <https://pdfs.semanticscholar.org/7781/50f03e46d8a6a4532e1a01556761aaf01eb2.pdf>.

Bennett, J. (2010). *Television personalities: Stardom and the small screen*. Routledge.

Bennett, N. J., Whitty, T. S., Finkbeiner, E., Pittman, J., Bassett, H., Gelcich, S., & Allison, E. H. (2018). Environmental stewardship: A conceptual review and analytical framework. *Environmental Management, 61*(4), 597–614.

Benton, T. G., Bieg, C., Harwhatt, H., Pudasaini, R., & Wellesley, L. (2021). *Food system impacts on biodiversity loss: Three levers for food system transformation in support of nature*. Chatham House. <https://www.chathamhouse.org/sites/default/files/2021-02/2021-02-03-food-system-biodiversity-loss-benton-et-al_0.pdf>.

Berlin, J. (2019, January 16). Australia's beloved kangaroos are now controversial pests. *National Geographic*. <https://www.nationalgeographic.com/magazine/2019/02/australia-kangaroo-beloved-symbol-becomes-pest/>.

Berry, E. (2020, March 25). *Pets: The voiceless victims of the COVID-19 crisis*. UNSW Newsroom. <https://newsroom.unsw.edu.au/news/social-affairs/pets-voiceless-victims-covid-19-crisis>.

Berry, H. L., Botterill, L. C., Cockfield, G., & Ding, N. (2016). Identifying and measuring agrarian sentiment in regional Australia. *Agriculture and Human Values, 33*, 929–41.

Berry, K. (2019, July 31). *Allied Pinnacle to phase out cage eggs*. Food & Drink Business. <https://www.foodanddrinkbusiness.com.au/news/allied-pinnacle-to-phase-out-cage-eggs>.

Berry, R. J. (Ed.). (2006). *Environmental stewardship: Critical perspectives, past and present*. T & T Clark.

Bettles, C. (2013a, June 30). *Coles bags a boost for NFF*. Farm Online. <https://www.farmonline.com.au/story/3588905/coles-bags-a-boost-for-nff/>.

Bettles, C. (2013b, June 8). *Coles cops more criticism*. Farm Online. <https://www.farmonline.com.au/story/3589497/coles-cops-more-criticism/>.

Bettles, C. (2017, March 29). *Cattle Council: Productivity report high-jacked by an extreme vegan agenda*. Farm Online. <https://www.farmonline.com.au/story/4564008/cattle-council-productivity-report-high-jacked-by-an-extreme-vegan-agenda/>.

Biddle, N., Edwards, B., Herz, D., & Makkai, T. (2020). *Exposure and the impact on attitudes of the 2019–20 Australian bushfires*. ANU Centre for Social Research Methods. <https://csrm.cass.anu.edu.au/sites/default/files/docs/2020/2/Exposure_and_impact_on_attitudes_of_the_2019-20_Australian_Bushfires_publication.pdf>.

Bignell, J. (2004). *An introduction to television studies*. Routledge.

Bird Control Australia. (n.d.). *Visual scare feather crow*. Retrieved March 3, 2021, from <https://birdcontrolaustralia.com.au/shop/residential-bird-deterrent/visual-scare-feather-crow-lifelike-dead-crow-with-real-feathers/>.

Birdlife Australia. (n.d.). *Our threatened birds are in the grip of the extinction crisis – can you help, before they disappear forever?* Retrieved June 1, 2021, from <https://birdlife.org.au/current-appeal>.

Bittman, M. (2013). *VB6: Eat vegan before 6:00 to lose weight and restore your health... for good*. Clarkson Potter.

Black, H. (2013, September 1). Crows show off their social skills: New findings on crows' intelligence lend perspective on how social smarts evolve. *Scientific American*. <https://www.scientificamerican.com/article/crows-show-off-social-skills/>.

Blagrove, A. (2013). Red Dog: The Pilbara wanderer. In B. Boyd & R. Norman (Eds.), Coolabah. Placescape, placemaking, placemarking, placedness... geography and cultural production. [Special issue]. *Coolabah, 11*, 19–24. <https://revistes.ub.edu/index.php/coolabah/article/view/15544/18696>.

Blue, G. (2015). Multispecies publics in the Anthropocene: From symbolic exchange to material-discursive intra-action. In The Human Animal Research Network (Ed.), *Animals in the Anthropocene: Critical perspectives on non-human futures* (pp. 165–76). Sydney University Press.

Bogueva, D., & Marinova, D. (2020). Cultured meat and Australia's generation z. *Frontiers in Nutrition, 7*, 1–15.

Bohmann, U., & Montero, D. (2014). History, critique, social change and democracy. An interview with Charles Taylor. *Constellations, 21*(1), 3–15.

Bonner, F. (2011). *Personality presenters: Television's intermediaries with viewers.* Routledge.

Boom, K., Ben-Ami, D., Croft, D. B., Cushing, N., Ramp, D., & Boronyak, L. (2012). 'Pest' and resource: A legal history of Australia's kangaroos. *Animal Studies Journal, 1*(1), 17–40.

Boom, K., Ben-Ami, D., Boronyak, L., & Riley, S. (2013). The role of inspections in the commercial kangaroo industry. *International Journal of Rural Law and Policy, 1*, 1–19.

Botterill, L. C. (2005). Policy change and network termination: The role of farm groups in agricultural policy making in Australia. *Australian Journal of Political Science, 40*(2), 1–13.

Botterill, L. C. (2009). The role of agrarian sentiment in Australian rural policy. In F. Merlan & D. Raftery (Eds.), *Tracking rural change: Community, policy and technology in Australia, New Zealand and Europe* (pp. 59–78). ANU Press.

Bourdieu, P. (1984). *Distinction: A social critique of the judgement of taste.* (R. Nice, Trans.). Harvard University Press. (Original work published 1979).

Bourdieu, P. (1990). *The logic of practice.* (R. Nice, Trans.). Stanford University Press. (Original work published 1980).

Bourke, L. (2018, December 4). Live export industry imposes summer ban in wake of backlash over animal deaths. *The Sydney Morning Herald.* <https://www.smh.com.au/politics/federal/live-export-industry-imposes-summer-ban-in-wake-of-backlash-over-animal-deaths-20181204-p50k6g.html>.

Boyd, M., Chrulew, M., Degeling, C., Mrva-Montoya, A., Probyn-Rapsey, F., Savvides, N., & Wadiwel, D. (2015). Introduction. In The Human Animal Research Network (Ed.), *Animals in the Anthropocene: Critical perspectives on non-human futures* (pp. vii–xxiv). Sydney University Press.

Boylan, J. (Series Producer). (2013–16). *River cottage Australia.* [TV series]. ITV Studios Australia; Keo Films; Special Broadcasting Service (SBS).

Bradshaw, L. (2019, March 6). *Farmers need to control the narrative*. Farm Weekly. <https://www.farmweekly.com.au/story/5940516/farmers-need-to-control-the-narrative/>.
Bray, J., Johns, N., & Kilburn, D. (2011). An exploratory study into the factors impeding ethical consumption. *Journal of Business Ethics*, *98*(4), 597–608.
Braysher, M., O'Brien, P., & Bomford, M. (1996). Towards 'best practice' vertebrate pest management in Australia using virally-vectored immunocontraception. *Proceedings of the Vertebrate Pest Conference*, *17*(17), 18–23.
Brent, L. J., Chang, S.W., Gariepy, J. F., & Platt, M. L. (2014). The neuroethology of friendship. *Annals of the New York Academy of Sciences*, *1316*, 1–17.
Bribie Island Environmental Protection Association. (n.d.). *Ocean beach protection program*. Retrieved April 30, 2021, from <https://biepa.org/ocean-beach-protection-project/>.
Brisbane City Council. (2019, May 10). *Torresian crow*. <https://www.brisbane.qld.gov.au/clean-and-green/natural-environment-and-water/biodiversity-in-brisbane/wildlife-in-brisbane/living-with-wildlife/torresian-crow>.
Brisbane City Council. (2020, October 20). *Wildlife movement solutions*. <https://www.brisbane.qld.gov.au/clean-and-green/natural-environment-and-water/biodiversity-in-brisbane/wildlife-in-brisbane/wildlife-movement-solutions>.
Broad, G. M. (2020). Making meat, better: The metaphors of plant based and cell-based meat innovation. *Environmental Communication*, *14*(7), 919–32.
Brockmeier, J. (2009). Reaching for meaning: Human agency and the narrative imagination. *Theory & Psychology*, *19*(2), 213–33.
Bruce, A., & Faunce, T. (2017). Food production and animal welfare legislation in Australia: Failing both animals and the environment. In G. Steier & K. Patel (Eds.), *International farm animal, wildlife and food safety law* (pp. 359–94). Springer International Publishing.
Bryant, C., & Barnett, J. (2018). Consumer acceptance of cultured meat: A systematic review. *Meat Science*, *143*, 8–17.
Bryant, C. J. (2019). We can't keep meating like this: Attitudes towards vegetarian and vegan diets in the United Kingdom. *Sustainability*, *11*(23), 1–17. <https://doi.org/10.3390/su11236844>.
Buckmaster, L. (2015, February 6). Dot and the Kangaroo rewatched – tear-jerking Australian animation trailblazer. *The Guardian*. <https://www.theguardian.com/film/2015/feb/06/dot-and-the-kangaroo-rewatched-tear-jerking-australian-animation-trailblazer>.
Buddle, E. A., Bray, H. J., & Ankeny, R. A. (2018). 'I feel sorry for them': Australian meat consumers' perceptions about sheep and beef cattle transportation. *Animals*, *8*(10), 1–13.

Burdon, A. (2016, March 8). Where the wild horses are. *Australian Geographic*. <https://www.australiangeographic.com.au/topics/wildlife/2016/03/where-the-wild-horses-are/>.

Burgin, S., Mattila, M., McPhee, D., & Hundloe, T. (2015). Feral deer in the suburbs: An emerging issue for Australia? *Human Dimensions of Wildlife*, *20*(1), 65–80.

Burke, P., Eckert, C., & Davis. S. (2014). Segmenting consumers' reasons for and against ethical consumption. *European Journal of Marketing*, *48*(11/12), 2237–61.

Burns, G. (2014). Anthropomorphism and animals in the Anthropocene. In G. Burns & M. Paterson (Eds.), *Engaging with animals: Interpretations of a shared existence* (pp. 3–20). Sydney University Press.

Burns, G., & Paterson, M. (2014). Introduction. In G. Burns & M. Paterson (Eds.), *Engaging with animals: Interpretations of a shared existence* (pp. ix–xiv). Sydney University Press.

Burton, L., Barker, E., Prendergast, J., & Collins, A. (2018, September 19). *What would Australia look like without live exports?* ABC News. <https://www.abc.net.au/news/rural/2018-09-19/impact-on-australia-without-live-export/10254566>.

Bush Heritage. (2021). *Who we are*. <https://www.bushheritage.org.au/who-we-are>.

Byrnes, P. (n.d.). *Dot and the Kangaroo: The food of understanding (1977)*. National Film and Sound Archive of Australia. Retrieved December 1, 2021, from <https://www.nfsa.gov.au/collection/curated/dot-and-kangaroo-food-understanding-1977>.

Calarco, M. (2015). Foreword. In D. Wadiwel (Ed.), *The war against animals* (pp. ix–xi). Brill.

Callari, M. (2020, November 30). *Koala rescue: 1 orphaned joey, and her species' fight for survival*. Deutsche Welle (DW). <https://www.dw.com/en/koala-rescue-an-orphaned-joey-and-her-species-fight-for-survival/a-55768411>.

Callicott, J. B. (2013). *Thinking like a planet, the land ethic and the earth ethic*. Oxford University Press.

Campbell, A. (2008). *Paddock to plate: Food, farming and Victoria's progress to sustainability*. Australian Conservation Foundation. <https://www.researchgate.net/publication/236627322_Paddock_to_Plate_Food_farming_and_Victoria's_progress_to_sustainability>.

Cao, D. (2010). *Animal law in Australia and New Zealand*. Thompson Reuters.

Cao D. (2015). *Animal Law in Australia*. (2nd edn.). Thomson Reuters.

Carey, R., Parker, C., & Scrinis, G. (2020). How free is sow stall free? Incremental regulatory reform and industry co-optation of activism. *Law & Policy*, *42*(3), 284–309. <https://doi.org/10.1111/lapo.12154>.
Carrington, D. (2018, May 22). Humans just 0.01% of all life but have destroyed 83% of wild mammals – study. *The Guardian*. <https://www.theguardian.com/environment/2018/may/21/human-race-just-001-of-all-life-but-has-destroyed-over-80-of-wild-mammals-study>.
Carty, V., & Onyett, J. (2006). Protest, cyberactivism and new social movements: The re-emergence of the peace movement post 9/11. *Social Movement Studies*, *5*(3), 229–49.
Casey, B. (2015, February 17). Public outrage over live baiting in greyhound racing following ABC Four Corners investigation. *Herald Sun*. <https://www.heraldsun.com.au/news/victoria/public-outrage-over-live-baiting-in-greyhound-racing-following-abc-four-corners-investigation/news-story/8daf30d1cf6084eae6000be9d930e0f4>.
Cashin, P., & McDermott, C. J. (2002). 'Riding on the sheep's back': Examining Australia's dependence on wool exports. *Economic Record*, *78*(242), 249–63.
Caso, N. (2010). *Practicing memory in Central American literature*. Palgrave Macmillan.
Castoriadis, C. (1987). *The imaginary institution of society* (K. Blamey, Trans.). Polity Press. (Original work published 1975).
Celermajer, D. (2021). *Summertime: Reflections on a vanishing future*. Hamish Hamilton.
Centre for International Economics. (2010). *Evaluation of the LiveCorp, MLA and the Australian government and industry partnership*. Meat & Livestock Australia. <https://www.mla.com.au/contentassets/698fb9367a43436bb76862b5f00f7357/w.liv.0153_final_report.pdf>.
CGBR (Citizens of the Great Barrier Reef). (n.d.-a). *Citizens of the Great Barrier Reef*. Retrieved February 2, 2021, from <https://citizensgbr.org/>.
CGBR (Citizens of the Great Barrier Reef). (n.d.-b). *The Great Reef census*. Retrieved February 2, 2021, from <https://greatreefcensus.org/>.
Chakrabarty, D. (2012). Postcolonial studies and the challenge of climate change. *New Literary History*, *43*(1), 1–18.
Chapin, F. S., Carpenter, S. R., Kofinas, G. P., Folke, C., Abel, N., Clark, W. C., Olsson, P., Smith, D. M., Walker, B., Young, O. R., Berkes, F., Biggs, R., Grove, J. M., Naylor, R. L., Pinkerton, E., Steffen, W., & Swanson, F. J. (2010). Ecosystem stewardship: Sustainability strategies for a rapidly changing planet. *Trends in Ecology & Evolution*, *25*(4), 241–9.

Chapin, F. S., Kofinas, G. P., & Folke, C. (2009). *Principles of ecosystem stewardship: Resilience-based natural resource management in a changing world.* Springer.

Chapin, F. S., Mark, A. F., Mitchell, R. A., & Dickinson, K. J. M. (2012). Design principles for social-ecological transformation toward sustainability: Lessons from New Zealand sense of place. *Ecosphere, 3*(5), 1–22.

Chapman, B. (2017, November 1). *Why do shark bites seem to be more deadly in Australia than elsewhere?* The Conversation. <https://theconversation.com/why-do-shark-bites-seem-to-be-more-deadly-in-australia-than-elsewhere-85986>.

Chappell, J. (2006). Living with the trickster: Crows, ravens, and human culture. *PLoS Biology, 4*(1), 16–17.

Charles, N. (2015). 'Animals just love you as you are': Experiencing kinship across the species barrier. *Sociology, 48*(4), 715–30.

Charles, N., & Davies, C. (2008). My family and other animals: Pets as kin. *Sociological Research Online, 13*, 13–26.

Chattoo Borum, C., & Feldman, L. (2017). Storytelling for social change: Leveraging documentary and comedy for public engagement in global poverty. *Journal of Communication, 67*, 678–701.

Chauvet, D. (2018). Should cultured meat be refused in the name of animal dignity? *Ethical Theory and Moral Practice, 21*, 387–411.

Chen, P. J. (2016). *Animal welfare in Australia: Politics and policy.* Sydney University Press.

Chiew, F. (2015). The paradox of self-reference: Sociological reflections on agency and intervention in the Anthropocene. In The Human Animal Research Network (Ed.), *Animals in the Anthropocene: Critical perspectives on nonhuman futures* (pp. 1–18). Sydney University Press.

Čičigoj, K. (2019). Regenerating the future without reproducing it: Donna Haraway's nature-cultural, multi-species kinship. *Maska, 34*(196–197), 26–43.

City of Ballarat. (2019). *Responsible pet ownership.* <https://www.ballarat.vic.gov.au/me/pets/responsible-pet-ownership>.

City of Canning. (2018). *Local biodiversity strategy.* <https://www.canning.wa.gov.au/CanningWebsite/media/Files/Community/Conservation/City-of-Canning-Local-Biodiversity-Strategy-2018.PDF>.

City of Fremantle. (2020). *Greening Fremantle: Strategy 2020.* <https://www.fremantle.wa.gov.au/sites/default/files/sharepointdocs/Greening%20Fremantle%20Strategy%202020-C-000625.pdf>.

City of Melville. (n.d.). *Living with ravens.* [Information Sheet]. Retrieved February 24, 2021, from <https://www.melvillecity.com.au/CityOfMelville/media/Documents-and-PDF-s/raven-info-sheet.pdf>.

City of Parramatta. (2021, June 10). *Mural inspired by native wildlife breathes new life into Rydalmere*. [Media release]. <https://www.cityofparramatta.nsw.gov.au/about-parramatta/news/media-release/mural-inspired-by-native-wildlife-breathes-new-life-into>.

City of Rockingham. (2021). *Native plants giveaway*. <https://rockingham.wa.gov.au/your-city/news/2021/april/native-plants-giveaway>.

City of Sydney. (2020). *Urban ecology strategic action plan*. <https://www.cityofsydney.nsw.gov.au/strategies-action-plans/urban-ecology-strategic-action-plan>.

City of Townsville. (2022, March 17). *New mural to shine a light on native animal's plight*. [Media release]. <https://www.townsville.qld.gov.au/about-council/news-and-publications/media-releases/2022/march/new-mural-to-shine-a-light-on-native-animals-plight>.

Clark, C. (2013, June 16). *Animals Australia: A closer look at the organisation and what it stands for*. ABC News. <https://www.abc.net.au/news/2013-06-16/a-closer-look-at-animals-australia-landline-explainer/4755298>.

Clark, S. (2008). "I knew him by his voice": Can animals be our friends? *Philosophy Now: A Magazine of Ideas*, 67. <https://philosophynow.org/issues/67/I_knew_him_by_his_voice_Can_Animals_Be_Our_Friends>.

Clark, W. C., Crutzen, P. J., & Schellnhuber, H. J. (2005). *Science for global sustainability: Toward a new paradigm. Center for International Development working paper no. 120*. [Working paper series]. Center for International Development, Harvard University. <https://www.hks.harvard.edu/sites/default/files/centers/cid/files/publications/faculty-working-papers/120.pdf>.

Clarke, R., Heitlinger, S., Light, A., Forlano, L., Foth, M., & DiSalvo, C. (2019). More-than-human participation: Design for sustainable smart city futures. *ACM Interactions*, *26*(3), 60–3.

Clarkson, B. D., Wehi, P. M., & Brabyn, L. (2007). *Bringing back nature into cities: Urban land environments, indigenous cover and urban restoration* (Research, Report No. 52). Centre for Biodiversity and Ecology, University of Waikato, Hamilton, New Zealand. <https://hdl.handle.net/10289/3786>.

Clemons, R., & Day, K. (2017, May 12). Free-range eggs: What does 'free range' really mean, and are consumers being misled? *Choice*. <https://www.choice.com.au/food-and-drink/meat-fish-and-eggs/eggs/articles/free-range-eggs>.

Climate Council. (2018). *Lethal consequences: Climate change impacts on the Great Barrier Reef*. <https://www.climatecouncil.org.au/wp-content/uploads/2018/07/CC_MVSA0147-Report-Great-Barrier-Reef_V4-FA_Low-Res_Single-Pages.pdf>.

Climate Council of Australia. (2020). *Summer of crisis*. <https://www.climatecouncil.org.au/wp-content/uploads/2020/03/Crisis-Summer-Report-200311.pdf>.

Cobble, D. S. (2010). More intimate unions. In E. Boris & R. S. Parreñas (Eds.), *Intimate labors: Cultures, technologies, and the politics of care* (pp. 280–95). Stanford Social Sciences.

Cochrane, A. (2012). *Animal rights without liberation: Applied ethic and human obligations*. Columbia University Press.

Cochrane, A. (2013). From human rights to sentient rights. *Critical Review of International Social and Political Philosophy, 16*(5), 655–75.

Coghlan, S. (2012, November 12). *Opinion: Live animal export: A moral violation?* Animals Australia. <https://www.animalsaustralia.org/media/opinion.php?op=298>.

Cole, M. (2008). Asceticism and hedonism in research discourses of veg*anism. *British Food Journal, 110*(7), 706–16.

Cole, M., & Morgan, K. (2013). Engineering freedom? A critique of biotechnological routes to animal liberation. *Configurations, 21*(2), 201–29.

Coleman, G. (2007, May 18). Public perceptions of animal pain and animal welfare. In *Australian animal welfare strategy science summit on pain and pain management*. [Conference proceedings] (pp. 1–8). Department of Agriculture and Australian Veterinary Association Annual Conference, Melbourne. <https://faunalytics.org/wp-content/uploads/2015/05/Citation1332.pdf>.

Coleman G. (2008). Public Perceptions of animal pain and animal welfare. In D. Mellor, P. M. Thornber, D. Bayvel, & S. Kahn (Eds.), *Scientific assessment of management of animal pain* (pp. 26–37). OIE (World Organisation for Animal Health).

Coleman, G. J. (2018). Public animal welfare discussions and outlooks in Australia. *Animal Frontiers, 8*(1), 14–19.

Coleman, G. J., Rohlf, V., Toukhsati, S., & Blache, D. (2015). Public attitudes relevant to livestock animal welfare policy. *Farm Policy Journal, 12*, 45–57.

Coleman, G. J., Rohlf, V., Toukhsati, S., & Blache, D. (2017). Public attitudes predict community behaviours relevant to the pork industry. *Animal Production Science, 58*, 416–23.

Coles. (2016). *Responsible sourcing*. <https://www.coles.com.au/corporate-responsibility/responsible-sourcing/responsible-sourcing>.

Collard, S., O'Connor, P., & Prowse, T. (2019, May 9). *Fixing Australia's extinction crisis means thinking bigger than individual species*. The Conversation. <https://theconversation.com/fixing-australias-extinction-crisis-means-thinking-bigger-than-individual-species-115559>.

Collins, J., Klomp, N., & Birckhead, J. (1996). Aboriginal use of wildlife: Past, present, future. In M. Bomford & J. Caughley (Eds.), *Sustainable use of wildlife by Aboriginal peoples and Torres Strait Islanders* (pp. 14–36). Australian Government Publishing Service.

Commonwealth of Australia. (2017). *Australian pest animal strategy 2017–2027*. Invasive Plants and Animals Committee. <https://www.agriculture.gov.au/sites/default/files/sitecollectiondocuments/pests-diseases-weeds/consultation/apas-final.pdf>.

Commonwealth of Australia. (2019). *Interim report: Australia's faunal extinction crisis*. Environment and Communications References Committee. <https://www.aph.gov.au/Parliamentary_Business/Committees/Senate/Environment_and_Communications/Faunalextinction/Interim_report>.

Commonwealth of Australia. (2020). *Tackling the feral cat pandemic: A plan to save Australian wildlife*. House of Representatives Standing Committee on the Environment and Energy. <https://parlinfo.aph.gov.au/parlInfo/download/committees/reportrep/024580/toc_pdf/TacklingtheferalcatpandemicaplantosaveAustralianwildlife.pdf;fileType=application%2Fpdf>.

Commonwealth of Australia. (2021). *Threatened species strategy 2021–2031*. <https://apo.org.au/sites/default/files/resource-files/2021-05/apo-nid312450.pdf>.

Connolly, J. J. T., Svendsen, E. S., Fisher, D. R., & Campbell, L. K. (2014). Networked governance and the management of ecosystem services: The case of urban environmental stewardship in New York city. *Ecosystem Services*, *10*, 187–94.

Context Pty Ltd. (2015, December). *National cultural heritage values assessment & conflicting values report*: The wild horse population Kosciuszko national park. NSW National Parks and Wildlife Service. <https://www.environment.nsw.gov.au/-/media/OEH/Corporate-Site/Documents/Animals-and-plants/Pests-and-weeds/Kosciuszko-wild-horses/national-cultural-heritage-values-assessment-conflicting-values-report-2015.pdf?la=en&hash=BC256E1E20DDFED99FA078049C07CB81F691611C>.

Convention on Biological Diversity. (2020). *Global biodiversity outlook 5*. <https://www.cbd.int/gbo/gbo5/publication/gbo-5-en.pdf>.

Cooper, M. M. (2017). How dogs (and other nonhuman animals) become interesting. In K. Bjørkdahl & A. C. Parrish (Eds.), *Rhetorical animals: Boundaries of the human in the study of persuasion* (pp. 219–35). Lexington Books.

Coorey, L., & Coorey-Ewings, C. (2018). Animal victims of domestic and family violence: Raising youth awareness. *Animal Studies Journal*, *7*(1), 1–40.

Cornell, H. N., Marzluff, J. M., & Pecoraro, S. (2012). Social learning spreads knowledge about dangerous humans among American crows. *Proceedings of the Royal Society B*, *279*, 499–508.

Cosslett, T. (2016). *Talking animals in British children's fiction, 1786–1914*. Routledge.

Coughlan, M. (2019, September 12). *Vegan extremists in government sights as crackdown on farm invasions passes parliament*. 7 News. <https://7news.com.au/lifestyle/vegan/vegan-extremists-cop-parliamentary-pile-on-c-449203>.

Coulter, K. (2016a). *Animals, work, and the promise of interspecies solidarity*. Palgrave Macmillan.

Coulter, K. (2016b). Beyond human to humane: A multispecies analysis of care work, its repression, and its potential. *Studies in Social Justice, 10*(2), 199–219.

Cowan, M., Blythman, M., Angus, J., & Gibson, L. (2020). Post-release monitoring of western grey kangaroos (Macropus fuliginosus) relocated from an urban development site. *Animals, 10*(1914), 1–23.

Cox, L. (2021, January 13). Australia the only developed nation on world list of deforestation hotspots. *The Guardian*. <https://www.theguardian.com/environment/2021/jan/13/australia-the-only-developed-nation-on-world-list-of-deforestation-hotspots>.

Crane, E. (2015, September 25). Chased into an electric fence and shot at multiple times: Professional kangaroo culler slams the inhumane killing of more than 30 animals after they were killed just for fun. *Daily Mail*. <https://www.dailymail.co.uk/news/article-3248511/Professional-kangaroo-culler-slams-inhumane-killing-30-animals-killed-just-fun.html>.

Creed, B. (2017). *Stray: Human-animal ethics in the Anthropocene*. Power Publications.

Cresswell, I. D., & Murphy, H. T. (2017). *Australia state of the environment 2016: Biodiversity* (Independent report to the Australian Government Minister for the Environment and Energy). Australian Government Department of the Environment and Energy, Canberra. <https://soe.environment.gov.au/sites/g/files/net806/f/soe2016-biodiversity-launch-version2-24feb17.pdf?v=1488792935>.

Crimston, C. R., Hornsey, M. J., Bain, P. G., & Bastian, B. (2018). Toward a psychology of moral expansiveness. *Current Directions in Psychological Science, 27*(1), 14–19.

Crist, E. (2015). I walk in the world to love it. In G. Wuerthner, E. Crist, & T. Butler (Eds.), *Protecting the wild: Parks and wilderness, the foundation for conservation* (pp. 82–95). The Island Press.

Croft, D. B. (1999). When big is beautiful: Some consequences of bias in kangaroo culling. In M. Wilson (Ed.), *The kangaroo betrayed: World's largest wildlife slaughter* (pp. 70–3). Hill of Content Publishing Co.

Crook, W. (Executive Producers). (2018). *Larry the wonderpup*. [TV series]. Chocolate Liberation Front; ABC iView.

Crow Away. (2021, February 4). *Get rid of crows – guaranteed!* <https://crowaway.com.au/get-rid-of-crows-guaranteed/>.

Crutzen, P. J. (2002). Geology of mankind. *Nature, 415*, 23. <https://doi.org/10.1038/415023a>.

Crutzen, P. J., & Stoermer, E. F. (2000). The 'Anthropocene'. *Global Change Newsletter*, *41*, 17–18. <https://www.igbp.net/download/18.316f183213234701 77580001401/1376383088452/NL41.pdf>.
Cudworth, E. (2011). Walking the dog explorations and negotiations of species difference. *PAN: Philosophy, Activism, Nature*, *8*, 14–22.
Cudworth, E. (2015). Killing animals: Sociology, species relations and institutionalized violence. *The Sociological Review*, *63*, 1–18.
Cunningham, S. (2000). History, contexts, politics, policy. In G. Turner & S. Cunningham (Eds.), *The Australian TV book* (pp. 13–32). Allen & Unwin.
Curtain, F., & Grafenauer, S. (2019). Plant-based meat substitutes in the flexitarian age: An audit of products on supermarket shelves. *Nutrients*, *11*(2603), 1–14.
Cushing, N. (2019). To eat or not to eat kangaroo: Bargaining over food choice in the Anthropocene. *M/C Journal of Media and Culture*, *22*(2). <https://doi.org/10.5204/mcj.1508>.
Dalton, J. (2018, April 9). Australian sheep bound for the Middle East 'cooked alive in harrowing and deadly conditions' on live export ships. *Independent*. <https://www.independent.co.uk/news/world/australasia/sheep-middle-east-live-expo rts-australia-ship-conditions-animal-rights-ban-trade-a8296511.html>.
Dalziell, J., & Wadiwel, D. J. (2016). Live exports, animal advocacy, race and 'animal nationalism'. In A. Potts (Ed.), *Meat culture* (pp. 73–89). Brill.
Dant, T. (2012). *Television and the moral imaginary: Society through the small screen*. Palgrave Macmillan.
Davidson, J. (2020, January 20). *UN biodiversity chief: Humans risk living in an 'empty world' with 'catastrophic' consequences*. EcoWatch: Environmental News for a Healthier Planet and Life. <https://www.ecowatch.com/world-economic-forum-climate-crisis-empty-world-2644865096.html?rebellti tem=3#rebelltitem3>.
Davies, J., Campbell, D., Campbell, M., Douglas, J., Hueneke, H., LaFlamme, M., Pearson, D., Preuss, K., Walker, J., & Walsh, F. J. (2010). *Livelihoods inLand: Promoting health and wellbeing outcomes from desert Aboriginal land management*. Desert Knowledge CRC Report 78, Desert Knowledge Cooperative Research Centre. <https://nintione.com.au/resource/dkcrc-rep ort-78-livelihoods-in-land_health-and-wellbeing-from-desert-alm.pdf>.
Davies, J. E. (2002). *Stories of change: Narrative and social movements*. State University of New York Press.
Dawson, T. J. (2012). *Kangaroos*. (2nd edn.). CSIRO Publishing.
Day, J. C., & Heron, S. F. (2019, September 12). The Great Barrier Reef is in trouble. There are a whopping 45 reasons why. The Conversation. <https://theconve rsation.com/the-great-barrier-reef-is-in-trouble-there-are-a-whopping-45-reas ons-why-122930>.

De Brito, S. (2015, February 19). Opinion, poll: Greyhound racing and live baiting: Time to ban the 'sport of grubs'. *Central Western Daily*. <https://www.centralwesterndaily.com.au/story/2893371/opinion-poll-greyhound-racing-and-live-baiting-time-to-ban-the-sport-of-grubs/>.

Debus, S. (2012, January 27). Stone the crows! Could corvids be Australia's smartest export? The Conversation. <https://theconversation.com/stone-the-crows-could-corvids-be-australias-smartest-export-4346>.

De Fina, A., & Georgakopoulou, A. (2008). Analysing narratives as practices. *Qualitative Research*, *8*(3), 379–87.

de la Bellacasa, M. P. (2012). 'Nothing comes without its world': Thinking with care. *The Sociological Review*, *60*(2), 197–216.

Delforce, C. (Director). (2018). *Dominion*. [Film]. Dominion Movement. <https://www.dominionmovement.com/watch>.

Deloitte Access Economics. (2016). *What are our stories worth? Measuring the economic and cultural value of Australia's screen sector*. Deloitte Touche Tohmatsu. <https://www.screenaustralia.gov.au/getmedia/13dceb59-0a88-432f-adb3-958fcc04e6bb/Deloitte-Access-Economics-Screen-Currency.pdf>.

Deloitte Access Economics. (2017). *At what price? The economic, social and icon value of the Great Barrier Reef*. Report for the Great Barrier Reef Foundation. <https://www2.deloitte.com/content/dam/Deloitte/au/Documents/Economics/deloitte-au-economics-great-barrier-reef-230617.pdf>.

De Moor, J. (2016). Lifestyle politics and the concept of political participation. *Acta Politica*, *52*(2), 1–19.

Denholm, M. (2021, September 19). Animal welfare lobby splits over 'false impression' salmon certification. *The Australian*. <https://www.theaustralian.com.au/nation/animal-welfare-lobby-splits-over-false-impression-salmon-certification/news-story/13b71662768ec05e973e79c06872044d>.

Department of Biodiversity, Conservation and Attractions. (2017). *Raven – Damage prevention and control*. Government of Western Australia. <https://www.dpaw.wa.gov.au/images/documents/plants-animals/animals/living-with-wildlife/raven_fauna_note_2017.pdf>.

Department of Justice and Community Safety. (2013, January 31). *Beechworth correctional centre – Raptor rehabilitation program*. [Video]. YouTube. <https://www.youtube.com/watch?v=zeoqacC_JhA>.

Derrida, J. (1997). *Politics of friendship* (G. Collins, Trans.). Verso. (Original work published 1994).

Devadas, V., & Mummery, J. (2007). Community without community. *Borderlands*, *6*(1). <https://webarchive.nla.gov.au/awa/20070830081416/http://www.borderlandsejournal.adelaide.edu.au/vol6no1_2007/devadasmummery_intro.htm>.

De Villiers, J. H. (2017). Animal rights theory, animal welfarism and the 'new welfarist' amalgamation: A critical perspective. *Southern African Public Law*, *30*, 406–33.

De Waal, F. B. M., & Tyack, P. L. (Eds.). (2003). *Animal social complexity: Intelligence, culture, and individualized societies*. Harvard University Press.

Dickson-Smith, D. (2020, October 29). *The great reef census: Citizen science on the Great Barrier Reef*. Diveplanit. <https://www.diveplanit.com/marine-environment/great-reef-census/>.

Doherty, P. (2013, October 13). Apps for animal lovers. *Sydney Morning Herald*. <https://www.smh.com.au/technology/apps-for-animal-lovers-20131010-2v97g.html>.

Doherty, T. S., Driscoll, D. A., Nimmo, D. G., Ritchie, E. G., & Spencer, R. (2019). Conservation or politics? Australia's target to kill 2 million cats. *Conservation Letters: A Journal of the Society for Conservation Biology*, *12*(4), 1–16. <https://doi: 10.1111/conl.12633>.

Dole, N. (2019, October 18). *Animal cruelty probe launched after Meramist abattoir workers filmed mistreating racehorses destined for slaughter*. ABC News. <https://www.abc.net.au/news/2019-10-18/horse-racing-industry-responds-abc-investigation-animal-welfare/11615070>.

Donaldson, S., & Kymlicka, W. (2011). *Zoopolis: A political theory of animal rights*. Oxford University Press.

Donaldson, S., & Kymlicka, W. (2012, September 29–30). *Citizen canine: Agency for domesticated animals*. [Paper presentation]. Symposium on domesticity and beyond: Living and working with animals, Queen's University, Kingston, Canada. <https://christianebailey.com/wp-content/uploads/2012/12/CitizenCanine_Agency-for-Domesticated-Animals-Donaldson-and-Kymlicka-2013.pdf>.

Doyle, M. (Producer). (2011, May 30). (1961–present). A bloody business. *Four corners*. [TV series]. Australian Broadcasting Commission.

Dubois, E., & Gaffney, D. (2014). The multiple facets of influence: Identifying political influentials and opinion leaders on Twitter. *American Behavioral Scientist*, *58*(10), 1260–77.

Ducarme, F., Luque, G. M., & Courchamp, F. (2013, July). *What are 'charismatic species' for conservation biologists?* [BioSciences Masters Reviews], Ecole Normale Superieur de Lyon, France. <https://biologie.ens-lyon.fr/ressources/bibliographies/pdf/m1-11-12-biosci-reviews-ducarme-f-2c-m.pdf?lang=fr>.

Duffy, E. (2019, November 20). *Pet-friendly properties a top priority for home buyers looking for fur-ever home*. <https://www.savings.com.au/home-loans/pet-friendly-properties-a-top-priority-for-home-buyers-looking-for-fur-ever-home>.

Dyer, M., Newlands, M., Bradshaw, E., & Hernandez, S. (2020). Stewardship in the Great Barrier Reef: A review of concepts and definitions of stewardship in *the Great Barrier Reef applied to Reef health.* Great Barrier Reef Marine Park Authority. <https://elibrary.gbrmpa.gov.au/jspui/bitstream/11017/3781/3/Stewardship-for-the-Great-Barrier-Reef.pdf>.

Eckhardt, G., Belk, R., & Devinney, T. (2010). Why don't consumers consume ethically? *Journal of Consumer Behaviour, 9*(6), 426–36.

Ecological Society of Australia. (2020, August 26). *Scientists' declaration: Strong legislation needed to curb Australia's accelerating rate of land clearing.* <https://www.ecolsoc.org.au/working-groups/policy/scientists-declaration-on-land-clearing/>.

Eddie, R. (2016, July 15). 'If you have a cat keep it inside… or you may not have a cat for long': Vigilante group threatens to catch and kill pets – and posts graphic images of their prey online. *Daily Mail Australia.* <https://www.dailymail.co.uk/news/article-3690372/Cat-Busters-trap-pets-Sydney-Melbourne-shoot-ferals-rural-areas.html>.

Edgar's Mission. (2021). *About Edgar's Mission.* <https://www.edgarsmission.org.au/about-us/.

Eger, B. (2020). *Halt the extinction of the Greater Glider.* Change.org. https://www.change.org/p/robert-gordon-stokes-mp-nsw-minister-for-planning-halt-the-extinction-of-the-greater-glider>.

Elder, C. (2007). *Being Australian: Narratives of national identity.* Allen & Unwin.

Elhacham, E., Ben-Uri, L., Grozovski, J., Bar-On, Y. M., & Milo, R. (2020). Global human-made mass exceeds all living biomass. *Nature, 588,* 442–4.

Elkins, J. (2008). *Visual literacy.* Routledge.

Ellicott, J. (2018, March 15). *Little pay incentive for shooters to join kangaroo meat industry.* The Land. <https://www.theland.com.au/story/5285265/top-roo-shooter-says-harvesting-is-a-low-paid-job/>.

Ellis, E. J. (2010). Making sausages and law: The failure of animal welfare laws to protect both animals and fundamental tenets of Australia's legal system. *Australian Animal Protection Law Journal, 4,* 6–22.

Engelen, J. (2011, December 19). ROA in Australia – Urban Art meets the Outback. de de ce Blog. <https://www.dedeceblog.com/2011/12/19/roa-murals-in-australia/>.

Englefield, B., Blackman, S. A., Starling, M., & Mcgreevy, P. D. (2019). A review of Australian animal welfare legislation, regulation, codes of practice, and policy, and their influence on stakeholders caring for wildlife and the animals for whom they care. *Animals, 9*(6). <https://www.mdpi.com/2076-2615/9/6/335/htm>.

Englezos, E. (2018). Ag-gag laws in Australia: Activists under fire may not be out of the woods yet. *Griffith Journal of Law & Human Dignity, 6*(1), 272–93.

EPBC ACT 1999 (*Environmental Protection and Biodiversity Conservation Act* 1999) (Cth). (n.d.). About the EPBC Act. Retrieved January 10, 2021, from <https://www.environment.gov.au/epbc/about>.

Epstein, D., Farina, C., & Heidt, J. (2014). The value of words: Narrative as evidence in policy making. *Evidence & Policy*, *10*(2), 243–58.

Estok, S. (2021). Merchandizing veganism. In L. Wright (Ed.), *The Routledge handbook of vegan studies.* (pp. 333–42). Routledge.

Ethical Farmers. (2020, June 23). *How to eat nose to tail.* <https://www.ethicalfarmers.com.au/blogs/news/how-to-eat-nose-to-tail>.

European Food Safety Authority. (2012). Scientific opinion: Statement on the use of animal-based measures to assess the welfare of animals. *European Food Safety Authority Journal*, *10*, 2767–96.

Evans, B. L. (2018). *Animal cruelty, discourse, and power: A study of problematisations in the live export policy debates.* [Doctoral dissertation, Queensland University of Technology]. <https://eprints.qut.edu.au/122417/1/Brodie_Evans_Thesis.pdf>.

Evans, M. (2019, August 21). *The part of the meat industry no one likes to talk about.* SBS. <https://www.sbs.com.au/food/article/2019/08/21/part-meat-industry-no-one-likes-talk-about>.

Evans, N. (2011). *More than a sign on the fence? Teacher learning and the reef guardian schools program in Far North Queensland.* VDM Verlag Dr. Müller.

Eyers, R. (2016). *A regulatory study of the Australian animal welfare framework for Queensland saleyard animals.* [Doctoral dissertation, Griffith University]. <https://research-repository.griffith.edu.au/handle/10072/366332>.

FAO. (2006). *Livestock's long shadow: Environmental issues and options.* Food and Agriculture Organization of the United Nations (FAO). <https://www.fao.org/3/a0701e/a0701e.pdf>.

Farm Online. (2013, June 5). *Coles withdraws AA bags.* <https://www.farmonline.com.au/story/3589560/coles-withdraws-aa-bags/>.

Fernandes, A. M., de Souza Teixeira, O., Philippe, J., Revillion, P., & de Souza, Â. R. L. (2021). Panorama and ambiguities of cultured meat: An integrative approach. *Critical Reviews in Food Science and Nutrition.* <https://doi 10.1080/10408398.2021.1885006>.

Fernández, C. M. (2021). The loss of food sovereignty in synthetic meat transition: A critique from eco-republican justice. In H. Schübel & I. Wallimann-Helmer (Eds.), *Justice and food security in a changing climate* (pp. 103–8). Wageningen Academic Publishers.

Fernández, N., Navarro, L. M., & Pereira, H. M. (2017). Rewilding: A call for boosting ecological complexity in conservation. *Conservation Letters*, *10*, 276–8.

Finn, H. (2017, July 5). *Land clearing isn't just about trees – it's an animal welfare issue too*. The Conversation. <https://theconversation.com/land-clearing-isnt-just-about-trees-its-an-animal-welfare-issue-too-80398>.

Fisher, P., Warburton, B., Morgan, D., Cowan, P., & Duckworth, J. (2008, June 28–July 1). Animal welfare in vertebrate pest management and research in New Zealand. In P. Cragg, J. Keenan, & G. Sutherland (Eds.), *Blue sky to deep water: The reality and the promise*. [Conference proceedings] (pp. 89–94). Australian and New Zealand Council for the Care of Animals in Research and Teaching (ANZCCART), Wellington, New Zealand. <https://www.researchgate.net/profile/Penny_Fisher/publication/259312631_Animal_welfare_in_vertebrate_pest_management_and_research_in_New_Zealand/links/5546888c0cf24107d397ed09/Animal-welfare-in-vertebrate-pest-management-and-research-in-New-Zealand.pdf>.

FitzRoy, L. (n.d.). *Schools programs*. Retrieved January 19, 2021, from <https://www.frompaddocktoplate.com.au/school-programs/>.

Flanagan, R. (2021). *Toxic: The rotting underbelly of the Tasmanian salmon industry*. Penguin.

Flannery, T. (2006). Beautiful lies: Population and environment in Australia. *4 Classic quarterly essays on the Australian story* (pp. 219–97). Quarterly Essay.

Fleming, P. A., Wickham, S. L., Barnes, A. L., Miller, D. W., & Collins, T. (2020). Varying opinions about animal welfare in the Australian live export industry: A survey. *Animals: An Open Access Journal from MDPI, 10*(10), 1–18. <https://doi.org/10.3390/ani10101864>.

Flew, T., & Harrington, S. (2010). Television. In S. Cunningham & G. Turner (Eds.), *The media & communications in Australia* (pp.155–72). Allen & Unwin.

Flint, J. (2019, April 7). *WA bid to save death-row SA border collies goes viral*. Perth Now. <https://www.perthnow.com.au/news/animals/wa-bid-to-save-death-row-sa-dogs-goes-viral-ng-b881157368z>.

Flint, N. (2013, May 9). ABC loses its balance over animal welfare. *The Sydney Morning Herald*. <https://www.smh.com.au/opinion/abc-loses-its-balance-over-animal-welfare-20130508-2j7va.html>.

Foley, M. (2020, June 21). Koala recovery plan five years overdue as populations are 'smashed'. *The Sydney Morning Herald*. <https://www.smh.com.au/politics/federal/koala-recovery-plan-five-years-overdue-as-populations-are-smashed-20200619-p554eq.html>.

Food Frontier. (2020). *2020 State of the industry Australia's plant-based meat sector*. <https://www.foodfrontier.org/wp-content/uploads/dlm_uploads/2021/03/Food-Frontier-2020-State-of-the-Industry.pdf>.

Food Frontier & Life Health Foods. (2019, October). *Hungry for plant-based: Australian consumer insights*. <https://www.foodfrontier.org/wp-cont

ent/uploads/2019/10/Hungry-For-Plant-Based-Australian-Consumer-Insights-Oct-2019.pdf>.

Francione, G. L. (2008). *Animals as persons: Essays on the abolition of animal exploitation*. Columbia University Press.

Francione, G. L. (2010a). Animal welfare and the moral value of nonhuman animals. *Law, Culture and the Humanities*, *6*(1), 24–36.

Francione, G. L. (2010b, February 3). *Is every campaign a single-issue campaign?* Animal rights: The abolitionist approach. <https://www.abolitionistapproach.com/is-every-campaign-a-single-issue-campaign>.

Francione, G. L. (2010c, April 17). *A short note on abolitionist veganism as a single issue campaign*. Animal rights: The abolitionist approach. <https://www.abolitionistapproach.com/a-short-note-on-abolitionist-veganism-as-a-single-issue-campaign>.

Francione, G. L. (2010d, February 1). *Single-issue campaigns in human & nonhuman contexts*. Animal rights: The abolitionist approach. <https://www.abolitionistapproach.com/single-issue-campaigns-and-in-human-nonhuman-contexts/>.

Francione, G. L. (2012, July 12). *Sentience*. Animal rights: The abolitionist approach. <https://www.abolitionistapproach.com/sentience/>.

Francione, G. L., & Garner, R. (2010). *The animal rights debate*. Columbia University Press.

Franklin, A. (1996). Australian hunting and angling sports and the changing nature of human-animal relations in Australia. *Journal of Sociology*, *32*(3), 39–56.

Franklin, A. (2007). Human-nonhuman animal relationships in Australia: An overview of results from the first national survey and follow-up case studies 2000–2004. *Society & Animals*, *15*, 7–27.

Franklin, A. (2008). The 'animal question' and the 'consumption' of wildlife. In B. Lovelock (Ed.), *Tourism and the consumption of wildlife: Hunting, shooting and sport fishing* (pp. 31–44). Routledge.

Fraser, D. (2010). Animal welfare. In M. Bekof (Ed.), *Encyclopedia of animal rights and animal welfare* (Vol. 1, pp. 47–9). Greenwood Press.

Frew, W. (2019, December 5). *Caring for country should be a sustainability and green design priority*. The Fifth Estate. <https://thefifthestate.com.au/innovation/design/caring-for-country-should-be-a-sustainability-and-green-design-priority/>.

Futureye. (2018). *Australia's shifting mindset on farm animal welfare*. <https://www.outbreak.gov.au/sites/default/files/documents/farm-animal-welfare.pdf>.

Gabb, J. (2011). Family lives and relational living: Taking account of otherness. *Sociological Research Online*, *16*, 141–50.

Gall, S. (2020, January 6). *Zero grey kangaroo quota will harm western towns, Senators say*. Farm Online. <https://www.farmonline.com.au/story/6565014/qlds-nil-kangaroo-quota-to-bring-new-hardship/>.

Gallo-Cajiao, E., Archibald, C., Ritchie, E., Friedman, R., Fuller, R., Steven, R., & Morrison, T. (2018, May 30). *Crowdfunded campaigns are conserving the earth's environment*. The Conversation. <https://theconversation.com/crowdfunded-campaigns-are-conserving-the-earths-environment-97312>.

Gambetta, D. (1990). Can we trust trust? In D. Gambetta (Ed.), *Trust: Making and breaking cooperative relations* (pp. 213–37). John Wiley & Sons.

Gamborg, C., Palmer, C., & Sandoe, P. (2012). Ethics of wildlife management and conservation: What should we try to protect? *Nature Education Knowledge*, *3*(10), 1–9.

Gantz, A. (2019, December 10). *Australia foodservice company commits to cage-free eggs*. WATT Poultry. <https://www.wattagnet.com/articles/39277australia-foodservice-company-commits-to-cage-free-eggs>.

Gardening Australia. (2012, May 5). *Animal attraction: Series 23, episode 06*. [Fact sheet]. ABC. <https://www.abc.net.au/gardening/factsheets/animal-attraction/9433658>.

Gardening Australia. (2016, July 31). *Hollows for habitat: Series 27, episode 21*. [Fact sheet]. ABC. <https://www.abc.net.au/gardening/factsheets/hollows-for-habitat/9438022>.

Gardening Australia. (2018, May 25). *Frog bog: Series 29, episode 18*. [Fact sheet]. ABC. <https://www.abc.net.au/gardening/factsheets/frog-bog/9801490>.

Gardening Australia. (2019, June 28). *Backyard biodiversity: Series 30, episode 22*. [Fact sheet]. ABC. <https://www.abc.net.au/gardening/factsheets/backyard-biodiversity/11260610>.

Gardening Australia. (2020a, April 17). *Handmade habitat: Series 31, episode 11*. [Fact sheet]. ABC. <https://www.abc.net.au/gardening/factsheets/handmade-habitat/12155214>.

Gardening Australia. (2020b, March 6). *Frog pond fix-up: Series 31, episode 05*. [Fact sheet]. ABC. <https://www.abc.net.au/gardening/factsheets/frog-pond-fix-up/12030676>.

Gardening Australia. (2020c, June 12). *Gardens for wildlife: Series 31, episode 19*. [Fact sheet]. ABC. <https://www.abc.net.au/gardening/factsheets/gardens-for-wildlife/12346606>.

Gately, I., & Kemdall, B. (2020). *Indigenous world 2020: Australia*. The International Work Group for Indigenous Affairs (IWGIA). <https://www.iwgia.org/en/australia/3642-iw-2020-australia.html>.

Gatens, M., & Lloyd, G. (1999). *Collective imaginings*. Routledge.

Gay'Wu Group of Women. (2019). *Song spirals: Sharing women's wisdom of Country through songlines.* Allen & Unwin.

GBRF (Great Barrier Reef Foundation). (2021). *Help save the endangered green turtle.* <https://www.barrierreef.org/>.

GBRMPA (Great Barrier Reef Marine Park Authority). (2015). *Reef 2050 long-term sustainability plan.* <https://elibrary.gbrmpa.gov.au/jspui/bitstream/11017/2934/1/reef-2050-long-term-sustainability-plan.pdf>.

GBRMPA (Great Barrier Reef Marine Park Authority). (2018, July). *Reef 2050 long-term sustainability plan.* <https://www.environment.gov.au/marine/gbr/publications/reef-2050-long-term-sustainability-plan-2018>.

GBRMPA (Great Barrier Reef Marine Park Authority). (2019). *Great Barrier Reef outlook report 2019.* <https://www.gbrmpa.gov.au/our-work/outlook-report-2019>.

GBRMPA (Great Barrier Reef Marine Park Authority). (2020). *Reef 2050 long-term sustainability plan public consultation draft.* <https://haveyoursay.awe.gov.au/reef-2050-plan>.

GBRMPA (Great Barrier Reef Marine Park Authority). (2021a). *Our story.* <https://www.gbrmpa.gov.au/about-us/about-us>.

GBRMPA (Great Barrier Reef Marine Park Authority). (2021b). *Reef guardians.* <https://www.gbrmpa.gov.au/our-work/our-programs-and-projects/reef-guardians>.

GBRMPA (Great Barrier Reef Marine Park Authority). (2021c). *Traditional owners of the Great Barrier Reef.* <https://www.gbrmpa.gov.au/our-partners/traditional-owners/traditional-owners-of-the-great-barrier-reef>.

Gelber, K., & O'Sullivan, S. (2021). Cat got your tongue? Free speech, democracy and Australia's 'ag-gag' laws. *Australian Journal of Political Science, 56*(1), 19–34.

Germano, J. M., Field, K. J., Griffiths, R. A., Clulow, S., Foster, J., Harding, G., & Swaisgood, R. R. (2015). Mitigation-driven translocations: Are we moving wildlife in the right direction? *Frontiers in Ecology and the Environment, 13*(2), 100–5.

Gertenbach, L., Lamla, J., & Laser, S. (2021). Eating ourselves out of industrial excess? Degrowth, multi-species conviviality and the micro-politics of cultured meat. *Anthropological Theory, 21*(3), 386–408.

Geysen, T. L., & White, S. (2009). The role of law in addressing the interests of animals. *Queensland Law Society, 29*, 24–6.

Gibbs, L. (n.d.). *Evaluation of the Stephanie Alexander kitchen garden program.* Retrieved January 19, 2021, from <https://mspgh.unimelb.edu.au/centres-institutes/centre-for-health-equity/research-group/jack-brockhoff-child-health-wellbeing-program/research/previous-projects/evaluation-of-the-stephanie-alexander-kitchen-garden-program>.

Gibbs, S. (2020, January 15). Where bushfire donations really go: How to make sure your money gets to where it is most desperately needed – and why millions of dollars will still be gobbled up in 'support costs'. *Daily Mail Australia*. <https://www.dailymail.co.uk/news/article-7880659/Where-bushfire-donations-really-charities-spend-millions-dollars-support-costs.html>.

Glasgow, D. (2008). The law of the jungle: Advocating for animals in Australia. *Deakin Law Review*, *13*(1), 181–210.

Glenn, E. N. (2000). Creating a caring society. *Contemporary Sociology*, *29*(1), 84–94.

Glynn-McDonald, R., & Sinclair, R. (2021, January 15). *Connection to animals and country*. Common Ground. <https://www.commonground.org.au/learn/connection-to-animals-and-country>.

Godin, S. (2017). Tricked! In B. Kateman (Ed.), *The reducetarian solution* (pp. 18–20). TarcherPerigee.

Goffman, E. (1974). *Frame analysis*. Harper & Row.

Gore, J. G. (2016, October 18). Looking back: Australia's emu wars. *Australian Geographic*. <https://www.australiangeographic.com.au/topics/wildlife/2016/10/on-this-day-the-emu-wars-begin/>.

Government of South Australia. (2016). *Creating a wildlife friendly garden*. Natural Resources Centre Eastwood. <https://www.naturalresources.sa.gov.au/files/sharedassets/adelaide_and_mt_lofty_ranges/plants_and_animals/urban_biodiversity/creating-wildlife-friendly-garden.pdf>.

Great Ape Project. (n.d.). *Mission and vision*. Retrieved March 30, 2021, from <https://www.projetogap.org.br/en/mission-and-vision/>.

Green, M. C., & Brock, T. C. (2000). The role of transportation in the persuasiveness of public narratives. *Journal of Personality and Social Psychology*, *79*(5), 701–21. <https://doi.org/10.1037/0022-3514.79.5.701>.

Green, M. C., & Brock, T. C. (2002). In the mind's eye: Transportation-imagery model of narrative persuasion. In M. C. Green, J. J. Strange, & T. C. Brock (Eds.), *Narrative impact: Social and cognitive foundations* (pp. 315–41). Lawrence Erlbaum Associates Publishers.

Green, M. C., Brock, T. C., & Kaufman, G. F. (2004). Understanding media enjoyment: The role of transportation into narrative worlds. *Communication Theory*, *14*(4), 311–27.

Green-Barber, J. M., & Old, J. M. (2019). What influences road mortality rates of eastern grey kangaroos in a semi-rural area? *BMC Zoology*, *4*(11), 1–10. <https://doi.org/10.1186/s40850-019-0047-8>.

Greengarten, L. (2010, April 6). Farm fresh down under. *The New York Times*. <https://www.nytimes.com/2010/04/11/travel/11headsup.html>.

Greenpeace. (n.d.). *Greenpeace*. Retrieved March 3, 2021, from <https://www.greenpeace.org.au/>.

Greer, A. (2013, August 9). 'Akin to terrorism': The war on animal activists. Overland. <https://overland.org.au/2013/08/akin-to-terrorism-the-war-on-animal-activists/>.

Gressier, C. (2016). Going feral: Wild meat consumption and the uncanny in Melbourne, Australia. *The Australian Journal of Anthropology, 27*(1), 49–65.

Gross. Y. (Director). (1977). *Dot and the kangaroo*. [Film]. Yoram Gross Films; The Australian Film Commission.

Gross, Y. (Director), (1984). *The camel boy*. [Film]. Yoram Gross Studios.

Gruen, L. (2014). *Entangled empathy: An alternative ethic for our relationships with animals*. Lantern Books.

Gunderson, R. (2013). From cattle to capital: Exchange value, animal commodification, and barbarism. *Critical Sociology, 39*(2), 259–75.

Hall, S. (1996). Ethnicity: Identity and difference. In G. Eley, G. Suny, & L. Freedman (Eds.), *Becoming national: A reader* (pp. 339–51). Oxford University Press.

Halpin, D. (2004). Transitions between formations and organisations: An historical perspective on the political representation of Australian farmers. *Australian Journal of Politics and History, 50*(4), 469–90.

Ham, A. (2021, March 17). Australia's cats kill two billion animals annually. Here's how the government is responding to the crisis. *Smithsonian Magazine*. <https://www.smithsonianmag.com/science-nature/australias-cats-kill-two-billion-animals-annually-180977235/>.

Hamilton, L. (2014, November 24). *Australia's weird obsession with cooking shows*. Vagabond Journey. <https://www.vagabondjourney.com/australias-weird-obsession-with-cooking-shows/>.

Hansson, N., & Jacobsson, K. (2014). Learning to be affected: Subjectivity, sense, and sensibility in animal rights activism. *Society & Animals, 22*, 262–88.

Haraway, D. (2003). *The companion species manifesto*. Prickly Paradigm Press.

Haraway, D. (2008). *When species meet*. University of Minnesota Press.

Haraway, D. (2016). *Staying with the trouble: Making kin in the chthulucene*. Duke University Press.

Haraway, D. (2018). Staying with the trouble for multispecies environmental justice. *Dialogues in Human Geography, 8*(1), 102–5.

Hariman, R., & Lucaites, J. L. (2016). *The public image: Photography and civic spectatorship*. University of Chicago Press.

Harris, R., Bowman, D., & Beaumont, L. (2018, July 5). *Ecosystems across Australia are collapsing under climate change*. The Conversation. <https://theconversation.com/ecosystems-across-australia-are-collapsing-under-climate-change-99367>.

Hastie, H. (2018, July 26). *No humans allowed: Main Roads building WA's first animal bridge*. WA Today. <https://www.watoday.com.au/national/west

ern-australia/no-humans-allowed-main-roads-building-wa-s-first-animal-bridge-20180726-p4ztre.html>.
Hayes, W., & Sahu, S. (2020). The human microbiome: History and future. *Journal of Pharmacy and Pharmaceutical Sciences, 23*, 404–11.
Hayward, T. (1997). Anthropocentrism: A misunderstood problem. *Environmental Values, 6*(1), 49–63.
Heathcote, A. (2019, March 4). Sharks and the media: How should we report? *Australian Geographic*. <https://www.australiangeographic.com.au/topics/science-environment/2019/03/sharks-and-the-media-how-should-we-report/>.
Henderson, E. (2017, May 5). Flexitarianism: Is eating less meat really better for us and the environment? *Independent*. <https://www.independent.co.uk/life-style/food-and-drink/flexitarianism-diet-eat-less-meat-better-health-environment-vegetarian-food-a7712711.html>.
Hens, K. (2009). Ethical responsibilities towards dogs: An inquiry into the dog-human relationship. *Journal of Agricultural and Environmental Ethics, 22*(1), 3–14.
Hepburn, J. (2015, December 22). *Cats came to Australia with European settlers not 17th-Century shipwrecks, two studies say*. ABC News. <https://www.abc.net.au/news/2015-12-22/cats-came-to-australia-with-european-settlers/7049308>.
Herd, P., & Meyer, M. H. (2002). Care work: Invisible civic engagement. *Gender & Society, 16*(5), 665-88.
Herman, D. (2009). *Basics of narrative*. John Wiley & Sons.
Hill, K. (2017, January 10). *Kelpie lovers beware, your own Red Dog comes with hard work*. ABC News South East SA. <https://www.abc.net.au/news/2017-01-10/red-dog-not-suitable-for-everyone/8172568>.
Hill, R., Grant, C., George, M., Robinson, C. J., Jackson, S., & Abel, N. (2012). A typology of Indigenous engagement in Australian environmental management: Implications for knowledge integration and social-ecological system sustainability. *Ecology and Society, 17*(1), 1–18. <https://dx.doi.org/10.5751/ES-04587-170123>.
Hill, R., Pert, P.L., Davies, J., Robinson, C. J., Walsh, F., & Falco-Mammone, F. (2013). *Indigenous land management in Australia: Extent, scope, diversity, barriers and success factors*. CSIRO Ecosystem Sciences. <https://www.agriculture.gov.au/sites/default/files/sitecollectiondocuments/natural-resources/landcare/submissions/ilm-report.pdf>.
Hilton, C. (Executive Producer) (2010–present). *Gourmet farmer* [TV series]. Essential Media & Entertainment Productions; Special Broadcasting Service (SBS).

Hollingsworth, J. (2019, April 30). *The case against cats: Why Australia has declared war on feral felines.* CNN. <https://edition.cnn.com/2019/04/26/asia/feral-cats-australia-intl/index.html>.

Hose, N., & Deacon, B. (2018, September 16). *Could Aboriginal cat-hunting be the key to saving the bilby?* ABC News. <https://www.abc.net.au/news/2018-09-16/feral-cat-hunting-in-kiwirrkurra-wa-to-protect-endangered-bilby/10241554>.

Hosking, C. (2020, July 2). *Stopping koala extinction is agonisingly simple. But here's why I'm not optimistic.* The Conversation. <https://theconversation.com/stopping-koala-extinction-is-agonisingly-simple-but-heres-why-im-not-opt imistic-141696>.

Houlihan, R. (2019, December 10). *'People are dumping their dogs': Shelter lists grow as owners abandon pets.* The Age. <https://www.theage.com.au/natio nal/victoria/people-are-dumping-their-dogs-shelter-lists-grow-as-owners-abandon-pets-20191208-p53hy6.html>.

Howell, L. G., & Witt, R. R. (2020, September 17). *Environment Minister Sussan Ley faces a critical test: Will she let a mine destroy koala breeding grounds?* The Conversation. <https://theconversation.com/environment-minister-sussan-ley-faces-a-critical-test-will-she-let-a-mine-destroy-koala-breeding-grounds-145839>.

Hughes, L. (2020, March). *The milk of human genius.* The Monthly. <https://www.themonthly.com.au/issue/2020/march/1582981200/lesley-hughes/milk-human-genius#mtr>.

IbisWorld. (2021). *Level of urbanisation.* Business Environment Profiles – Australia. <https://www.ibisworld.com/au/bed/level-of-urbanisation/25029/>.

Idato, M. (2009, September 10). From casting pouch to TV queen. *The Age.* <https://www.theage.com.au/entertainment/from-casting-pouch-to-tv-queen-20090 910-ge8342.html>.

IPBES. (2019). *Summary for policymakers of the global assessment report on biodiversity and ecosystem services of the intergovernmental science-policy platform on biodiversity and ecosystem services.* <https://ipbes.net/sites/default/files/2020-02/ipbes_global_assessment_report_summary_for_policymakers_en.pdf>.

IPCC (Intergovernmental Panel on Climate Change). (2018). *Global warming of 1.5°C: An IPCC special report on the impacts of global warming of 1.5°C above pre-industrial levels and related global greenhouse gas emission pathways, in the context of strengthening the global response to the threat of climate change, sustainable development, and efforts to eradicate poverty.* Cambridge University Press.

Ivin, G. (Director). (2020). *Penguin bloom.* [Film]. Made Up Stories; Penguin Bloom Films.

Jackson, S., & Altman, J. C. (2009). Indigenous rights and water policy: Perspectives from tropical northern Australia. *Australian Indigenous Law Review*, *13*(1), 27–48.

Jackson, W. J., Argent, R. M., Bax, N. J., Bui, E., Clark, G. F., Coleman, S., Cresswell, I. D., Emmerson, K. M., Evans, K., Hibberd, M. F., Johnston, E. L., Keywood, M. D., Klekociuk, A., Mackay, R., Metcalfe, D., Murphy, H., Rankin, A., Smith, D. C., & Wienecke, B. (2016). *Overview: Invasive species are a potent, persistent and widespread threat to Australia's environment. Australia state of the environment 2016*. Australian Government Department of the Environment and Energy. <https://soe.environment.gov.au/theme/overview/topic/invasive-species-are-potent-persistent-and-widespread-threat-australias>.

Jacobs, J. M. (1996). *Edge of empire: Postcolonialism and the city*. Routledge.

Jagot, J. (2017, October 20). *The rule of law and reconciliation* [Opening address]. Law society of New South Wales young lawyers' conference, Federal Court of Australia, Sydney, NSW, Australia. <https://www.fedcourt.gov.au/digital-law-library/judges-speeches/justice-jagot/jagot-j-20171020>.

James, L. (2019, July 30). With plant-based eggs and lab grown meat, JUST Inc. is transforming food. *Forbes*. <https://www.forbes.com/sites/laurajames/2019/07/30/with-plant-based-eggs-and-lab-grown-meat-just-inc-is-transforming-food/#4621ea4a74d7>.

Janssen, J. (2014). On the relationship between animal victimization and stigmatization of ethnic groups: The case of ritual slaughter. In T. Spapens, R. White, & M. Kluin (Eds.), *Environmental crime and its victims: Perspectives within green criminology* (pp. 205–17). Routledge.

Jarvie, K., Evans, J., & McKemmish, S. (2021). Radical appraisal in support of archival autonomy for animal rights activism. *Archival Science*, *21*, 353–72.

Jarvis, D., Stoeckl, N., Hill, R., & Pert, P. (2018). Indigenous land and sea management programs: Can they promote regional development and help 'close the (income) gap'? *Australian Journal of Social Issues*, *53*(3), 283–303.

Jasper, J. M. (1998). The emotions of protest: Affective and reactive emotions in and around social movements. *Sociological Forum*, *13*(3), 397–424.

Jasper, J. M., & Nelkin, D. (1992). *The animal rights crusade: The growth of a moral protest*. Free Press.

Jasper, J. M., & Poulsen, J. (1995). Recruiting strangers and friends: Moral shocks and social networks in animal rights and anti-nuclear protests. *Social Problems*, *42*(4), 493–512.

Jena, N. P. (2017). Animal welfare and animal rights: An examination of some ethical problems. *Journal of Academic Ethics*, *15*(4), 377–95.

Jensen, M. P., Allen, C. D., Eguchi, T., Bell, I. P., LaCasella, E. L., Hilton, W. A., Hof, C. A. M., & Dutton, P. H. (2018). Environmental warming and feminization of one of the largest sea turtle populations in the world. *Current Biology*, *28*(1), 154–59.

Johnson, R. (2017, September 15). Agriculture is now the powerhouse driving economic growth in Australia. *Agricultural Appointments*. <https://www.agri.com.au/agriculture-is-now-the-powerhouse-driving-economic-growth-in-australia/>.

Jones, B. (2011). *The slaughter of Australian cattle in Indonesia: An observational study*. RSPCA Australia. <https://kb.rspca.org.au/wp-content/uploads/2019/03/The-slaughter-of-Australian-cattle-in-Indonesia-RSPCA-Research-Report-2011.pdf>.

Jones, B., & Davies, J. (2016). *Backlash: Australia's conflict of values over live export*. Finlay Lloyd.

Jones, D., & Everding, S. (1993). Crows in suburbia. *Cumberland Bird Observers Club Inc. Newsletter 14*(4), 1–3.

Jooste, J. (2016, February 15). *Animals Australia ready to launch new advertisements calling for ban on live exports, after complaints about previous campaign dismissed*. ABC News. <https://www.abc.net.au/news/rural/2016-02-15/live-export-animals-australia-advertising-complaint-dismissed/7168534>.

Joshi, V. K., & Kumar, S. (2015). Meat analogues: Plant based alternatives to meat products – A review. *International Journal of Food and Fermentation Technology*, *5*(2), 107–19.

Joy, M. (2008). *Strategic action for animals: A handbook on strategic movement building, organizing, and activism for animal liberation*. Lantern Books.

Kangaroo Industry Association of Australia. (2019). *Addressing the differences between commercial and non-commercial harvesting of kangaroo*. [Policy paper #3]. Kangaroo Industry Association of Australia. <https://www.kangarooindustry.com/wp-content/uploads/2019/06/Commercial-and-non-commercial-harvesting.pdf>.

Kao, G. (2018). My life with Morris: A feminist account of friendship and conversion in a bicultural context. In T. Bechtel, M. Eaton, & T. Harvie (Eds.), *Encountering earth: Thinking theologically with a more than human world* (pp. 75–87). Wipf and Stock Publishers.

Kateman, B. (2017). Introduction. In B. Kateman (Ed.), *The reducetarian solution* (pp. xv–xviii). TarcherPerigee.

Katz, E., & Lazarsfeld, P. F. (2006). *Personal influence: The part played by people in the flow of mass communications*. Transaction Publishers.

Keck, M. (2020, July 22). *Australia's nature laws are 'ineffective' and fail to protect threatened species: Report*. Global Citizen. <https://www.globalcitizen.org/en/content/australias-nature-laws-are-ineffective-report/>.

Keefe, L. M. (2018). #FakeMeat: How big a deal will animal meat analogs ultimately be? *Animal Frontiers, 8*(3), 33–7.
Keogh, M. (2013, May 22). *Truth the first casualty in live exports war.* [Online forum post]. Australian Farm Institute. <https://www.farminstitute.org.au/ag-forum/truth-the-first-casualty-in-live-exports-war>.
Keogh, M., Henry, M., & Day, N. (2016). *Enhancing the competitiveness of the Australian livestock export industry.* Australian Farm Institute. <https://www.mla.com.au/research-and-development/search-rd-reports/final-report-details/Enhancing-the-competitiveness-of-the-Australian-livestock-export-industry/3853>.
Kestenbaum, R. (2018, November 27). The biggest trends in the pet industry. *Forbes.* <https://www.forbes.com/sites/richardkestenbaum/2018/11/27/the-biggest-trends-in-the-pet-industry/?sh=52799e68f099>.
Kharroub, T., & Bas, O. (2016). Social media and protests: An examination of Twitter images of the 2011 Egyptian revolution. *New Media & Society, 18*(9), 1973–92.
Khorana, S. (2020). Diverse Australians on television: From nostalgic whiteness to aspirational multiculturalism. *Media International Australia, 174*(1), 29–38.
Kilbourne, W. E., & Polonsky, M. J. (2005). Environmental attitudes and their relation to the dominant social paradigm among university students in New Zealand and Australia. *Australasian Marketing Journal, 13*(2), 37–48.
Kilvert, N. (2020a, October 8). *Land clearing in Australia: How does your state (or territory) compare?* ABC News. <https://www.abc.net.au/news/science/2020-10-08/deforestation-land-clearing-australia-state-by-state/12535438>.
Kilvert, N. (2020b, February 16). *Think Australia's bushfires killed a lot of animals? Weak environmental laws threaten the lives of more.* ABC News. <https://www.abc.net.au/news/science/2020-02-16/bushfire-wildlife-extinction-offsets/9622980>.
Kim, S. (2020, October 8). *Makeshift research fleet undertakes first Great Barrier Reef 'census' citizen science project.* ABC News. <https://www.abc.net.au/news/2020-10-08/citizens-of-the-gbr-launch-great-reef-census/12711870>.
King, B. (2013). *How animals grieve.* The University of Chicago Press.
King, B. (2021). *Animals' best friends: Putting compassion to work for animals in captivity and in the wild.* University of Chicago Press.
Kirby, M. (2011, August 13). Stand up and speak up for animals that cannot. *The Sydney Morning Herald.* <https://www.smh.com.au/environment/animals/stand-up-and-speak-up-for-animals-that-cannot-20110813-1irp5.html>.
Kirjner, D. (2015). Painfully, from the first-person singular to first-person plural: The role of feminism in the study of the Anthropocene. In the Human-Animal Research Network Editorial Collective (Ed.), *Animals in the*

Anthropocene: Critical perspectives on non-human futures (pp. 135–50). Sydney University Press.

Knaus, C. (2017, March 17). Greyhound racing: Euthanasia of healthy dogs sparks call for stronger penalties. *The Guardian*. <https://www.theguardian.com/sport/2017/mar/17/greyhound-racing-euthanasia-of-healthy-dogs-sparks-call-for-stronger-penalties>.

Knaus, C. (2021, June 29). High court to hear bid to overturn New South Wales hidden camera laws. *The Guardian*. <https://www.theguardian.com/australia-news/2021/jun/29/high-court-to-hear-bid-to-overturn-new-south-wales-ag-gag-laws>.

Kopnina, H., Washington, H., Taylor, B., & Piccolo, J. (2018). Anthropocentrism: More than just a misunderstood problem. *Journal of Agricultural and Environmental Ethics*, *31*(1), 109–27.

Kotzmann, J. (2019, October 3). *ACT's new animal sentience law recognises an animal's psychological pain and pleasure, and may lead to better protections*. The Conversation. <https://theconversation.com/acts-new-animal-sentience-law-recognises-an-animals-psychological-pain-and-pleasure-and-may-lead-to-better-protections-124577>.

KPMG. (2018). *Talking 2030: Growing agriculture into a $100 billion industry*. <https://docs.wixstatic.com/ugd/f0cfd1_26dbb49eea91458d8b1606a0006ec20e.pdf>.

Kurki, V. (2021). Legal personhood and animal rights. *Journal of Animal Ethics*, *11*(1), 47–62.

Kwaymullina, A. (2005). Seeing the light: Aboriginal law, learning and sustainable living in country. *Indigenous Law Bulletin*, *6*(11). <https://www6.austlii.edu.au/cgi-bin/viewdoc/au/journals/IndigLawB/2005/27.html>.

La Canna, X. (2017, September 12). *Australia's introduced animals: Eradication programs under the spotlight*. ABC News. <https://www.abc.net.au/news/2017-09-12/should-australia-rethink-eradication-programs-of-feral-animals/8830998>.

La Fontaine, M. (2006). *New legend: A story of law and culture and the fight for self-determination in the Kimberley, Australia*. Kimberley Aboriginal Law and Culture Centre.

Lane, A., Wallis, K., & Phillips, S. (2020). *A review of the conservation status of New South Wales populations of the koala (Phascolarctos cinereus) leading up to and including part of the 2019/20 fire event*. (Report to International Fund for Animal Welfare). Biolink Ecological Consultants. <https://d1jyxxz9imt9yb.cloudfront.net/resource/348/attachment/original/file-1389175135551e339cf3296d6f24ed522.pdf>.

Larson, S., Stoeckl, N., Jarvis, D., Addison, J., Grainger, D., Watkin Lui, F., Walalakoo Aboriginal Corporation, Bunuba Dawangarri Aboriginal Corporation Rntbc, Ewamian Aboriginal Corporation Rntbc, & Yanunijarra Aboriginal Corporation Rntbc. (2020). Indigenous Land and Sea Management Programs (ILSMPs) enhance the wellbeing of Indigenous Australians. *International Journal of Environmental Research and Public Health, 17*(1), 1–15.

Latour, B. (2003). The promise of constructivism. In E. Selinger (Ed.), *Chasing technoscience: Matrix for materiality* (pp. 27–46). Indiana University Press.

Lattouf, A. (2016, August 25). Changing the whitewash channel on Australian television. ABC News. <https://www.abc.net.au/news/2016-08-25/whitewash-channel-australian-tv-diversity/7783428>.

Lauder, S. (2015, July 16). *The war on feral cats begins*. [Interview]. ABC AM. <https://www.abc.net.au/am/content/2015/s4274581.htm>.

Lawrence, G. (1987). *Capitalism and the countryside: The rural crisis in Australia*. Pluto Press.

Lawyers for Animals. (2021). *The law*. <https://lawyersforanimals.org.au/information/the-law/>.

Leahy, G. (Director). (2016). *Baxter and me*. [Film]. Gecko Films.

Leahy, S. (2019, May 6). One million species at risk of extinction, UN report warns. *National Geographic*. <https://www.nationalgeographic.com/environment/2019/05/ipbes-un-biodiversity-report-warns-one-million-species-at-risk/>.

Lee, J. R., Maggini, R., Taylor, M. F. J., & Fuller, R. A. (2015). Mapping the drivers of climate change vulnerability for Australia's threatened species. *PLoS ONE, 10*(5), e0124766. <https://doi.org/10.1371/journal.pone.0124766>.

Lefebvre, H. (1991). *The production of space*. (D. Nicolson-Smith, Trans.). Blackwell. (Original work published 1974).

Lefebvre, H. (2004). *Rhythmanalysis: Space, time and everyday life*. (S. Elden & G. Moore, Trans). Continuum. (Original work published 1992).

Leopold, A. (1949). *A sand county almanac*. Oxford University Press.

Leopold, A. (1987). *A sand county almanac and sketches here and there*. Oxford University Press.

Lerner, S. (2017). Meatless Monday: One day a week, cut out meat. In B. Kateman (Ed.), *The reducetarian solution* (pp. 141–2). TarcherPerigee.

Leroy, F., & Praet, I. (2017). Animal killing and postdomestic meat production. *Journal of Agricultural Environmental Ethics, 30*, 67–86.

Letnic, M., & Feit, B. (2019, April 4). *Like cats and dogs: Dingoes can keep feral cats in check*. The Conversation. <https://theconversation.com/like-cats-and-dogs-dingoes-can-keep-feral-cats-in-check-114748>.

Lewis, J. (2014, July 21). *Netflix – friend or foe?* BARB. <https://www.barb.co.uk/news/netflix-friend-or-foe/>.

Lezy-Bruno, L. (2017). *Let go of some urban domestication: How would you convince the mayor to re-wild the city?* The Nature of Cities. <https://www.thenatureofcities.com/2017/11/13/re-wilding-make-cities-better-just-wilder/>.

Lindell, J. (2021, May 14). More than 1500 kangaroos to be killed in Canberra nature reserves as annual cull returns. *The Canberra Times*. <https://www.canberratimes.com.au/story/7252883/more-than-1500-kangaroos-to-be-killed-as-annual-cull-returns/>.

Lindenmayer, D., & Burgman, M. (2005). *Practical conservation biology*. CSIRO Publishing.

LiveCorp. (2019). *Annual Report 2018/19*. <https://livecorp.com.au/article/2irWvEVpR3XeeBTcDe3cnb>.

Locke, S. (2017, January 5). *Caged egg sales trend lower as demand for free-range increases*. ABC Rural. <https://www.abc.net.au/news/rural/2017-01-05/caged-egg-market-trending-down-in-response-to-free-range/8164004>.

Lockie, S. (2015). *Australia's agricultural future: The social and political context*. (Report to SAF07). Australia's Agricultural Future Project, Australian Council of Learned Academies, Melbourne. <https://acola.org.au/wp/PDF/SAF07/social%20and%20political%20context.pdf>.

Lomas, G. (Executive Producer 2015–21). (1990–present). *Gardening Australia*. [TV series]. Australian Broadcasting Corporation (ABC).

Loretto, M. C., Fraser, O. N., & Bugnyar, T. (2012). Ontogeny of social relations and coalition formation in common ravens (*Corvus corax*). *International Journal of Comparative Psychology*, 25(3), 180–94.

Lowe, B. M. (2006). *Emerging moral vocabularies: The creation and establishment of new forms of moral and ethical meanings*. Lexington Books.

Lowrey, T. (2018, May 20). *Culling of Kosciuszko brumbies to be banned under plan to protect 'national icons'*. ABC News. <https://www.abc.net.au/news/2018-05-20/culling-kosciousko-brumbies-banned-under-plan-national-icon/9780558>.

Lucas, P., Webb, T., Valentine, P., Marsh, H., Great Barrier Reef Marine Park Authority, & Environment Australia. (1997). *The outstanding universal value of The Great Barrier Reef World Heritage Area*. Great Barrier Reef Marine Park Authority. <https://elibrary.gbrmpa.gov.au/jspui/handle/11017/301>.

Luke, B. (2007). Justice, caring, an animal liberation. In J. Donovan & C. J. Adams (Eds.), *The feminist care tradition in animal ethics* (pp. 125–52). Columbia University Press.

Luna, T. (2017). The element of surprise. In B. Kateman (Ed.), *The reducetarian solution* (pp. 6–9). TarcherPerigee.

Lynn, W. S. (2015, October 7). *Australia's war on feral cats: Shaky science, missing ethics*. The Conversation. <https://theconversation.com/australias-war-on-feral-cats-shaky-science-missing-ethics-47444>.
Mackay Conservation Group (n.d.). *We must safeguard landscapes like this forever*. Retrieved March 3, 2021, from <https://www.mackayconservationgroup.org.au/>.
Małecki, W., Sorokowski, P., Pawłowski, B., & Cieński, M. (2019). *Human minds and animal stories: How narratives make us care about other species*. Routledge.
Maller, C., Murnaw, L., & Cooke, B. (2019). Health and social benefits of living with 'wild' nature. In N. Pettorelli, S. Durant, & J. Du Toit (Eds.), *Rewilding*. (165–81). Cambridge University Press.
Mann, C. (2013, October 19–20). *Communicating animal social justice*. [Paper presentation]. Animal Activists Forum, Melbourne, Australia.
Manning, J., Power, D., & Cosby, A. (2021). Legal complexities of animal welfare in Australia: Do on-animal sensors offer a future option? *Animals, 11*(1). <https://www.ncbi.nlm.nih.gov/pmc/articles/PMC7825130/pdf/animals-11-00091.pdf>.
Mar, R. A., & Oatley, K. (2008). The function of fiction is the abstraction and simulation of social experience. *Perspectives on Psychological Science, 3*(3), 173–92.
Marks, C. A. (1999). Ethical issues in vertebrate pest control: Can we balance the welfare of individuals and ecosystems? In D. J. Mellor & V. Monamy (Eds.), *The use of wildlife in research* (pp. 79–89). Australian and New Zealand Council for the Care of Animals in Research and Teaching.
Markwell, K. (2020, January 7). *Koalas are the face of Australian tourism. What now after the fires?* The Conversation. <https://theconversation.com/koalas-are-the-face-of-australian-tourism-what-now-after-the-fires-129347>.
Markwell, K., & Cushing, N. (2014). Animals and Australian identity. *On Line Opinion: Australia's e-Journal of Social and Political Debates*. <https://www.onlineopinion.com.au/view.asp?article=16804>.
Marshall, A. (2019, September 13). *Why farmers fear fake meat activists more than fake meat*. Farm Online. <https://www.farmonline.com.au/story/6384291/farmers-dont-fret-about-fake-meat-just-fake-labels/>.
Marx, K., & Engels, F. (1955). *The Communist manifesto*. Appleton-Century-Crofts.
Masterman-Smith, H., Ragusa, A., & Crampton, A. (2014, November 24–27). Reproducing speciesism: A content analysis of Australian media representations of veganism. In B. West (Ed.), *Proceedings of The Australian Sociological Association (TASA) conference: Challenging identities, institutions and communities*. [Conference proceedings]. (pp. 1–13). The Australian Sociological Association (TASA), Adelaide, Australia.

Mata, L., Ramalho, C. E., Kennedy, J. E., Parris, K. M., Valentine, L., Miller, M., Bekessy, S., Hurley, S., & Cumpston, Z. (2020). Bringing nature back into cities. *People and Nature*, *2*(2), 350–68.

Mathews, F. (1997). Living with animals. *Animal Issues*, *1*(1), 4–16.

Mathews, F. (2020, March 9). *Koala makes us Australian: Reflections on the great fires*. [Opinion]. ABC Religion and Ethics. <https://www.abc.net.au/religion/koala-makes-us-australian/12039676>.

Matthews, K. (2020, January 24). *Rewilding cities: How bringing nature back to cities is good for our health*. RMIT. <https://www.rmit.edu.au/news/all-news/2020/jan/rewilding-cities>.

Mayhall, T. A. (2019). The meat of the matter: Regulating laboratory-grown alternative. *Food and Drug Law Journal*, *74*(1), 151–69.

Mazur, N. A. (2006). *Social research to support the Australian animal welfare strategy: A synthesis report*. ENVision Environmental Consulting.

McAlpine, C. A., Syktus, J., Ryan, J. G., Deo, R. C., McKeon, G. M., McGowan, H. A., & Phinn, S. R. (2009). A continent under stress: Interactions, feedbacks and risks associated with impact of modified land cover on Australia's climate. *Global Change Biology*, *15*, 2206–23.

McCann, A., & Pearce, D. (Director & Writer). (2019). *Koko: The red dog story*. [Film]. Woss Group Film Productions.

McCarthy, M., & Henderson, A. (2018, June 16). *What is 'mince'? Supermarkets and farmers clash over Funky Fields plant product*. ABC News. <https://www.abc.net.au/news/rural/2018-06-16/meaning-of-mince-sparks-clash-between-supermarket-and-farmers/9875242>.

McCausland, C. (2014). A utilitarian argument against animal exploitation. In G. Burns & M. Paterson (Eds.), *Engaging with animals: Interpretations of a shared existence* (pp. 205–23). Sydney University Press.

McDonald, S. (Director). (2015). *Oddball*. [Film]. The Film Company; Practical Pictures Kmunications.

McElligott, A. G., O'Keeffe, K. H., & Green, A. C. (2020). Kangaroos display gazing and gaze alternations during an unsolvable problem task. *Biology Letters*, *16*(12), 1–4. <https://doi.org/10.1098/rsbl.2020.0607>.

McEwan, A. (2020, September 25). *Greyhound pups must be tracked from birth to death, so we know how many are killed*. The Conversation. <https://theconversation.com/greyhound-pups-must-be-tracked-from-birth-to-death-so-we-know-how-many-are-killed-144868>.

McGhee, K. (2020, February 27). Our koalas: Post-bushfire recovery and future challenges. *Australian Geographic*. <https://www.australiangeographic.com.au/topics/wildlife/2020/02/our-koalas-post-bushfire-recovery-and-future-challenges/>.

McGowan, M. (2019, October 18). Prosecutions 'should occur' after footage reveals racehorse slaughter and cruelty. *The Guardian*. <https://www.theguardian.com/world/2019/oct/18/prosecutions-footage-racehorse-cruelty-slaughter>.

McGrail, S., Halamish, E., Teh-White, K., & Clark, M. (2013). Diagnosing and anticipating social issue maturation: Introducing a new diagnostic framework. *Futures, 46*, 50–61.

McGrath, M. (2018, October 11). *'Flexitarian' diets key to feeding people in a warming world*. BBC News. <https://www.bbc.com/news/science-environment-45814659>.

McLagan, M., & McKee, Y. (2012). Introduction. In M. McLagan & Y. McKee (Eds.), *Sensible politics: The visual culture of nongovernmental politics* (pp. 9–26). Zone Books.

McLennan, C. (2021, February 8). *Duck season outrage – no one is happy*. Farm Online. <https://www.farmonline.com.au/story/7116636/duck-season-outrage-no-one-is-happy/>.

McLeod, S. (1996). *The foraging behaviour of the arid zone herbivores the red kangaroo (Macropus rufus) and the sheep (Ovis aries) and its role on their competitive interaction, populations dynamics and life-history strategies*. [Doctoral dissertation, University of New South Wales]. <https://unsworks.unsw.edu.au/fapi/datastream/unsworks:49487/SOURCE01?view=true>.

Meat & Livestock Australia. (2020). *State of the industry report 2019: The Australian red meat and livestock industry*. <https://www.mla.com.au/globalassets/mla-corporate/prices--markets/documents/trends--analysis/soti-report/mla-state-of-industry-report-2020.pdf>.

Meijer, E. (2019). *When animals speak: Toward an interspecies democracy*. New York University Press.

Memphis Meats. (n.d). *About Memphis Meats*. Retrieved January 19, 2021, from <https://www.memphismeats.com/about>.

Menz, C., & Sharp, A. (2018). *Rewilding Southern Yorke Peninsula community engagement report*. Natural Resources Northern & Yorke. <https://www.naturalresources.sa.gov.au/files/sharedassets/northern_and_yorke/land/great_southern_ark_community_engagement_report.pdf>.

Metcalf, J. (2013). Meet shmeat: Food system ethics, biotechnology and re-worlding technoscience. *Parallax, 19*(1), 74–87.

Mika, M. (2006). Framing the issue: Religion, secular ethics and the case of animal rights mobilization. *Social Forces, 85*(2), 915–41.

Mills, B. (2017). *Animals on television: The cultural making of the non-human*. Palgrave Macmillan.

Milman, O. (2014, April 2). Koalas may disappear in areas affected by offset scheme, says foundation. *The Guardian*. <https://www.theguardian.com/environm

ent/2014/apr/02/koalas-may-disappear-in-areas-affected-by-offset-scheme-says-foundation>.

Milroy, G., & Milroy, J. (2008). Different ways of knowing: Trees are our families too. In S. Morgan, T. Mia, & B. Kwaymullina (Eds.), *Heartsick for country: Stories of love, spirit and creation* (pp. 22–42). Fremantle Press.

Mitchell, W. J. T. (2013). Image, space, revolution: The arts of occupation. In W. J. T. Mitchell, B. E. Harcourt, & M. Taussig (Eds.), *Occupy: Three inquiries in disobedience* (pp. 93–130). University of Chicago Press.

Molloy, C. (2011). *Popular media and animals*. Palgrave Macmillan.

Monahan, C. (Director). (2014). *Healing*. [Film]. Pointblank Pictures; Screen Australia.

Monastersky, R. (2015). Anthropocene: The human age. *Nature, 519*(7542), 144–7.

Moon, D. L. (2005). *A study of the abundance, distribution and daily activities of the Australian raven (Corvus coronoides) in urban wetland parks* [Unpublished honours dissertation]. Edith Cowan University. <https://ro.ecu.edu.au/cgi/viewcontent.cgi?article=1966&context=theses_hons>.

Moore, T. (2019, November 18). Ministers assess protection of 'very poor' Great Barrier Reef ahead of UN scrutiny. *The Sydney Morning Herald*. <https://www.smh.com.au/environment/climate-change/ministers-assess-protection-of-very-poor-great-barrier-reef-ahead-of-un-scrutiny-20191118-p53bqa.html>.

Moore, T. (2020, August 19). UNESCO study produces fresh fears over Great Barrier Reef's health. *The Sydney Morning Herald*. <https://www.smh.com.au/environment/climate-change/unesco-study-produces-fresh-fears-over-great-barrier-reef-s-health-20200819-p55n6s.html>.

Moraro, P. (2019, July 22). *It isn't clear how the new bill against animal rights activists will protect farmers*. The Conversation. <https://theconversation.com/it-isnt-clear-how-the-new-bill-against-animal-rights-activists-will-protect-farmers-120588>.

Moret, G. (2019). *Animal rights group creates online map showing farm locations and contact details*. ABC News. <https://www.abc.net.au/news/2019-01-21/animal-rightsgroup-aussie-farms-online-map-farmers-backlash/10731560>.

Moreton Bay Regional Council. (n.d.). *Crows*. Retrieved March 15, 2021, from <https://www.moretonbay.qld.gov.au/Services/Environment/Local-Wildlife/Crows>.

Morton, J. (1991). Black and white totemism: Conservation, animal symbolism, and human identification in Australia. In D. B. Croft (Ed.), *Australian people and animals in today's dreamtime* (pp. 21–51). Praeger Publishers.

Morton, R., Hebart, M. L., Ankeny, R. A., & Whittaker, A. L. (2021). Assessing the uniformity in Australian animal protection law: A statutory comparison.

Animals, *11*(1). <https://www.ncbi.nlm.nih.gov/pmc/articles/PMC7824303/pdf/animals-11-00035.pdf>.

Morton, R., Hebart, M. L., & Whittaker, A. L. (2020). Explaining the gap between the ambitious goals and practical reality of animal welfare law enforcement: A review of the enforcement gap in Australia. *Animals*, *10*(3). <https://www.mdpi.com/2076-2615/10/3/482/htm>.

Moss, P. (2018). *Review of the regulatory capability and culture of the Department of Agriculture and Water Resources in the regulation of live animal exports*. <https://www.agriculture.gov.au/animal/welfare/export-trade/independent-review-of-regulation>.

Muller, S. (2003). Towards decolonisation of Australia's protected area management: The Nantawarrina Indigenous protected area experience. *Australian Geographical Studies*, *41*(1), 29–43.

Muller, S. (2008). Indigenous payment for environmental service (PES) opportunities in the Northern Territory: Negotiating with customs. *Australian Geographer*, *39*(2), 149–70.

Mummery, J. (2017). *Radicalizing democracy for the twenty-first century*. Routledge.

Mummery, J., & Rodan, D. (2019a). Digitising kids with chooks to supercharge one online activism campaign. In L. Green, D. Holloway, K. Stevenson, & K. Jaunzems (Eds.), *Digitising early childhood* (pp. 319–36). Cambridge Scholars Press.

Mummery, J., & Rodan, D. (2019b). The multiple modes of protesting live exports in Australia. *Contention: The Multidisciplinary Journal of Social Protest*, *7*(1), 49–65.

Mundt, J. (1993). Externalities: Uncalculated outcomes of exchange. *Journal of Macromarketing*, *13*(2), 46–53.

Munro, L. (2004). Animals, 'nature' and human interests. In R. White (Ed.), *Controversies in environmental sociology* (pp. 61–76). Cambridge University Press.

Munro, L. (2015). The live animal export controversy in Australia: A moral crusade made for mass media. *Social Movement Studies: Journal of Social, Cultural and Political Protest*, *14*(2), 214–29.

Murphy, K. (2018, June 9). Live export opponents should check their moral compass, Minister says. *The Guardian*. <https://www.theguardian.com/australia-news/2018/jun/09/live-export-opponents-should-check-their-moral-compass-minister-says>.

Nabi, R. (2009). The effect of disgust-eliciting visuals on attitudes toward animal experimentation. *Communication Quarterly*, *46*(4), 472–84.

NACCHO (National Aboriginal and Community Controlled Health Organisation). (n.d.). *Aboriginal Community Controlled Health Organisations*

(ACCHOs). Retrieved October 14, 2021, from <https://www.naccho.org.au/acchos?hsCtaTracking=8583a101-0de0-4928-ab27-4678e2202539%7C0e1aaf2b-f754-4d70-83f7-1cb7bbb6186a>.

Nancy, J. L. (1991). *The inoperative community*. (P. Connon, et al., Trans.). University of Minnesota Press.

Nason, J. (2013, June 4). *Coles bagged over Animals Australia campaign*. Beef Central. <https://www.beefcentral.com/news/coles-bagged-over-animals-australia-campaign/>.

NatureLink Perth. (n.d.). *NatureLink Perth: Linking people, perspectives and knowledge to connect nature to the city*. Retrieved March 3, 2021, from <https://www.naturelinkperth.org/?mc_cid=3b0e39e50b&mc_eid=0c9b41ac95>.

News Chant. (2021, April 18). *How MasterChef changed the way we eat*. <https://au.newschant.com/entertainment/how-masterchef-changed-the-way-we-eat/>.

NFF (National Farmers' Federation). (2019). *National Farmers' Federation: Submission to ACCC inquiry into water markets in the Murray-Darling Basin issues paper*. <https://www.accc.gov.au/system/files/Water%20Inquiry%20-%20Submission%20-%20NFF%20-%206%20December%202019.pdf>.

Nibert, D. A. (2013). *Animal oppression and human violence: Domesecration, capitalism, and global conflict*. Columbia University Press.

Noble, F. (2020, January 14). *Government set to revise total number of hectares destroyed during bushfire season to 17 million*. 9 News. <https://www.9news.com.au/national/australian-bushfires-17-million-hectares-burnt-more-than-previously-thought/b8249781-5c86-4167-b191-b9f628bdd164>.

Nonhuman Rights Project. (2021). *Litigation: Challenging the legal thinghood of autonomous nonhuman animals*. <https://www.nonhumanrights.org/litigation/>.

North Queensland Conservation Council. (n.d.). Retrieved March 3, 2021, from <https://www.nqcc.org.au/>.

NSW Government. (2019, June 19). *Wild horses*. [Media release]. NSW Department of Planning, Industry and Environment. <https://www.environment.nsw.gov.au/topics/animals-and-plants/pest-animals-and-weeds/pest-animals/wild-horses>.

NSW Government. (2020, January 12). *Aerial food drops for endangered wildlife*. [Media release]. <https://www.environment.nsw.gov.au/news/aerial-food-drops-for-endangered-wildlife>.

Nulsen, R. A. (2012). Changes in soil properties. In R. J. Hobbs & D. A. Saunders (Eds.), *Reintegrating fragmented landscapes: Towards sustainable production and nature conservation* (pp. 107–45). Springer-Verlag.

Nurse, A. (2016). Beyond the property debate: Animal welfare as a public good. *Contemporary Justice Review*, *19*(2), 174–87.

Nursey-Bray, M. (2009). A Guugu Yimmithir Bam Wii: Ngawiya and Girrbithi: Hunting, planning and management along the Great Barrier Reef, Australia. *Geoforum*, *40*(3), 442–53.

Nursey-Bray, M., & Rist, P. (2009). Co-management and protected area management: Achieving effective management of a contested site, lessons from the Great Barrier Reef World Heritage Area (GBRWHA). *Marine Policy*, *33*, 118–27.

Oatley, K., Djikic, M., & Mar, R. (2016). The inwardness of James Joyce's story, 'The dead'. *Readings – A Journal for Scholars and Readers*, *2*(1), 1–14.

O'Brien, K. (2018, May 17). *Feral cat impact on native animal populations leads to construction of world's largest fence*. ABC News. <https://www.abc.net.au/news/2018-05-17/feral-cat-proof-fence-to-be-built-in-australia/9766830>.

OECD. (2015). *Meat consumption (indicator)*. <https://data.oecd.org/agroutput/meat-consumption.htm>.

Ogden, R. (2021). *Save Bribie's turtles!* Change.org. <https://www.change.org/p/qld-minister-for-environment-science-save-bribie-s-turtles>.

Oh, J-C., & Yoon, S-J. (2014). Theory-based approach to factors affecting ethical consumption. *International Journal of Consumer Studies*, *38*(3), 278–88.

Okoye, C. A. (2012). The problem of climate change: A study of anthropocentric environmental ethics. *Research on Humanities and Social Sciences*, *2*(9), 134–39.

Oliver, J. (2019). *Veg: Easy and delicious meals for everyone*. Penguin Random House.

Oliver, S. (Director) (2009). *Skippy – Australia's first superstar*. [Film]. Electric Pictures (Australia); Brook Lapping Productions (UK).

Olsberg.SPI. (2016, November 11). *Measuring the cultural value of Australia's screen sector*. Screen Australia. <https://www.screenaustralia.gov.au/getmedia/1dce395e-a482-42d1-b5a9-47bb6307f868/Screen-Currency-Olsberg-SPI-Nov2016.pdf>.

Olwig, K. F. (2002). The ethnographic field revisited: Towards a study of common and not so common fields of belonging. In V. Amit (Ed.), *Realizing community: Concepts, social relationships and sentiments* (pp. 124–45). Routledge.

Osborne, H., & van der Zee, B. (2020, January 20). Live export: Animals at risk in giant global industry. *The Guardian*. <https://www.theguardian.com/environment/2020/jan/20/live-export-animals-at-risk-as-giant-global-industry-goes-unchecked>.

O'Sullivan, S. (2015, February 17). New laws could stop revelations of animal abuse. *The Sydney Morning Herald*. <https://www.smh.com.au/opinion/new-laws-could-stop-revelations-of-animal-abuse-20150217-13g04u.html>.

O'Sullivan, S., McCausland, C., & Brenton, S. (2017). Animal activists, civil disobedience and global responses to transnational injustice. *Res Publica*, *23*(3), 261–80.

Pallotta, N. (2005). *Becoming an animal rights activist: An exploration of culture, socialization and identity transformation* [Doctoral dissertation, University of Georgia]. <https://getd.libs.uga.edu/pdfs/pallotta_nicole_r_200505_phd.pdf>.

Palmer, C. (2006). Stewardship: A case study in environmental ethics. In R. J. Berry (Ed.), *Environmental stewardship: Critical perspectives, past and present* (pp. 63–75). T & T Clark.

Parbery, P., & Wilkinson, R. (2012). *Victorians' attitudes to farming*. (Research papers 2012.3). Department of Environment and Primary Industries. <https://www.vgls.vic.gov.au/client/en_AU/search/asset/1018252/0>.

Parker, C., & Bromberg, L. (2021, July 5). *National plan to allow battery cages until 2036 favours cheap eggs over animal welfare*. The Conversation. <https://theconversation.com/national-plan-to-allow-battery-cages-until-2036-favours-cheap-eggs-over-animal-welfare-163552>.

Parker, R. (Series Producers). (2013–14). *Paddock to plate*. [TV series]. ITV Studios Australia Productions; Foxtel.

Parkes-Hupton, H. (2022, February 18). *Fatal shark attack in Sydney's Little Bay 'rare and uncommon', experts say*. ABC News. <https://www.abc.net.au/news/2022-02-18/little-bay-sydney-shark-attack-rare-and-uncommon/100840136>.

Paulson, S. (2019, December 6). Making kin: An interview with Donna Haraway. *Los Angeles Review of Books*. <https://lareviewofbooks.org/article/making-kin-an-interview-with-donna-haraway/>.

Paytas, T. (2018, July 8). *Vegan activists using direct action should rethink their approach*. ABC News. <https://www.abc.net.au/news/2018-07-08/vegans-turning-aussies-off-veganism/9944244>.

Pecl, G., & Hobday, A. (2016, April 6). Species everywhere are on the move. *Ecos*, *218*. <https://ecos.csiro.au/1459-2/>.

Pedersen, C., & White, R. (2021). Discourses of discord: Animal activism and moral judgement. *International Criminology*. <https://doi.org/10.1007/s43576-021-00027-w>.

Pedley, E. C. (2013). *Dot and the Kangaroo*. (F. P. Mahony, Illus.). Dodo Press. (Original work published 1899).

Pepin-Neff, C., & Wynter, T. (2018a). Reducing fear to influence policy preferences: An experiment with sharks and beach safety policy options. *Marine Policy*, *88*, 222–9.

Pepin-Neff, C., & Wynter, T. (2018b). Shark bites and shark conservation: An analysis of human attitudes following shark bite incidents in two locations in

Australia. *Conservation Letters: A Journal of the Society for Conservation*, *11*(2), 1–8. <https://doi.org/10.1111/conl.12407>.
Perth NRM. (2021). *Rewild Perth*. <https://www.perthnrm.com/project/rewild/>.
Petcare Information Advisory Service. (2010). *Pets in the city*. <https://www.ideas.org.au/uploads/resources/363/pets_in_the_city.pdf>.
Petrie, C. (2016). *Live export: A chronology*. (Research Paper Series, 2016–17). Parliament of Australia. <https://www.aph.gov.au/About_Parliament/Parliamentary_Departments/Parliamentary_Library/pubs/rp/rp1920/Chronologies/LiveExport>.
Petrie, C. (2018, September 4). *Export legislation amendment (live-stock) bill 2018*. [Bills digest no. 20, 2018–19]. Parliament of Australia. <https://www.aph.gov.au/Parliamentary_Business/Bills_Legislation/bd/bd1819a/19bd020>.
Pettorelli, N., Barlow, J., Stephens, P., Durant, S., Connor, B., Bühne, H. S., Sandom, C., Wentworth, J., & Toit, J. D. (2018). Making rewilding fit for policy. *Journal of Applied Ecology*, *55*, 1114–25.
Petty, M. (2019, September 22). *Let nature takes its course*. Medium. <https://medium.com/natural-world/let-nature-take-its-course-c57e88c1e02>.
Phillipov, M. (2017). *Media and food industries: The new politics of food*. Palgrave Macmillan.
Phillipov, M., & Gale, F. (2018). Celebrity chefs, consumption politics and food labelling: Exploring the contradictions. *Journal of Consumer Culture*, *20*(4), 400–18. <https://doi.org/10.1177%2F1469540518773831>.
Phoenix, J. (2020, February 12). *The speech that broke the Internet: Joaquin Phoenix Oscar speech*. [Video]. YouTube. <https://www.youtube.com/watch?v=LUllIT8rCj4>.
Pines, M., Petherick, C., Gaughan, J. B., & Phillips, C. J. C. (2007). Stakeholders' assessment of welfare indicators for sheep and cattle exported by sea from Australia. *Animal Welfare*, *16*(4), 489–98.
Piper, H. (2016). Broadcast drama and the problem of television aesthetics: Home, nation, universe. *Screen*, *57*(2), 163–93.
Plumwood, V. (2010). Nature in the active voice. In R. Irwin (Ed.), *Climate change and philosophy: Transformational possibilities* (pp. 32–47). Continuum.
Pobjie, B. (2016, May 25). *If you're looking for TV that reflects the real, multicultural Australia – watch a singing competition or a cooking show*. SBS. <https://www.sbs.com.au/guide/article/2016/05/25/why-reality-tv-so-much-more-diverse-scripted>.
Pointing, C. (2019, September 5). *Jamie Oliver gives up meat to live longer*. Live Kindly. <https://www.livekindly.co/jamie-oliver-meat-live-longer/>.
Pollan, M. (2008). *In the defense of food: An eater's manifesto*. Penguin.

Post, M. (2013, August 6). *Meet the new meat: A TEDx talk to pair with the first lab-grown hamburger.* [Video]. TEDxHarlem Conferences. <https://blog.ted.com/meet-the-new-meat-a-tedx-talk-to-pair-with-the-first-lab-grown-hamburger/>.
Predovnik, R. (2015, February 20). *WA government and Stockland handball costs to WA volunteers.* WA Today. <https://www.watoday.com.au/national/western-australia/wa-government-and-stockland-handball-costs-to-wa-volunteers-20150219-13j0ov.html>.
Productivity Commission. (2016, November 15). *Regulation of Australian agriculture: Productivity Commission inquiry report.* <https://www.pc.gov.au/inquiries/completed/agriculture/report/agriculture.pdf>.
Pryor, S. (2018, June 21). One of Canberra's largest ever kangaroo culls has finished. *The Canberra Times.* <https://www.canberratimes.com.au/story/6015818/one-of-canberras-largest-ever-kangaroo-culls-has-finished/>.
Purtill, J. (2019, February 20). *An Australian rodent has become the first climate change mammal extinction.* ABC Triple J Hack. <https://www.abc.net.au/triplej/programs/hack/bramble-cay-melomys-first-climate-change-mammal-extinction/10830080>.
Quail, G. (Executive Producer). (2011–present). *Send in the dogs Australia.* [TV series]. Quail Television; Nine Network.
Rahbek, U. (2007). Revisiting Dot and the Kangaroo: Finding a way in the Australian bush. *Australian Humanities Review, 41.* <https://australianhumanitiesreview.org/2007/02/01/revisiting-dot-and-the-kangaroo-finding-a-way-in-the-australian-bush/>.
Ratelle, A. (2014). *Animality and children's literature and film.* Palgrave Macmillan.
Rawls, J. (1971). *A theory of justice.* Harvard University Press.
Readfearn, G. (2019, December 2). Great Barrier Reef world heritage values damaged by climate change, government admits. *The Guardian.* <https://www.theguardian.com/environment/2019/dec/02/great-barrier-reef-world-heritage-values-damaged-climate-change-government-admits>.
Readfearn, G. (2020, February 3). Koala 'massacre': Scores of animals found dead or injured after plantation logging. *The Guardian.* <https://www.theguardian.com/environment/2020/feb/03/koala-massacre-animals-reported-starving-or-dead-after-plantation-logging>.
Readfearn, G. (2021, May 21). New threatened species strategy won't overcome Australia's appalling record, campaigners say. *The Guardian.* <https://www.theguardian.com/environment/2021/may/21/new-threatened-species-strategy-wont-overcome-australias-appalling-record-campaigners-say>.
Reeves, M. (2019, August 7). *MOOving on: Consumers are switching to soy, almond and pea protein milks.* [Press release]. Ibis*World* Industry Insider. <https://

www.ibisworld.com/industry-insider/press-releases/mooving-on-consumers-are-switching-to-soy-almond-and-pea-protein-milks/>.

Regan, T. (1986). A case for animal rights. In M. W. Fox & L. D. Mickley (Eds.), *Advances in animal welfare science 1986/87* (pp. 179–89). The Humane Society of the United States.

Rendall, J. (2020, July 22). *Plan to build in core koala habitat sparks local anger as developer challenges council rejection*. ABC News. <https://www.abc.net.au/news/2020-07-22/queensland-koala-habitat-under-threat-development/12476108>.

Rhodes, J., McAlpine, C., Lunney, D., Wilson, K., & Santika, T. (2016, July 6). *Koalas are feeling the heat, and we need to make some tough choices to save our furry friends*. The Conversation. <https://theconversation.com/koalas-are-feeling-the-heat-and-we-need-to-make-some-tough-choices-to-save-our-furry-friends-61306>.

Rhodes, J. R., Beyer, H. L., Preece, H. J., & McAlpine, C. A. (2015). *South East Queensland koala population modelling study*. UniQuest. <https://environment.des.qld.gov.au/__data/assets/pdf_file/0029/88913/seq-koala-population-modelling-study.pdf>.

Rhodes, J. R., Hood, A., Melzer, A., & Mucci, A. (2017). *Queensland koala expert panel: A new direction for the conservation of koalas in Queensland*. (A report to the Minister for Environment and Heritage Protection). Queensland Government. <https://environment.des.qld.gov.au/__data/assets/pdf_file/0031/88582/qld-koala-expert-panel-report-2017.pdf>.

Riber, A. B., Weerd, H. A. V. D., Jong, I. C. D., & Steenfeldt, S. (2017). Review of environmental enrichment for broiler chickens. *Poultry Science*, *97*(2), 378–96.

Ritchie, E., Coetsee, A., Rendall, A., Doherty, T., & Miritis, V. (2020, June 16). *Cats wreak havoc on native wildlife, but we've found one adorable species outsmarting them*. The Conversation. <https://theconversation.com/cats-wreak-havoc-on-native-wildlife-but-weve-found-one-adorable-species-outsmarting-them-132265>.

Ritchie, E., Tulloch, A., Driscoll, D., Evans, M. C., & Doherty, T. (2021, January 21). *It's not too late to save them: 5 ways to improve the government's plan to protect threatened wildlife*. The Conversation. <https://theconversation.com/its-not-too-late-to-save-them-5-ways-to-improve-the-governments-plan-to-protect-threatened-wildlife-147669>.

Robinson, A. (2020, March 17). 'Dogs have a magic effect': How pets can improve our mental health. *The Guardian*. <https://www.theguardian.com/society/2020/mar/17/dogs-have-a-magic-effect-the-power-of-pets-on-our-mental-health>.

Robinson, L., & Hill, D. (Series Producer). (1967–68). *Skippy the bush kangaroo*. [TV series]. Norfolk International Production; Nine Network.

Rockström, J., Steffen, W., Noone, K., Persson, A., Chapin, F. S. III., Lambin, E., Lenton, T. M., Scheffer, M., Folke, C., Schellnhuber, H. J., Nykvist, B., de Wit, C. A., Hughes, T., van der Leeuw, S., Rodhe, H., Sorlin, S., Snyder, P. K., Costanza, R., Svedin, U., ... Foley. J. (2009). Planetary boundaries: Exploring the safe operating space for humanity. *Ecology and Society*, *14*(2), 1–33. <https://www.ecologyandsociety.org/vol14/iss2/art32/>.

Rodan, D., & Mummery, J. (2014). The 'Make it Possible' multi-media campaign: Generating a new 'everyday' in animal welfare. *Media International Australia*, *153*, 78–87.

Rodan, D., & Mummery, J. (2016). Doing animal welfare activism everyday: Questions of identity. *Continuum: Journal of Media & Cultural Studies*, *30*(4), 381–96.

Rodan, D., & Mummery, J. (2018). *Activism and digital culture in Australia*. Rowman & Littlefield.

Rodan, D., & Mummery, J. (2019). Animals Australia and the challenges of vegan stereotyping. *M/C Journal of Media and Culture*, *22*(2). <https://doi.org/10.5204/mcj.1510>.

Rodrigues, U. M. (2020, August 17). *Whitewash on the box: How a lack of diversity on Australian television damages us all*. The Conversation. <https://theconversation.com/whitewash-on-the-box-how-a-lack-of-diversity-on-australian-television-damages-us-all-143434>.

Romolini, M., Grove, J. M., Ventriss, C. L., Koliba, C. J., & Krymkowski, D. H. (2016). Toward an understanding of citywide urban environmental governance: An examination of stewardship networks in Baltimore and Seattle. *Environmental Management*, *58*(2), 254–67.

Rosane, O. (2020, September 16). *World failed to meet a single goal to save nature: UN biodiversity report*. EcoWatch: Environmental News for a Healthier Planet and Life. <https://www.ecowatch.com/nature-destruction-biodiversity-loss-un-2647682550.html>.

Rose, D., James, D., & Watson, C. (2003). *Indigenous kinship with the natural world in New South Wales*. New South Wales National Parks and Wildlife Service. <https://www.heritage.nsw.gov.au/assets/Uploads/publications/518/indigenous-kinship-with-the-natural-world-new-south-wales.pdf>.

Rose, D. B. (1996). *Nourishing terrains: Australian Aboriginal views of landscape and wilderness*. Australian Heritage Commission.

Rose, D. B. (2007). Recursive epistemologies and an ethics of attention. In B. Miller (Ed.), *Extraordinary anthropology: Transformations in the field* (pp. 88–102). University of Nebraska Press.

Rose, G. (2016). *Visual methodologies: An introduction to researching with visual materials.* (4th edn.). Sage Publications.
Rosenbloom, C. (2016, July 8). Can't do vegetarian? How about flexitarian? *The Washington Post.* <https://www.washingtonpost.com/lifestyle/wellness/cant-do-vegetarian-how-about-flexitarian/2016/07/07/9d2610aa-3d57-11e6-80bc-d06711fd2125_story.html>.
Ross, H., Grant, C., Robinson, C. J., Izurieta, A., Smyth, D., & Rist, P. (2009). Co-management and Indigenous protected areas in Australia: Achievements and ways forward. *Australasian Journal of Environmental Management, 16*(4), 242–52.
Roudavski, S. (2020). Multispecies cohabitation and future design. In S. Boess, M. Cheung, & R. Cain (Eds.), *Proceedings of Design Research Society (DRS) 2020 International Conference: Synergy* (pp. 731–50). Design Research Society.
Roughley, A., & Williams, S. (2007). *The engagement of Indigenous Australians in natural resource management: Key findings and outcomes from Land & Water Australia funded research and the broader literature.* Land & Water Australia. <https://parlinfo.aph.gov.au/parlInfo/search/display/display.w3p;query=Id%3A%22library%2Fcatalog%2F00146992%22;src1=sm1>
Rowlands, M. (2002). *Animals like us.* Verso.
Roy Morgan. (2019). *Rise in vegetarianism not halting the march of obesity.* <https://www.roymorgan.com/findings/7944-vegetarianism-in-2018-april-2018-201904120608>.
RSPCA. (n.d.). *Choose wisely: Putting humane food on the menu.* Retrieved January 19, 2021, from <https://www.choosewisely.org.au/about/>.
RSPCA. (2017). *The Productivity Commission's recommendations for improving animal welfare governance in Australia.* [Information paper]. <https://kb.rspca.org.au/wp-content/uploads/2019/01/An-Australian-Commission-for-Animal-Welfare-RSPCA-Information-Paper-May-2017.pdf>.
RSPCA. (2018, April 27). *New poll finds 3 in 4 Australians want live export to end, greatest concern over standards in rural and country areas.* [Media release and statements]. <https://www.rspca.org.au/media-centre/news/2018/new-poll-finds-3-4-australians-want-live-export-end-greatest-concern-over>.
RSPCA. (2019a, March 31). *Fact check: RSPCA South Australia's seizure of 10 adult border collie dogs.* RSPCA South Australia. <https://www.rspcasa.org.au/fact-check-rspca-border-collie-seizure/>.
RSPCA. (2019b, September 5). *Good animal welfare.* <https://kb.rspca.org.au/knowledge-base/good-animal-welfare/>.

RSPCA. (2019c, May 1). *Is there a need to kill kangaroos or wallabies?* <https://kb.rspca.org.au/knowledge-base/is-there-a-need-to-kill-kangaroos-or-wallabies/>.
RSPCA. (2019d, July 31). *The RSPCA approved farming scheme – 22 years and a better life for more than 2 billion animals.* <https://www.rspca.org.au/media-centre/news/2019/rspca-approved-farming-scheme-%E2%80%93-22-years-and-better-life-more-2-billion>.
RSPCA. (2019e, May 2). *RSPCA Australia mission statement, vision and objectives.* <https://kb.rspca.org.au/knowledge-base/rspca-australia-mission-statement-vision-and-objectives/>.
RSPCA. (2019f, April 3). *RSPCA SA v Ross and Fitzpatrick – get the facts.* RSPCA South Australia. <https://www.rspcasa.org.au/wp-content/uploads/2019/04/RPSCA-v-Ross-and-Fitzpatrick-get-the-facts-3-April-2019.pdf?fbclid=IwAR17wJYP22QyQiqmjdxhZGpAn5GNMhPzfGwhEbxyGogcoDK7XFyQlpRtGkY>.
RSPCA. (2019g, May 1). *What is the RSPCA doing to get hens out of battery cages?* <https://kb.rspca.org.au/knowledge-base/what-is-the-rspca-doing-to-get-hens-out-of-battery-cages/>.
RSPCA. (2019h, May 2). *What are ag-gag laws and how would they affect transparency and trust in animal production?* <https://kb.rspca.org.au/knowledge-base/what-are-ag-gag-laws-and-how-would-they-affect-transparency-and-trust-in-animal-production/>.
RSPCA. (2019i, August 8). *What is the RSPCA approved farming scheme?* <https://kb.rspca.org.au/knowledge-base/what-is-the-rspca-approved-farming-scheme/>.
RSPCA. (2019j, October 8). *What are the five domains and how do they differ from the five freedoms?* <https://kb.rspca.org.au/knowledge-base/what-are-the-five-domains-and-how-do-they-differ-from-the-five-freedoms/>.
RSPCA. (2019k, October 8). *What are the five freedoms of animal welfare?* <https://kb.rspca.org.au/knowledge-base/what-are-the-five-freedoms-of-animal-welfare/>.
RSPCA. (2020, March 12). *Australia slips further down international animal welfare ranking with 'D' scorecard.* <https://www.rspca.org.au/media-centre/news/2020/australia-slips-further-down-international-animal-welfare-ranking-%E2%80%98d%E2%80%99>.
RSPCA. (2021a). *About us.* <https://www.rspca.org.au/about-us>.
RSPCA. (2021b, March 3). *How are national farm animal welfare standards developed?* <https://kb.rspca.org.au/knowledge-base/how-are-national-farm-animal-welfare-standards-developed/>.

Rubbo, L., & Wellauer, K. (2020, March 7). *Koala losses from recent NSW bushfires 'One of the most significant biodiversity impacts in our history'*. ABC News. <https://www.abc.net.au/news/2020-03-07/koalas-losses-post-bushfires-bigger-than-modelled/12033834>.

Russell, D., & Singer, P. (1997). An interview with Professor Peter Singer, *Animal Issues*, *1*(1), 37–44.

Saalfeld, W. K., Delaney, R., Fukuda, Y., & Fisher, A. J. (2014). *Management program for the saltwater crocodile in the Northern Territory of Australia, 2014-2015*. Northern Territory Department of Land Resource Management.

Sacre, H. (2018, April 8) (Producer). (1979–present). Sheep, ships and videotape. *60 Minutes* Australia. 2018. [TV series]. Channel Nine, Nine Network.

Safran, H. (Director). (1976). *Storm boy*. [Film]. South Australian Film Company.

Sanda, D. (2018, March 11). Australia leads on extinction rate: Report. *The Sydney Morning Herald*. <https://www.smh.com.au/environment/conservation/australia-leads-on-extinction-rate-report-20180311-p4z3vn.html#:~:text=Australia%20'leading%20the%20world'%20on%20extinction%20rate>.

Sanders, B. (2020). *Global animal slaughter statistics & charts: 2020 update*. Faunalytics. <https://faunalytics.org/global-animal-slaughter-statistics-and-charts-2020-update/>.

SBS News. (2019, April 8). *Vegans versus farmers in national stoush*. <https://www.sbs.com.au/news/vegans-versus-farmers-in-national-stoush>.

Schawbel, D. (2017, April 18). Brian Kateman: Why and how he launched the Reducetarian Movement. *Forbes*. <https://www.forbes.com/sites/danschawbel/2017/04/18/brian-kateman-why-and-how-he-launched-the-reducetarian-movement/#77e4e7a3de6d>.

Schedvin, C. B. (1979). Midas and the merino: A perspective on Australian economic historiography. *Economic History Review*, *32*(4), 542–56.

Schloegl, C., Kotrschal, K., & Bugnyar, T. (2007). Gaze following in common ravens, *Corvus corax*: Ontogeny and habituation. *Animal Behaviour*, *74*, 769–78.

Scholz, S. J. (2008). *Political solidarity*. Pennsylvania State University Press.

Screen Australia. (2016a). *Screen currency: Valuing our screen industry*. <https://www.screenaustralia.gov.au/getmedia/1b1312e5-89ad-4f02-abad-daeee601b739/screencurrency-sa-report.pdf>.

Screen Australia. (2016b). *Seeing ourselves: Reflections on diversity in Australian TV drama*. <https://www.screenaustralia.gov.au/fact-finders/reports-and-key-issues/reports-and-discussion-papers/seeing-ourselves>.

Screen Australia & Ipsos Australia. (2013). *Hearts & minds: How local screen stories capture the hearts & minds of Australians*. Screen Australia. <https://www.screenaustralia.gov.au/getmedia/b2dc80e7-ebb7-4341-9a20-8225b00064bb/Report-hearts-and-minds.pdf?ext=.pdf>.

Searchinger, T., Waite, R., Hanson, C., Ranganathan, J., Dumas, P., & Matthews, E. (2018). *Creating a sustainable food future: A menu of solutions to feed nearly 10 billion people by 2050*. Synthesis report. World Resource Institute. <https://www.undp.org/content/dam/undp/library/Sustainable%20Development/Creating-a-sustainable-food-future.pdf>.

Sebo, J. (2018). The ethics and politics of plant-based and cultured meat. *The Ethics Forum*, *13*(1), 159–83.

Sefcovic, E., & Condit, C. M. (2001). Narrative and social change: A case study of the Wagner Act of 1935. *Communication Studies*, *52*(4), 284–301.

Sekhri, A. (2013, January 18). UN Secretary General Ban Ki-Moon addresses Mali, Syria, women's rights at Stanford. *The Stanford Daily*. <https://www.stanforddaily.com/2013/01/18/ban-urges-international-cooperation-to-solve-global-challenges/>.

Senate Select Committee on the Encouragement of Australian Productions for Television. (1963). *Report from the select committee on the encouragement of Australian productions for television: Part 1*. [Vincent committee]. Government of the Commonwealth of Australia.

Serna, J., & Rust, S. (2020, January 14). An Australia in flames tries to cope with an 'animal apocalypse.' Could California be next? *Los Angeles Times*. <https://www.latimes.com/environment/story/2020-01-14/australia-fires-killed-millions-of-animals-kangaroo-island>.

Seymour, F. (2020, July 15). They're advised not to shoot mums in Canberra – but they do, here's how. *The District Bulletin*. <https://districtbulletin.com.au/theyre-advised-not-shoot-mums-canberra-heres/>

Shafer, M. E., & Taddicken, M. (2015). Mediatized opinion leaders: New patterns of opinion leadership in new media environments? *International Journal of Communication*, *9*, 960–81.

Sharman, K. (2009). Farm animals and welfare law: An unhappy union. In P. J. Sankoff & S. W. White (Eds.), *Animal law in Australasia: A new dialogue* (pp. 35–56). Federation Press.

Shaw, D., & Shiu, E. M. K. (2002). An assessment of ethical obligation and self-identity in ethical consumer decision-making: A structural equation modelling approach. *International Journal of Consumer Studies*, *26*(4), 286–93.

Shea, M. (2015). Punishing animal rights activists for animal abuse: Rapid reporting and the new wave of Ag-Gag Laws. *Columbia Journal of Law and Social Problems*, *48*(3), 338–71.

Shotwell, A. (2011). *Knowing otherwise: Race, gender and implicit understanding*. Pennsylvania State University Press.

Shyam, G. (2018). Is the classification of animals as property consistent with modern community attitudes? *The University of New South Wales Law Journal*, *41*(4), 1418–44.

Silvers, A., & Francis, L. (2005). Justice through trust: Disability and the 'outlier problem' in social contract theory. *Ethics*, *116*(1), 40–76.

Simo, F. (2020, January 10). *More than $50 million raised for Australia wildfire relief efforts*. [Facebook app]. Facebook. <https://about.fb.com/news/2020/01/australia-wildfire-relief/>.

Sinclair, M., Derkley, T., Fryer, C., & Phillips, C. (2018). Australian public opinions regarding the live export trade before and after an animal welfare media exposé. *Animals: An Open Access Journal from MDPI*, *8*(7), 1–12. <https://doi.org/10.3390/ani8070106>.

Singer, P. (1974). All animals are equal. *Philosophic Exchange*, *5*(1), 103–16. <https://digitalcommons.brockport.edu/phil_ex/vol5/iss1/6/>.

Singer, P. (1990). *Animal liberation*. Random House. (Original work published 1975).

Singer, P. (2011). *The expanding circle: Ethics, evolution, and moral progress*. Princeton University Press. (Original work published 1981).

Singer, P. (2018, May 18). *The live export trade is unethical: It puts money ahead of animals' pain*. The Conversation. <https://theconversation.com/the-live-export-trade-is-unethical-it-puts-money-ahead-of-animals-pain-96849>.

Smallacombe, S. (2020, November 2). Australia's Indigenous culture has always had a place for dogs. They are part of our dreaming. *The Guardian*. <https://www.theguardian.com/commentisfree/2020/nov/02/australias-indigenous-culture-has-always-had-a-place-for-dogs-they-are-part-of-our-dreaming>.

Smee, B. (2019, November 2). Queensland shark attack follows a familiar pattern: First horror, then a media feeding frenzy. *The Guardian*. <https://www.theguardian.com/environment/2019/nov/02/blood-in-the-water-sharks-drum-lines-and-the-media-hysteria-that-surrounds-them>.

Smil, V. (2011). Harvesting the biosphere: The human impact. *Population and Development Review*, *37*, 613–36.

Smith, A. (1994). *An inquiry into the nature and causes of the wealth of nations*. The Modern Library.

Smith, B. (2012). The 'pet effect': Health related aspects of companion animal ownership. *Australian Family Physician*, *41*(6), 439–42.

Smith, M. (2009). Against ecological sovereignty: Agamben, politics, and globalization. *Environmental Politics*, *18*(1), 99–116.

Smith Maguire, J., & Matthews, J. (2010). Cultural intermediaries and the media. *Sociology Compass*, *4*(7), 405–16.

Smith Maguire, J., & Matthews, J. (2012). Are we all cultural intermediaries now? An introduction to cultural intermediaries in context. *European Journal of Cultural Studies*, *15*(5), 551–62.

Snow, D. A., & Benford, R. D. (1988). Ideology, frame resonance and participant mobilization. In B. Klandermans, H. Kriesi, & S. Tarrow (Eds.), *From structure to action: Comparing social movement research across cultures* (pp. 197–217). JAI Press.

Sookram, R. (2013). Environmental attitudes and environmental stewardship: Implications for sustainability. *The Journal of Values-Based Leadership*, *6*(2), 1–11.

Sools, A. (2012). 'To see a world in a grain of sand': Towards future-oriented what-if analysis in narrative research. *Narrative Works: Issues, Investigations and Interventions*, *2*(1), 83–105.

Sorenson, J. (2009). Constructing terrorists: Propaganda about animal rights. *Critical Studies on Terrorism*, *2*(2), 237–56.

Sorenson, J. (2016). *Constructing ecoterrorism: Capitalism, speciesism and animal rights*. Fernwood Publishing.

Southwell, A., Bessey, A., & Barker, B. (2006). *Attitudes towards animal welfare: A research report*. TNS Social Research.

Spencer, A., Gill, J., & Schmahmann, L. (2015, December 9–11). *Urban or suburban? Examining the density of Australian cities in a global context* [Paper presentation]. 7th State of Australian Cities Conference. Gold Coast, Queensland, Australia. <https://apo.org.au/node/63334>.

Spring, A., & Earl, C. (2019, May 15). Australia's biodiversity at breaking point – a picture essay. *The Guardian*. <https://www.theguardian.com/environment/2019/may/15/australias-biodiversity-at-breaking-point-a-picture-essay>.

Springmann, M., Clark, M., Mason-D'Croz, D., Wiebe, K., Bodirsky, B. L., Lassaletta, L., De Vries, W., Vermeulen, S. J., Herrero, M., Carlson, K. M., Jonell, M., Troell, M., Declerck, F., Gordon, L. J., Zurayk, R., Scarborough, P., Rayner, M., Loken, B., Fanzo, J., … Willett, W. (2018). Options for keeping the food system within environmental limits. *Nature*, *562*, 519–25.

Squire, C., Andrews, M., & Tamboukou, M. (2008). Introduction: What is narrative research? In M. Andrews, C. Squire, & M. Tamboukou, *Doing narrative research* (pp. 1–21). Sage Publications.

Stadler, J. (2003). Narrative, understanding and identification in *Steps for the future* HIV/AIDS documentaries. *Visual Anthropology Review*, *19*(1, 2), 86–101.

Stamps, J. A., & Swaisgood, R. R. (2007). Someplace like home: Experience, habitat selection and conservation biology. *Applied Animal Behaviour Science*, *102*(3-4), 392–409.

Stănescu, V. (2010). 'Green' eggs and ham? The myth of sustainable meat and the danger of the local. *Journal for Critical Animal Studies, VIII*(1/2), 8–32.
Starling, M., & McGreevy, P. (2018). *Making dogs happy*. Murdoch Books.
Starling, M., & McGreevy, P. (2020, February 6). *8 things we do that really confuse our dogs*. The Conversation. <https://theconversation.com/8-things-we-do-that-really-confuse-our-dogs-122616>.
Stenders, K. (Director). (2011). *Red dog*. [Film]. Woss Group Film Productions.
Stenders, K. (Director). (2016). *Red dog: True blue*. [Film]. Woss Group Film Productions.
Stephens, N., Di Silvioc, L., Dunsfordb, I., Ellisd, M., Glencrosse, A., & Sexton, A. (2018). Bringing cultured meat to market: Technical, socio-political, and regulatory challenges in cellular agriculture. *Trends in Food Science & Technology, 78*, 155–66.
Stephany, R. W. (2010). Hormonal growth promoting agents in food producing animals. In D. Thieme & P. Hemmersbach (Eds.), *Doping in sports: Handbook of experimental pharmacology* (pp. 355–67). Springer-Verlag.
Stewart, P. J. (1997). *Some aspects of the ecology of an urban corvid: The Australian raven (Corvus coronoides) in metropolitan Perth* [Unpublished honours dissertation]. Edith Cowan University. <https://ro.ecu.edu.au/cgi/viewcontent.cgi?article=1294&context=theses_hons>.
Stobo-Wilson, A., Murphy, B., Gillespie, G., Dielenberg, J., & Woinarski, J. (2020, July 29). *The mystery of the top end's vanishing wildlife, and the unexpected culprits*. The Conversation. <https://theconversation.com/the-mystery-of-the-top-ends-vanishing-wildlife-and-the-unexpected-culprits-143268>.
Stuparyk, B., Horn, C. J., Karabatsos, S., & Arteaga-Torres, J. (2018). A meta-analysis of animal survival following translocations: Comparisons between conflicts and conservation efforts. *Canadian Wildlife Biology and Management, 7*(1), 3–17.
Sullivan, K. (2019, April 11). *Scott Morrison's Government fired up over vegan trespasses as federal election looms*. ABC News. <https://www.abc.net.au/news/2019-04-11/government-response-to-a-vegan-debate/10990540>.
Sullivan, K., & Verley, A. (2021, April 14). *Australian live cattle exports will lift with NZ banning live trade, industry analyst*. ABC News. <https://www.abc.net.au/news/2021-04-14/nz-decision-to-ban-exports-of-live-cattle/100068068>.
Suncorp. (2020, April 29). *House hunting: Who comes first, pets or friends?* <https://www.suncorp.com.au/learn-about/buying-a-home/pets-or-friends.html>.
Sustainable Table. (n.d.). *Are you game? A guide to ethical meat eating*. Retrieved March 30, 2021, from <https://sustainabletable.org.au/wp-content/uploads/Are_You_Game.pdf>.

Sutton, M. (2019, October 26). *Vegans a 1 per cent minority in a country of meat eaters, survey finds*. ABC News. <https://www.abc.net.au/news/2019-10-26/vegans-comprise-just-1-per-cent-of-the-population-survey-finds/11635306>.

Sutton, Z. (2021). Researching towards a critically posthumanist future: On the political 'doing' of critical research for companion animal liberation. *International Journal of Sociology & Social Policy*, 41(3/4), 376–90.

Swan Valley Wildlife and Environment Advocacy Group. (2021). *Save the wildlife from property developers in the Swan Valley*. Change.org. <https://www.change.org/p/mayor-and-councillors-city-of-swan-save-the-wildlife-from-property-developers-in-the-swan-valley>.

Sweeney, O. F., Turnbull, J., Jones, M., Letnic, M., Newsome, T. M., & Sharp, A. (2019a). An Australian perspective on rewilding. *Conservation Biology*, 33(4), 812–20.

Sweeney, O. F., Turnbull, J., Jones, M., Letnic, M., & Newsome, T. M. (2019b). *We can 'rewild' swathes of Australia by focusing on what makes it unique*. The Conversation. <https://theconversation.com/we-can-rewild-swathes-of-australia-by-focusing-on-what-makes-it-unique-111749>.

Tait, A. (2017, July 9). Are you a reducetarian? Meat eating is no longer all or nothing. *New Statesman America*. <https://www.newstatesman.com/culture/observations/2017/07/are-you-reducetarian-meat-eating-no-longer-all-or-nothing>.

Tal-Or, N., & Cohen, J. (2016) Unpacking engagement: Convergence and divergence in transportation and identification. *Annals of the International Communication Association*, 40(1), 33–66. <https://doi.org/10.1080/23808985.2015.11735255>.

Taplin, L. (2017). *Review of a trial harvest of estuarine crocodile eggs in the Pormpuraaw deed of grant in trust lands and recommendations as to an experimental commercial harvest*. (Report to Department of Environment and Science). <https://environment.des.qld.gov.au/__data/assets/pdf_file/0016/90133/review-trial-harvest-est-croc-eggs-pormpuraaw-dgtl.pdf>.

Tasmanian Times. (2021, January 4). *Interest in vegan products surges 51% in Australia*. <https://www.tasmaniantimes.com/2021/01/interest-in-vegan-products-surges-51-in-australia/>.

Taylor, A. (2014). Settler children, kangaroos and the cultural politics of Australian national belonging. *Global Studies of Childhood*, 4(3), 169–82.

Taylor, C. (2004). *Modern social imaginaries*. Duke University Press.

Taylor, M. (2017, October 18). CSIRO independent analysis: No support for kangaroo research assumptions. *The District Bulletin*. <https://districtbulletin.com.au/csiro-independent-analysis-no-support-kangaroo-research-assumptions-2/>.

Taylor, M. (2018, June 27). Us and them – the end game? *The District Bulletin*. <https://districtbulletin.com.au/us-end-game/>.

Taylor, N., & Signal, T. D. (2009). Willingness to pay: Australian consumers and "on the farm" welfare. *Journal of Applied Animal Welfare Science, 12*, 345–59.

Taylor, P. (2015, October 26). Pintubi cat hunters to take skills across Australia. *The Australian*. <https://www.theaustralian.com.au/news/pintubi-cat-hunters-to-take-skills-across-australia/news-story/e80f111d790ac0f47c2134bb6a63b40d>.

Telfer, E. (1991). Friendship. In M. Pakaluk (Ed.), *Other selves: Philosophers on friendship* (pp. 250–67). Hackett.

Textor, M. (2011, June 4). Blood should flow after this anaemic response. *The Sydney Morning Herald*. <https://www.smh.com.au>.

The Age. (2004, November 11). A sheep in wolf's clothing? *The Age*. <https://www.theage.com.au/national/a-sheep-in-wolfs-clothing-20041111-gdyyy9.html>.

The Livestock Collective. (2021). *Live export*. <https://thelivestockcollective.com.au/live-export/>.

The Nationals. (n.d.). *Our history*. Retrieved January 15, 2021, from <https://nationals.org.au/about/our-history/>.

The Nature Conservancy Australia. (n.d.). *Why are we so afraid of sharks? Understanding the ocean's top predators*. Retrieved January 11, 2021, from <https://www.natureaustralia.org.au/what-we-do/our-priorities/wildlife/wildlife-stories/sharks/>.

The New Shorter Oxford English Dictionary (1993). Speciesism. In L. Brown (Ed.), *The new shorter oxford English dictionary* (p. 2972). Oxford University Press.

Thomas, I. (Director). (2018). *Emu runner*. [Film]. Imogen Thomas Films.

Thompson, J. (2019, January 27). *Exploring the strange, dangerous charm of keeping a saltwater crocodile as a pet*. ABC News. <https://www.abc.net.au/news/2019-01-27/crocs-as-pets-in-the-northern-territory/10441662>.

Ting, I. (2013, April 15). *Hold the red, pass the white – meat, that is*. Good Food. <https://www.goodfood.com.au/eat-out/news/hold-the-red-pass-the-white--meat-that-is-20130415-2hw6m>.

Topolski, R., Weaver, N., Martin, Z., & McCoy, J. (2013). Choosing between the emotional dog and the rational pal: A moral dilemma with a tail. *Anthrozoös, 26*, 253–63.

Torres, B. (2007). *Making a killing: The political economy of animal rights*. AK Press.

Towell, N., & Ilanbey, S. (2020). Mercy mission: Food drops to help save animals in Victoria's bushfire zones. *The Age*. <https://www.theage.com.au/national/victoria/mercy-mission-food-drops-to-help-save-animals-in-victoria-s-bushfire-zones-20200121-p53tdo.html>.

Townley, C. (2010). Animals as friends. *Between the species, 13*(10), 45–59.

Trading Economics. (2021). *Australia – urban population (% of total)*. <https://tradingeconomics.com/australia/urban-population-percent-of-total-wb-data.html>.
Trebeck, D. (1990). Farmer organisations. In D. B. Williams (Ed.), *Agriculture in the Australian Economy* (pp. 127–43). Sydney University Press.
Tronto, J. C. (1993). *Moral boundaries: A political argument for an ethic of care*. Routledge.
Tronto, J. C. (2013). *Caring democracy: Markets, equality, and justice*. New York University Press.
Tsing, A. L. (2011). Arts of inclusion, or, how to love a mushroom. *Australian Humanities Review*, *50*, 5–22.
Tsing, A. L. (2012). Unruly edges: Mushrooms as companion species. *Environmental Humanities*, *1*, 141–54.
Tulloch, A. I. T., Barnes, M. D., Ringma, J., Fuller, R. A., & Watson, J. E. M. (2016). Understanding the importance of small patches of habitat for conservation. *Journal of Applied Ecology*, *53*, 418–29.
Turow, J. (2017). *Media today: Mass communication in a converging world*. (6th edn.). Routledge.
Tutty, J. (2018, August 20). *Gardening shows prove a hit in the Friday TV ratings*. Mumbrella. <https://mumbrella.com.au/gardening-shows-prove-a-hit-in-the-friday-tv-ratings-536061>.
UNESCO. (n.d). *Great Barrier Reef*. Retrieved March 3, 2021, from <https://whc.unesco.org/en/list/154/>.
UNESCO. (2016). *World heritage marine sites: Best practice guide*. <https://whc.unesco.org/en/activities/868/>.
Union of Concerned Scientists. (n.d.). *Our work and from our blog*. Retrieved March 3, 2021, from <https://www.ucsusa.org/>.
Ursell, L. K., Metcalf, J. L., Parfrey, L. W., & Knight, R. (2012). Defining the human microbiome. *Nutrition Reviews*, *70*(Suppl 1), S38–S44. <https://doi.org/10.1111/j.1753-4887.2012.00493.x>.
van der Weele, C., Feindt, P., van der Goot, A. J., van Mierlo, B., & van Boekel, M. (2019). Meat alternatives: An integrative comparison. *Trends in Food Science & Technology*, *88*, 505–12.
van Dooren, T. (2011). Invasive species in penguin worlds: An ethical taxonomy of killing for conservation. *Conservation and Society*, *9*(4), 286–98.
van Dooren. T. (2019). *The wake of crows: Living and dying in shared worlds*. Columbia University Press.
van Dooren, T., Kirksey, E., & Münster, U. (2016). Multispecies studies: Cultivating arts of attentiveness. *Environmental Humanities*, *8*(1), 1–23.

Van Gurp, M. (2012, November 4). Factory farming the musical. *Osocio*. <https://osocio.org/message/factory-farming-the-musical/>.

Van Metter, J. E., Harriger, M.D., & Bolen, R. H. (2008). Environmental enrichment utilizing stimulus objects for African lions (*Panthera leo leo*), and Sumatran tigers (*Panthera tigris sumatrae*). *Bios*, *79*(1), 7–16.

Vegan First. (n.d.). *Did you know that chef Jamie Oliver endorsed a vegan diet?* Vegan First Daily. Retrieved January 20, 2021, from <https://www.veganfirst.com/article/did-you-know-that-chef-jamie-oliver-endorsed-a-vegan-diet>.

Vegan Society. (2008). *Memorandum and articles of association*. <https://www.vegansociety.com/about-us/further-information/memorandum-and-articles-association>.

Verbeke, W., Marcu, A., Rutsaert, P., Gaspar, R., Seibt, B, Fletcher, D., & Barnett, J. (2015a). 'Would you eat cultured meat?': Consumers' reactions and attitude formation in Belgium, Portugal and the United Kingdom. *Meat Science*, *102*, 49–58.

Verbeke, W., Sans, P., & Van Loo, E. J. (2015b). Challenges and prospects for consumer acceptance of cultured meat. *Journal of Integrative Agriculture*, *14*(2), 285–94.

Victorian Aboriginal Heritage Council. (2020). *Taking care of culture*. State of Victoria's Aboriginal Cultural Heritage. <https://content.vic.gov.au/sites/default/files/2021-01/Taking%20Care%20of%20Culture%20Discussion%20Paper_04012021_2.pdf>.

Viggers, K. L., & Hearn, J. P. (2005). The kangaroo conundrum: Home range studies and implications for land management. *Journal of Applied Ecology*, *42*(1), 99–107.

Villanueva, G. (2012, November 7). *Mainstream crusade – how the animal rights movement boomed*. The Conversation. <https://theconversation.com/mainstream-crusade-how-the-animal-rights-movement-boomed-10087>.

Vivian, L., & Godfree, R. (2014). *Relationships between vegetation condition and kangaroo density in lowland grassy ecosystems of the northern Australian Capital Territory: Analysis of data 2009, 2010 and 2013*. CSIRO. <https://www.environment.act.gov.au/__data/assets/pdf_file/0013/1225201/CSIRO-Rel-btw-vegetation-condition-and-kangaroo-density-2014.pdf>.

Voiceless. (n.d.). *About us*. Retrieved January 18, 2021, from <https://www.voiceless.org.au/about-us>.

Voiceless. (2019). *Live export*. <https://www.voiceless.org.au/hot-topics/live-export>.

Wadiwel, D. (2015). *The war against animals*. Brill.

Wagstaff, J. (2015, February 4). *Animals Australia launches new campaign against live exports*. The Weekly Times. <https://www.weeklytimesnow.com.au/agrib

usiness/animals-australia-launches-new-campaign-against-live-exports/news-story/8a6e61d75696de3d39e02e92bb93b4d7>.
Wahlen, S., & Laamanen, M. (2015). Consumption, lifestyle and social movements. *International Journal of Consumer Studies, 39*(5), 397–403.
Wahlquist, C. (2018, June 18). Not mincing words: Nationals denounce vegetarian product in meat aisle. *The Guardian*. <https://www.theguardian.com/lifeandstyle/2018/jun/18/not-mincing-words-nationals-denounce-vegetarian-product-in-meat-aisle>.
Wahlquist, C. (2019a, October 13). Australia's kangaroo meat trade could be the most sustainable in the world, despite welfare concerns. *The Guardian*. <https://www.theguardian.com/world/2019/oct/13/australias-kangaroo-cull-humane-and-sustainable-or-exercise-in-cruelty>.
Wahlquist, C. (2019b, July 20). 'It just didn't make any sense': Why Australians are turning away from meat. *The Guardian*. <https://www.theguardian.com/lifeandstyle/2019/jul/20/it-just-didnt-make-any-sense-why-australians-are-turning-away-from-meat>.
Waldron, D., & Townsend, S. (2012). *Snarls from the tea tree: Big cat folklore*. Arcadia Press.
Wallis, K., Lane, A., & Phillips, S. (2020). *A review of the conservation status of Queensland populations of the koala (Phascolarctos cinereus) arising from events leading up to and including the 2019/20 fire event.* (Report to World Wide Fund for Nature (WWF) Australia). Biolink Ecological Consultants. <https://www.wwf.org.au/ArticleDocuments/353/A%20Review%20of%20the%20Conservation%20Status%20of%20QLD%20Koalas.pdf.aspx>.
Wan, L. (2018, April 25). *Fact not fad: Why the vegan market is going from strength to strength in Australia*. Foodnavigator-Asia. <https://www.foodnavigator-asia.com/Article/2018/04/25/Fact-not-fad-Why-the-vegan-market-is-going-from-strength-to-strength-in-Australia>.
Wangan & Jagalingou Family Council (n.d.). *Our fight: Stop Adani destroying our land and culture*. Retrieved May 27, 2021, from <https://wanganjagalingou.com.au/our-fight/>.
Ward, R. (1966). *The Australian legend*. Oxford University Press.
Warkentin, T. (2010). Interspecies etiquette: An ethics of paying attention to animals. *Ethics & the Environment, 15*(1), 101–21.
Warren, M. A. (1997). *Moral status: Obligations to persons and other living things*. Oxford University Press.
Warren, R., Price, J., VanDerWal, J., Cornelius, S., & Sohl, H. (2018). The implications of the United Nations Paris Agreement on climate change for globally significant biodiversity areas. *Climatic Change, 147*(6), 395–409.

Waters, C. (2018, April 23). Paddock to plate: Farmers cut out the middle man. *The Sydney Morning Herald*. <https://www.smh.com.au/business/small-business/paddock-to-plate-farmers-cut-out-the-middle-man-20180419-p4zakj.html>.

Watson, E. M., Evans, T., Venter, O., Williams, B., Tulloch, A., Stewart, C., Thompson, I., Ray, J. C., Murray, K., Salazar, A., McAlpine, C., Potapov, P., Walston, J., Robinson, J. G., Painter, M., Wilkie, D., Filardi, C., Laurance, W. F., Houghton, R. A., ... Lindenmayer, D. (2018). The exceptional value of intact forest ecosystems. *Nature Ecology & Evolution*, 2(4), 599–610.

Wear, R. (1991). *Agrarian myth: Its history and use by the Australian Country Party*. [Masters dissertation, University of New England]. <https://rune.une.edu.au/web/handle/1959.11/18693>.

Weir, J. K., Stacey, C., & Youngetob, K. (2011). *The benefits associated with caring for country: Literature review*. Australian Institute of Aboriginal and Torres Strait Islander Studies. <https://aiatsis.gov.au/sites/default/files/research_pub/benefits-cfc_0_2.pdf>.

Welchman, J. (2012). A defence of environmental stewardship. *Environmental Values*, 21(3), 297–316.

Welin, S. (2013). Introducing the new meat. Problems and prospects. *Etikk i praksis – Nordic Journal of Applied Ethics*, 7(1), 24–37.

Whetham, B. (2020, February 14). *'Oddball effect' sees hundreds of maremma dogs needing rehoming after success of 2015 film*. ABC News South East SA. <https://www.abc.net.au/news/2020-02-14/maremmas-needed-rehoming-following-success-of-film-oddball/11951066>.

White, S. (2008). Regulation of animal welfare in Australia and the emergent Commonwealth: Entrenching the traditional approach of the states and territories or laying the ground for reform? *Federal Law Review*, 35(3), 347–74.

White, S. (2016). Animal protection law in Australia: Bound by history. In S. White & D. Cao (Eds.), *Animal law and welfare – international perspectives* (pp. 109–30). Springer.

WildArk. (n.d.). *Rewilding Australia one quoll at a time: Interview with Rewilding Australia founder, Robert Brewster*. Retrieved March 3, 2021, from <https://wildark.org/journals/rewilding-australia-one-quoll-time/>.

Wildcare. (2020, January 6). Want to help fire-affected wildlife? Regional network set up. *The District Bulletin*. <https://districtbulletin.com.au/want-to-help-fire-affected-wildlife-regional-network-set-up/>.

Wilderness Society. (n.d.). Retrieved March 3, 2021, from <https://www.wilderness.org.au/>.

Wilderness Society. (2021, February 25). *10 Facts about deforestation in Australia*. <https://www.wilderness.org.au/news-events/10-facts-about-deforestation-in-australia>.

Wildlife News Headlines. (2021, September 1). Nine News. <https://www.9news.com.au/wildlife/4>.
Wildlife Preservation Society of Queensland. (n.d.). Retrieved March 3, 2021, from <https://wildlife.org.au/>.
Wilkinson, T. (2016, May 24). To rescue or not, that is the question with distressed animals. *National Geographic*. <https://www.nationalgeographic.com/animals/article/160523-when-to-rescue-wild-animals>.
Wilks, M., & Phillips, C. J. C. (2017). Attitudes to in vitro meat: A survey of potential consumers in the United States. *PLoS ONE, 12*(2), Article e0171904. <https://doi.org/10.1371/journal.pone.0171904>.
Willett, W., Rockström, J., Loken, B., Springmann, M., Lang, T., Vermeulen, S., Garnett, T., Tilman, D., DeClerck, F., Wood, A., Jonell, M., Clark, M., Gordon, L. J., Fanzo, J., Hawkes, C., Zurayk, R., Rivera, J. A., De Vries, W., Sibanda, L. M., … Murray, C. J. L. (2019). Food in the Anthropocene: The EAT–Lancet Commission on healthy diets from sustainable food systems. *The Lancet, 393*(10170), 447–92. <https://doi.org/10.1016/S0140-6736(18)31788-4>.
Williams, D. K., Archer, C. J., & O'Mahony, L. (2021). Calm the farm or incite a riot? Animal activists and the news media: A public relations case study in agenda-setting and framing. *Public Relations Inquiry*, 1–23. <https://doi: 10.1177/2046147X21105519>.
Wilson, G. (2018, January/February). Kangaroos can be an asset rather than a pest. *Australasian Science Magazine, 39*(1). <https://www.australasianscience.com.au/category/magazine-issue/janfeb-2018>.
Wilson, G. R., & Edwards, M. (2019). Professional kangaroo population control leads to better animal welfare, conservation outcomes and avoids waste. *Zoologist, 40*(1), 181–202.
Winskell, K., & Enger, D. (2014). Storytelling for social change. In K. Gwinn Wilkins, T. Tufte, & R. Obregon (Eds.), *The handbook of development communication and social change* (pp. 189–206). John Wiley & Sons.
Wintle, B., & Bekessy, S. (2017, October 17). *Let's get this straight, habitat loss is the number-one threat to Australia's species*. The Conversation. <https://theconversation.com/lets-get-this-straight-habitat-loss-is-the-number-one-threat-to-australias-species-85674>.
Wintle, B. A., Legge, S., & Woinarski, J. C. Z. (2020). After the megafires: What next for Australian wildlife? *Trends in Ecology & Evolution, 35*(9), 753–57. <https://www.cell.com/trends/ecology-evolution/pdf/S0169-5347(20)30171-3.pdf>.
Winton, T. (2018, July 11). *Predator or prey?* [Opinion]. Australian Marine Conservation Society. <https://www.marineconservation.org.au/planet-shark-predator-or-prey-tim-winton/>.

Witherell, C., & Noddings, N. (1991). Prologue: An invitation to our readers. In C. Witherell & N. Noddings (Eds.), *Stories lives tell: Narrative and dialogue in education* (pp. 1–12). Teachers College Press.

Wittmayer, J., Backhaus, J., Avelino, F., Pel, B., Strasser, T., & Kunze, I. (2015). *Narratives of change: How social innovation initiatives engage with their transformative ambitions.* TRANSIT working paper 4, 1–25. [Working paper series]. Transformative Social Innovation Resource Hub. <https://www.transitsocialinnovation.eu/content/original/Book%20covers/Local%20PDFs/181%20TRANSIT_WorkingPaper4_Narratives%20of%20Change_Wittmayer%20et%20al_October2015_2.pdf>.

Woinarski, J. (2016). Weighing inequality in a less natural world. *Wildlife Australia*, 53(3), 23–6.

Woinarski, J., Burbidge, A., & Harrison, P. (2014). *The action plan for Australian mammals 2012.* CSIRO Publishing.

Woinarski, J. C. Z. (2014). The illusion of nature: Perception and the reality of natural landscapes, as illustrated by vertebrate fauna in the Northern Territory, Australia. *Ecological Management & Restoration*, 15, 30–3.

Woodhouse, J. (2019a, September 4). *From human rights to sentient rights: The next generation of rights thinking.* Open Global Rights. <https://www.openglobalrights.org/from-human-rights-to-sentient-rights-the-next-generation-of-rights-thinking/>.

Woodhouse, J. (2019b, May 17). *Universal Declaration of Sentient Rights.* Medium. <https://jamie-woodhouse.medium.com/universal-declaration-of-sentient-rights-4b43d428c590>.

Woodward, E., Hill, R., Harkness, P., & Archer, R. (Eds.). (2020). *Our knowledge our way in caring for Country: Indigenous-led approaches to strengthening and sharing our knowledge for land and sea management. Best practice guidelines from Australian experiences.* NAILSMA and CSIRO. <https://www.nespnorthern.edu.au/wp-content/uploads/2020/11/Our-Knowledge-Our-Way-Guidelines.pdf>.

Workman, A. (2021, January 22). Eyes wide shut: 'Latest lamb ads should come with a warning'. *The Australian.* <https://www.theaustralian.com.au/commentary/strewth/eyes-wide-shut/news-story/e0841de876bb95e7c208a95408ffbc7c>.

World Animal Protection. (2016). *Advance Australian animal welfare: The urgent need to re-establish national frameworks.* <https://www.worldanimalprotection.org.au/sites/default/files/media/au_files/advance_aw_report_lr.pdf>.

World Animal Protection. (2020). *Animal Protection Index (API) 2020 Commonwealth of Australia: Ranking D.* <https://api.worldanimalprotection.org/sites/default/files/api_2020_-_australia.pdf>.

Worrell, R., & Appleby, M. C. (2000). Stewardship of natural resources: Definition, ethical and practical aspects. *Journal of Agricultural and Environmental Ethics*, *12*(3), 263–77.
Wrenn, C. (2017). Toward a vegan feminist theory of the State. In D. Nibert (Ed.), *Animal oppression and capitalism* (pp. 201–30). Praeger Press.
Wrenn, C. L. (2013). Resonance of moral shocks in abolitionist animal rights advocacy: Overcoming contextual constraints. *Society & Animals*, *21*(4), 379–94.
Wright, L. (2021). Framing vegan studies: Vegetarianism, veganism, animal studies, ecofeminism. In L. Wright (Ed.), *The Routledge handbook of vegan studies* (pp. 3–14). Routledge.
WWF (World Wide Fund for Nature). (n.d.). Retrieved March 3, 2021, from <https://www.worldwildlife.org/>.
WWF (World Wide Fund for Nature). (2018a, May 24). *Backyard barometer 2018: Environmental concern on the rise*. <https://www.wwf.org.au/news/news/2018/backyard-barometer-2018-environmental-concern-on-the-rise#gs.dnwmc5>.
WWF (World Wide Fund for Nature). (2018b). *Backyard barometer summary report: Australian attitudes to nature*. <https://www.wwf.org.au/ArticleDocuments/353/pub-summary-backyard-barometer-australian-attitudes-to-nature-23may18.pdf.aspx?Embed=Y>.
WWF (World Wide Fund for Nature). (2018c). *Fight for the reef*. <https://www.wwf.org.au/what-we-do/oceans/great-barrier-reef/fight-for-the-reef#gs.ruin90>.
WWF (World Wide Fund for Nature). (2018d). *Great Barrier Reef*. <https://www.wwf.org.au/what-we-do/oceans/great-barrier-reef#gs.rxu451>.
WWF (World Wide Fund for Nature). (2018e). *Koala*. <https://www.wwf.org.au/what-we-do/species/koala#gs.talkau>.
WWF (World Wide Fund for Nature). (2018f). *Wildlife in a warming world: The effects of climate change on biodiversity in WWF's priority places*. <https://www.earthhour.org.au/ArticleDocuments/634/pub-report-wildlife-in-a-warming-world-13mar18.pdf.aspx>.
WWF (World Wide Fund for Nature). (2019, January 31). *Australia launches largest-ever rewilding project*. <https://www.wwf.org.au/news/news/2019/australia-launches-largest-ever-rewilding-project#gs.u0ayg4>.
WWF (World Wide Fund for Nature). (2020a, April 5). *Following bushfires, call to list koalas as endangered in QLD, NSW & ACT*. <https://www.wwf.org.au/news/news/2020/following-bushfires-call-to-list-koalas-as-endangered-in-qld-nsw-act#gs.du2aqe>.

WWF (World Wide Fund for Nature). (2020b). *Living planet report 2020: Bending the curve of biodiversity loss.* <https://c402277.ssl.cf1.rackcdn.com/publications/1371/files/original/ENGLISH-FULL.pdf?1599693362>.

WWF (World Wide Fund for Nature). (2020c, July 28). *New WWF report: 3 billion animals impacted by Australia's bushfire crisis.* <https://www.wwf.org.au/news/news/2020/3-billion-animals-impacted-by-australia-bushfire-crisis#gs.d81k0y>.

WWF (World Wide Fund for Nature). (2021, January 28). *Report urging fix to failing nature laws puts government to the ultimate test.* <https://www.wwf.org.au/news/news/2021/report-urging-fix-to-failing-nature-laws-puts-government-to-the-ultimate-test#gs.solic4>.

Wynne, E. (2021, August 11). *Could keeping native animals as pets help conservation of threatened species?* ABC News. <https://www.abc.net.au/news/2021-08-11/keeping-australian-native-animals-as-pets/100362834>.

Young, E. A., Ross, H., Johnson, J., & Kesteven, J. (1991). *Caring for Country: Aborigines and land management.* Australian National Parks and Wildlife Service.

Yu, A. (2016, March 7). *TV cooking shows inspire Australian home cooking.* Euromonitor International. <https://www.euromonitor.com/article/tv-cooking-shows-inspire-australian-home-cooking>.

Zhou, J. (2016). Boomerangs versus javelins: How polarization constrains communication on climate change? *Environmental Politics, 25*(5), 788–811.

Zoethout, C. M. (2013). Ritual slaughter and the freedom of religion: Some reflections on a stunning matter. *Human Rights Quarterly, 35*(3), 651–72.

Zuolo, F. (2020). *Animals, political liberalism and public reason.* Springer International Publishing AG.

Zwartz, H. (2018, October 12). *Tasmania's salmon farms shooting thousands of non-lethal 'beanbag' rounds at seals.* ABC News. <https://www.abc.net.au/news/2018-10-12/seals-being-shot-with-thousands-of-beanbag-bullets-to-protect-s/10366006>.

Index

Aboriginal and Torres Strait Islander Peoples *see* First Nations Peoples
Ag-gag laws 40, 43, 53, 70, 71
animal activism
    moral shock 62, 90–2, 268
    priorities 9, 12, 13, 16, 17–8, 39–40, 43, 45–6, 49–54, 56, 57–8, 60, 62–5, 67, 70, 73, 75, 78, 81, 82, 89-90, 92, 93–6, 104–5, 112, 134, 161, 162, 188, 192, 201, 207, 208, 236, 245, 254, 263, 266, 268, 269–70
    vegan agenda, accusations 51, 102
    see also Ag-gag laws
animal agriculture
    economic value 28–9, 32, 38, 41, 45–9, 50–5, 61, 64–5, 67, 70, 71, 72–3, 77, 79, 82, 86, 89, 102, 111, 172, 173, 177, 189–90, 192, 214, 259
    industrialization 11, 21–2, 23, 25, 28–9, 47, 56–9, 61, 77–9, 84, 86, 93–4, 96, 98, 99–101, 235, 259
    live animal export 13, 50–2, 56, 57–64, 71-3, 75, 78, 90, 134, 255
    livestock welfare 13, 29–30, 33, 36–9, 46–9, 50–4, 60–7, 71, 72–3, 75, 77–9, 96–7, 99–101, 155–6
    Model Codes of Practice 36-7, 48-9
    see also animal rights; animal welfare; five domains; five freedoms; Meat & Livestock Australia (MLA); National Farmers' Federation (NFF); Royal Society for the Prevention of Cruelty to Animals (RSPCA)

animal rights 16, 34, 43, 53, 69–70, 76, 77, 79–81, 87, 89, 90–1, 93, 104, 150, 232, 259, 269
    see also sentience
animal welfare 2, 13, 16, 29–30, 33–40, 41–3, 45–52, 53–4, 56–70, 72–3, 75, 77–9, 80, 83, 84, 86, 88, 90, 92–8, 99–106, 109, 111, 112, 115, 145, 155–7, 163–6, 168, 176, 182, 184, 186, 187, 190, 191, 192, 207, 227, 232, 259, 263, 265, 266
    see also animal agriculture; sentience
animal welfare legislation 16, 30, 34–9, 41–3, 48–9, 52, 72
    see also animal agriculture; animal welfare; sentience
animals
    attitudes regarding 25–38, 39–40, 41–2, 50–4, 59–66, 67, 69, 70, 75, 78–9, 92, 100, 101, 103, 149, 166, 191, 192, 232, 263–4
    charisma 2, 150–1, 252
    cruelty to 19, 29–30, 35–6, 38, 39–40, 42, 48–50, 51–8, 60–1, 62, 72–3, 77–8, 83, 86, 88, 89, 90, 93–4, 98, 104, 150, 162, 175, 180, 188, 208, 263, 268, 270
    economic value 9–10, 26, 28–9, 32, 34, 45–7, 49, 51–5, 67, 68–70, 71, 77, 79, 82, 94, 111, 112, 149, 153, 177–8, 189–92, 205–7, 212, 214, 222, 223, 252, 256, 258–9, 267
    endangered and/or threatened species 15, 19, 158, 159, 160, 163, 164, 170,

183–4, 185, 188, 201, 205, 212, 213, 217, 218, 222, 235, 244–5
enrichment 49, 68, 71, 73
ethical and/or moral considerations 14, 16, 22–3, 25–7, 33–4, 41, 46, 49–50, 54, 59, 60, 61–2, 68–70, 75, 76, 83–4, 86–7, 89–92, 98, 99, 106, 108, 145, 162, 194, 209–11, 219, 222, 229, 230, 236, 253, 268, 269
iconic 2, 15, 174, 177–8, 180–1, 182, 195, 207
invasive 179, 265, 268
management of 2–3, 9, 19, 20, 28–30, 33, 34, 42, 47, 85, 154, 161, 173–4, 179–80, 182, 185, 187–8, 189–91, 212, 229, 239, 242–4, 259
property 16, 30, 36, 39, 42, 48, 168
totem 137, 146, 259
see also animal activism; animal agriculture; companion animals; feral animals; livestock animals; native animals; pest animals; sentience; war on animals
Animals Australia
Make it Possible 56–7, 59, 61, 77, 84–5, 89, 90, 93–4, 100, 102, 108
priorities 45–6, 51, 55–7, 60–1, 62, 63, 64, 73, 77, 78, 84–5, 89, 92, 95–6, 98, 102, 104, 105, 108, 111, 166, 266
Somewhere 56, 77, 93–4, 100
see also White, Lyn
Anthropocene 12, 20–5, 27, 31, 199
anthropocentrism 11, 12, 15, 20–30, 33–4, 39, 40, 45, 46–9, 54, 67, 68–70, 80, 87, 94, 104, 105, 106, 111, 112, 115, 116, 122, 123, 124, 134–5, 137, 139–40, 144–6, 150, 153–7, 161, 167, 171, 173, 176–7, 178, 183, 185, 190–1, 192, 197–9, 201, 207, 209, 210, 211, 212, 222–3, 227, 228, 229–30, 231, 232, 234–5, 236, 240, 242–3, 251–2, 253, 254–6, 257, 258, 260, 263–7
see also human exceptionalism; human superiority; speciesism
anthropomorphism 57, 124, 208
anti-cruelty legislation 35–7, 48–9
attentiveness 145, 233–7, 238, 241, 255, 256–7, 264, 267, 269, 271
see also empathy; friendship; kinship
Australian Animal Welfare Standards and Guidelines 37, 48, 73
Australian Animal Welfare Strategy 41–2
Australian Conservation Foundation (ACF) 158
Australian Marine Conservation Society (AMCS) 97–9, 188, 217, 218, 219

backfire effect 208
Baxter and Me 118, 119, 130, 143, 145
biodiversity
conservation initiatives 211–2, 214, 219–21, 243–50
legislation 158, 202–2, 205–6
loss 1–2, 9, 15, 21–5, 29, 158–9, 161, 198, 200–1, 223, 227, 243, 254, 256, 258, 266, 268, 270
offsetting schemes 202
significance of Australia 31, 41, 168, 213–4, 223
threats 169, 243, 249
see also biomass; habitat; native animals; rewilding
biomass 21–2
Black Summer fires 11, 32, 157, 162–6, 169–70, 172, 191, 197, 203, 206, 253, 255, 271
see also habitat; native animals
'Bloody Business, A' 58, 59, 62, 78, 90

Index

Camel Boy, The 119, 121, 130, 147, 192
capitalism, problem of 12, 20, 25, 27–9
chefs, high-profile 93, 98–102, 104, 105, 106, 108–9
  see also meat-eating; paddock to plate movement
climate change 2, 20, 23–4, 159, 171, 185, 192, 193, 203, 210, 215–6, 217, 223, 251, 256, 257, 270
community
  anthropocentric 229–31, 240, 242, 243
  definition 230
  ecological 199, 201, 204, 206, 207, 210, 211, 212, 213, 221, 222, 223–4, 225, 227, 228, 230, 235, 242, 243, 251, 254, 256, 257, 263, 266, 267, 270
  inclusive design 232, 256–7
  multispecies 11–2, 14, 24, 146, 200, 204, 224, 228–9, 230–5, 236–8, 241–3, 250–8, 260, 263, 266–8, 269, 271
  see also rewilding
companion animals 1, 10, 11, 13, 21, 29, 30, 32–3, 48, 81, 111–5, 118, 121–3, 126, 128–9, 133–4, 136, 141, 144, 146, 149, 150, 153, 155, 156, 157, 193, 194, 197, 229, 231, 235, 236–8, 249, 250, 253, 256, 265, 268
  see also animals; feral animals
Commonwealth Scientific and Industrial Research Organisation (CSIRO) 175
conservation culls 29, 173–4, 175, 180, 261, 270
  see also biodiversity; feral animals; native animals; pest animals
Country
  care for Country 161, 199, 211–2, 222, 224–5, 228, 231, 259–60, 263, 268, 269

definition 135, 160–1
  see also First Nations Peoples
cultural intermediaries 9, 76, 92, 99, 104–5, 269
cultured meat *see* meat-eating, alternatives

deforestation *see* land clearing
Dominion 77–8, 89, 90, 100
  see also Phoenix, Joaquin
Dot and the Kangaroo 118, 119, 137, 139–40, 142, 143, 145, 147, 149–50, 234,

ecosystems
  collapse 21, 23–4, 29, 159, 167, 169, 174–5, 182, 202, 204, 215, 216, 227, 251, 256
  ecological processes 23–4, 180, 223, 244–5
  natural balance 168, 188, 202, 213, 242–3, 250–1
  regulation 29, 30, 34, 45, 97, 115, 155, 173–5, 179–80, 182, 184, 185, 190, 191, 192, 198, 202, 209, 219, 242–4, 250, 251, 258
  see also biodiversity; climate change; conservation culls; habitat
empathy 7, 8, 46, 56–7, 60, 61, 62, 78, 79, 90, 111, 116, 117, 132, 237, 254
  see also attentiveness; friendship; kinship
Emu Runner 118, 119, 137–8, 139, 142, 145, 147, 148, 233
Environmental Protection and Biodiversity Conservation Act 1999 (EPBC Act) 158, 162, 171, 191, 201
extinction crisis 1–2, 20, 23, 25, 158–9, 161, 162, 171, 172, 183, 184, 185, 189, 191, 197, 200, 204, 227, 268

feral animals
    brumbies 13, 116, 154, 179, 180–3, 185, 186, 191, 195, 265
    definition 154, 167, 178–9
    feral cats 179, 183–5, 191, 200, 227, 258, 261
    problem 179–80, 181–2, 183–4, 200, 212, 229, 243, 249, 257, 258, 261, 265
    status 10, 11, 37, 154, 167, 190, 191, 192, 193, 229, 231, 236, 238, 243, 261, 265, 266
    see also animals; conservation culls; pest animals
First Nations Peoples 11–2, 15, 16, 17, 31, 71, 116, 125, 127, 135, 137, 139, 146, 147, 148, 149–50, 160–1, 168, 184, 189, 193–4, 195, 199, 211–2, 213–4, 224–5, 228, 230–2, 244, 259–60, 263, 269
    see also *Country*
five domains 67–8, 78
    see also Royal Society for the Prevention of Cruelty to Animals (RSPCA)
five freedoms 67–8, 78
    see also Royal Society for the Prevention of Cruelty to Animals (RSPCA)
flexitarianism *see* reducetarianism
food norms 11, 13, 75, 76, 93, 95, 99, 101, 102, 103, 106, 107, 109
    see also meat-eating; veganism; vegetarianism
friendship
    companionate relationships 10, 13, 112, 115, 117, 118, 121–2, 130, 131, 136–7, 146, 148, 149
    human-animal 13, 112, 122–32, 135, 137, 142–3, 144, 145, 147, 231, 233, 237, 253, 268
    other-regarding 124, 126, 129, 131, 132, 145, 150, 241
    properties of 122–4, 131–2, 133, 136, 139, 141–3, 145, 162, 231, 233
    stories of 112, 113, 124–32, 137, 144, 145, 237, 271
    see also attentiveness; empathy; kinship
Futureye report 33, 62, 63, 65, 66–7, 68, 70, 92, 106, 253, 264

Gardening Australia 246–8
GoodFish: Australia's Sustainable Seafood Guide 98–9
Great Barrier Reef 11, 13, 117, 155, 193, 199, 212–22,

habitat
    conservation 139, 149, 158–9, 172, 174, 175, 179, 180, 189, 201, 202, 204, 206, 209–11, 214, 243–6, 250
    fragmentation 171, 173, 175, 184, 191, 203, 206
    loss 11, 158, 169, 171–2, 175, 184–5, 191, 200, 202, 205, 206, 207, 223, 245, 266
    restoration 210, 211, 245, 249, 250
    see also biodiversity; conservation culls; land clearing; native animals; rewilding; stewardship, environmental
Haraway, Donna 25, 133–5, 136, 139, 231
    see also kinship
Healing 119, 138–9, 145, 147, 149
human chauvinism *see* human exceptionalism
human exceptionalism 12, 20, 25–7, 29–30, 107, 115, 134
    see also anthropocentrism; human superiority; speciesism
human superiority 22–3, 26, 54

Index 343

see also anthropocentrism; human exceptionalism; speciesism
hunting 2, 9, 16, 29, 155, 180, 184, 189, 212, 270

interspecies solidarity 254–6, 257, 259, 263, 266, 267, 268

Jones, Bidda 46–7, 51, 58, 59, 62, 72, 78
    see also Royal Society for the Prevention of Cruelty to Animals (RSPCA)

kangatarianism *see* meat-eating
kinship
    familial 133, 137, 139, 142
    First Nations kinship systems 116, 135, 137, 146, 160–1, 195, 212, 224, 225
    Haraway on kinship 133–5, 137, 138, 139
    making kin 134, 137, 142–3, 170, 227, 235, 236, 253, 255, 266, 268, 269
    properties 132–3, 135–6, 138, 140, 142–3, 145, 149–50, 162, 231, 233
    stories 112, 113, 122, 137–42, 144, 170, 171, 237, 260, 271
    see also attentiveness; empathy; friendship; Haraway, Donna
Kirby, Michael 59

land clearing 22, 28, 160, 171, 172, 175, 185, 191, 200–2, 205, 206, 245, 251, 257, 270
    see also habitat
Larry the Wonderpup 112, 118, 120, 124, 128–30, 131
livestock animals 9–11, 13, 21, 25, 26, 28, 30, 32, 33, 36–9, 40, 41, 45–66, 67, 70, 71–3, 75–9, 84–6, 89–90, 93–4, 96–7, 100–1, 102, 104, 105–7, 111, 112, 154, 155–6, 157, 168, 173, 175, 177, 182, 185, 189, 192, 193, 197, 227, 229, 231, 235, 236, 253, 255, 256–7, 260, 264, 265, 266, 268
    see also animal agriculture; animal welfare; animal welfare legislation; animals

Meat & Livestock Australia (MLA) 93, 103, 104–5
    see also animal agriculture; livestock animals; National Farmers' Federation
meat-eating
    alternatives 82, 83–9
    culture 13, 75, 79, 82, 84, 87, 104, 105, 106–7, 258
    kangatarianism 85–6
    meat paradox 91, 207
    nose-to-tail philosophy 85, 107–8
    see also chefs, high profile; food norms; paddock to plate movement; reducetarianism; veganism; vegetarianism
moral shock *see* animal activism

National Farmers' Federation (NFF) 17, 51, 71, 82, 93, 95, 102, 104–5
    see also animal agriculture; livestock animals; Meat & Livestock Australia
native animals
    attitudes regarding 21, 31–2, 157–60, 161, 162–7, 168, 206–7, 235
    crows 238–43
    dingoes 26, 185, 227, 252, 261
    emus 31, 125, 137–8, 261
    kangaroos 13, 26, 31, 85–6, 121, 126, 149, 150, 154, 172–8, 179, 180, 182–3, 185, 190, 191, 195, 203, 207, 227, 243, 251, 252, 261, 270

koalas 31, 117, 125, 157, 169–72, 175, 177–8, 185, 191, 200, 202, 203, 205–6, 249
native wildlife 1, 2, 10, 11, 15, 19, 21, 29, 31–2, 37, 41, 97, 98, 125, 139, 148, 149, 154, 157–60, 161, 162–7, 168–80, 182, 183–5, 191–2, 193, 194, 197, 198, 200, 202–4, 205–7, 227, 229, 231, 235, 238–45, 246–58, 261, 263, 265, 268
saltwater crocodiles 13, 154, 186, 188–90, 191
sharks 1, 2, 154, 186–90, 191
see also animals; biodiversity; feral animals; habitat; land clearing; pest animals; rewilding
nature, ideas of 27–8

paddock to plate movement 93, 99, 101
pest animals 10, 11, 29, 37, 173, 174, 177, 178, 179, 181, 182–4, 229, 242, 244, 248, 261,
see also biodiversity; conservation culls; feral animals; native animals
pets see companion animals
Phoenix, Joaquin 77–8, 79, 89
see also *Dominion*
plant-based meat see meat-eating, alternatives

Red Dog 112, 118–9, 120, 121, 130, 137, 141–2, 145, 147, 150
reducetarianism 75, 84–5, 86–7, 90, 93, 94, 96, 99, 101–2, 104, 105, 106, 109
Regan, Tom 16, 80
rewilding 11, 241, 243–50, 252, 257, 263
Royal Society for the Prevention of Cruelty to Animals (RSPCA)
analysis live animal export 51, 58, 59

priorities 35–6, 37–9, 45–6, 51, 53, 55, 57–8, 59, 63, 64, 65–6, 67–8, 80, 92, 95, 96–7, 100, 106, 111, 156, 175, 176
RSPCA Approved Farming Scheme 39, 96–7, 100, 106, 109
welfarism 38–9, 67, 156
see also animal agriculture; five domains; five freedoms; Jones, Bidda

Send in the Dogs 119, 121, 130, 192
sentience 11, 13, 41, 46, 49–50, 52, 53–5, 59, 61–70, 75–6, 77–81, 90, 92, 93–4, 96, 103–4, 105–6, 107, 111, 112, 115, 134, 153, 154, 170, 175–6, 180, 192, 197, 198–9, 222, 227, 231, 236, 241, 253, 255, 259, 260, 264, 266, 267, 270, 271
'Sheep, Ships and Videotape' 58–9
Singer, Peter 26, 49–50, 54, 55, 61, 79, 80, 194
Skippy the Bush Kangaroo 112, 119, 121, 124–6, 127–8, 129, 131, 132, 144, 145, 147, 148–9, 182, 233
social change 6–8, 63, 65–7, 75–6, 91–2, 106, 207–8, 231, 263, 268
social imaginary 3–8, 12, 14, 15, 17, 20, 40, 45, 70, 92, 105, 112, 113, 115, 124, 134, 137, 154, 161, 177, 212, 227, 230, 231, 232, 260, 264, 268, 271
social maturation 66, 92
speciesism 12, 20, 25, 26–7, 29–30, 50, 54, 69, 70, 80, 87, 94, 115, 145, 150, 154, 229, 233, 258, 260
see also anthropocentrism; human exceptionalism; human superiority
stewardship, environmental 11, 13, 199, 209–12, 214, 218, 220–25, 231, 243, 245, 246, 254, 263, 268, 269, 270, 271

Storm Boy 119, 121, 124, 127–8, 131, 145, 147, 148, 233
sustainability 13, 17, 65, 83, 84, 93, 95, 97, 98, 99, 100–1, 106, 174, 188, 199, 209–12, 217–8, 220–2, 223–4, 243, 244–5, 248

vegan before six 75, 84
veganism 13, 51, 75, 81, 82–3, 84, 86–7, 91, 93, 96, 100, 102–3, 104, 106, 107, 108, 109, 259, 268
vegetarianism 75, 82–3, 84, 86, 91, 96, 97, 100, 103, 104, 108, 109

Voiceless 55, 59–60, 72, 96

war on animals 29, 144, 146, 184, 228, 261
White, Lyn 56, 105–6
 see also Animals Australia
Wildlife Information and Rescue Service (WIRES) 163, 166
World Wide Fund for Nature (WWF) 1–2, 22, 159, 160, 163, 166, 170, 197, 203, 206, 213, 218, 243, 244

AUSTRALIAN STUDIES: INTERDISCIPLINARY PERSPECTIVES

Series Editor
Anne Brewster
Associate Professor,
University of New South Wales

This interdisciplinary book series showcases dynamic, innovative research on contemporary and historical Australian culture. It aims to foster interventions in established debates on Australia as well as opening up new areas of enquiry that reflect the diversity of interests in the scholarly community. The series includes research in a range of fields across the humanities and social sciences, such as history, literature, media, philosophy, cultural studies, gender studies and politics. Proposals are encouraged in areas such as Indigenous studies, critical race and whiteness studies, women's studies, studies in colonialism and coloniality, multiculturalism, the experimental humanities and ecocriticism. Of particular interest is research that promotes the study of Australia in cross-cultural, transnational and comparative contexts. Cross-disciplinarity and new methodologies are welcomed.

The series will feature the work of leading authors but also invites proposals from emerging scholars. Proposals for monographs, biographies and high-quality edited volumes are welcomed. Proposals and manuscripts considered for the series will be subject to rigorous peer review and editorial attention. The series is affiliated with the International Australian Studies Association (*www.inasa.org*).

Vol. 1 Geoff Rodoreda
        The Mabo Turn in Australian Fiction
        276 pages. 2018. ISBN 978-1-78707-264-0

Vol. 2 Xu Daozhi
        Indigenous Cultural Capital: Postcolonial Narratives in
        Australian Children's Literature
        250 pages. 2018. ISBN 978-1-78707-077-6

Vol. 3 Matteo Dutto
        Legacies of Indigenous Resistance: Pemulwuy,
        Jandamarra and Yagan in Australian Indigenous
        Film, Theatre and Literature
        254 pages. 2019. ISBN 978-1-78874-541-3

Vol. 4 Radha O'Meara, Tessa Dwyer, Stayci Taylor and
        Craig Batty (eds)
        TV Transformations & Transgressive Women: From
        Prisoner: Cell Block H to *Wentworth*
        492 pages. 2022. ISBN 978-1-78997-506-2

Vol. 5 Jane Mummery and Debbie Rodan
        Imagining New Human-Animal Futures in Australia
        362 pages. 2022. ISBN 978-1-78997-314-3

www.ingramcontent.com/pod-product-compliance
Ingram Content Group UK Ltd.
Pitfield, Milton Keynes, MK11 3LW, UK
UKHW021827210426
5322IPUK00003B/64